MARINE MAMMAL RESEARCH

Conservation beyond Crisis

Edited by
JOHN E. REYNOLDS III
WILLIAM F. PERRIN
RANDALL R. REEVES
SUZANNE MONTGOMERY
TIMOTHY J. RAGEN

THE JOHNS HOPKINS UNIVERSITY PRESS
Baltimore

Printed in the United States of America on acid-free paper

9 8 7 6 5 4 3 2 1

The Johns Hopkins University Press
2715 North Charles Street
Baltimore, Maryland 21218-4363
www.press.jhu.edu

Library of Congress Cataloging-in-Publication Data

Marine mammal research : conservation beyond crisis / edited by John E. Reynolds . . . [et al.].
p. cm
Includes bibliographical references (p.) and index.
ISBN 0-8018-8255-9 (alk. paper)
1. Marine mammals—Conservation. 2. Marine mammals—Effect of human beings on. I. Reynolds, John Elliott, 1952–
QL713.2M32 2005
599.5′17—dc22 2005047496

A catalog record for this book is available from the British Library.

Photo credits: Frontispiece and chapter opening art courtesy of Minden Pictures. The photographers are Flip Nicklin (frontispiece, Chapters 1, 4, 7, and 11), Yva Momatiuk and John Eastcott (Chapter 2), Frans Lanting (Chapters 3 and 10), Matthias Breiter (Chapter 5), Tui De Roy (Chapters 6 and 12), Norbert Wu (Chapter 8), Michio Hoshino (Chapter 9).

Contents

Preface

OVER THE PAST SEVERAL DECADES, as interest in marine mammal conservation has grown and research into the biology and status of these animals has expanded, the conflicts between certain human activities and the welfare of marine mammals and their ecosystems have been drawn into ever-sharper relief. In response, in 2002 the U.S. Congress issued a directive to the Marine Mammal Commission to conduct an expert consultation on future directions in marine mammal research.

The Commission initiated its response by appointing a steering committee that focused the consultation on key threats to marine mammals and their habitat: direct fishery interactions, indirect fishery interactions, disease, contaminants, harmful algal blooms, anthropogenic sound, habitat transformation, long-term environmental change, inadequate definition of conservation units, and human population growth and demography. The committee then engaged scientists with expertise in these topics to write background papers for the consultation. Authors were asked not only to summarize current research efforts but also to identify the future research needed to fill anticipated gaps in our knowledge of marine mammals and their habitat as related to those threats.

The background papers were delivered at an international meeting of experts in Portland, Oregon, on 4–7 August 2003. Meeting participants were asked to evaluate the findings from the background documents and to develop recommendations that would guide Congress in future funding decisions. In the course of their deliberations, participants were expected to consider the effects of human population growth and associated economic development, the important subsistence and cultural uses of marine mammals, the values attributed to marine mammals apart from resource use, and the necessity for an interdisciplinary approach to research. In July 2004 the Commission delivered its report to Congress (Ragen and Reeves 2004) and issued a separate and more detailed report of the consultation meeting (Reeves and Ragen 2004).

From the outset, the Marine Mammal Commission was eager to make the background papers for the consultation available to a wider audience, including scientists, managers, and anyone else interested in marine mammal research and conservation. To that end, the authors modified their papers in response to the discussions at the consultation meeting and to comments from the editors and external referees. The final compilation of papers, published in this volume, represents an up-to-date, authoritative benchmark in marine mammal science and conservation.

The list of threats discussed in the consultation and described in this volume is not exhaustive. Other threats were considered by the steering committee or in the course of the consultation. The committee initially intended to include "dead zones" among the threats to be considered. Dead zones occur where human-related runoff (e.g., agricultural fertilizer) stimulates phytoplankton blooms that, when degraded by oxygen-consuming bacteria, lead to the creation of anoxic waters. The dead zone growing off the Mississippi River Delta may be the most familiar example of this phenomenon, but dead zones appear to represent a serious and growing threat to coastal ecosystems in many areas of the world. This topic was described only briefly in the background paper on habitat degradation after a fuller treatment of the subject was precluded by a scheduling conflict.

Two additional threats or issues were discussed at the consultation meeting itself but are not addressed explicitly in this volume. Ship strikes are a serious threat to marine mammals (e.g., the North Atlantic right whale) and, absent suitable regulation, are likely to increase in the future as a result of expected increases in commercial shipping. Moreover, human interactions with large and growing marine mammal populations (e.g., California sea lions, northern elephant seals) pose a dilemma that will likely require considerable management attention in the future. The consultation discussions pertaining to ship strikes and to human interactions with growing marine mammal populations are described in the workshop report (Reeves and Ragen 2004).

Finally, whaling and sealing—whether for commercial purposes or for "subsistence"—were also not considered in depth at the consultation, even though they have been a long-term source of concern in connection with marine mammal conservation. Hunting by Alaska Natives is treated to some extent in the chapter on contaminants because tainted marine mammal tissues could impair the health of consumers. In addition, the chapter on units to conserve addresses the issue of stock structure, which is an important aspect of the current debates on the impact of whaling and sealing (e.g., Japanese whaling of "J stock" minke whales in the western North Pacific and the subsistence hunting of local harbor seal stocks by Natives in Alaska).

Still, the absence of a chapter devoted explicitly to the subject of direct exploitation merits explanation. Some 35 years ago, the continuing decimation of great whales by the commercial whaling industry was one of the driving forces that compelled Congress to formulate and pass the Marine Mammal Protection Act. The whaling issue still receives sporadic media attention, especially around the time of the annual meeting of the International Whaling Commission (IWC), because, even with the global moratorium on commercial whaling in place, more than a thousand whales are taken each year. Most of the ongoing whaling is legal, either under the guise of scientific research or as a result of formal objections to the moratorium provisions. Both strategies are permitted within the IWC framework. The stocks affected by ongoing commercial whaling (with the important exception of the J stock of minke whales) are not thought to be in immediate danger. Moreover, the IWC appears close to agreement on a revised management scheme that would ensure that any future commercial whaling is closely monitored and that catch levels are adjusted if necessary. Most importantly, the IWC has itself considerably widened its scope of interest and responsibility over the years. The activities of its scientific committee extend to consideration of incidental mortality from ship strikes and entanglement in fishing gear, ecological interactions between fisheries and whale populations, and the effects of contaminants, noise, harmful algal blooms, and long-term environmental change. In fact, the research agenda of that body, originally focused entirely on estimating whale population numbers and setting catch quotas, is now much more diverse, encompassing most of the topics addressed in this book.

In many respects, then, the recent success of the IWC in addressing the direct exploitation of large whales reflects a shift toward a more precautionary approach, which is lacking for the other threats identified by the consultation steering committee. The committee's decision not to include a chapter on whaling and sealing should not be interpreted to mean that it viewed direct exploitation as irrelevant to marine mammal conservation. Rather, it means that the committee chose to focus attention on those threats for which a rigorous, forward-looking approach has not yet been developed, but for which such an approach is clearly needed.

For their essential roles in the consultation and in completing the chapters in this volume, the Marine Mammal Commission and the editors thank the authors and the reviewers of the background papers (Jay Barlow, Ian Boyd, Deborah Fauquier, Tim Gerrodette, John Kucklick, David Lavigne, Lloyd Lowry, David Mellinger, Derek Muir, Eugene Murphy, Douglas Nowacek, Richard Pierce, Andre Punt, Patty Rosel, Brian Smith, Karen Steidinger, Brent Stewart, Paul Wade, and Stephanie Wong); the meeting participants; the other members of the steering committee (David Cottingham, Tara Cox, Paul Dayton, Robert Mattlin, Teri Rowles, and Whitney Tilt); the facilitators for the Portland meeting (Dustin Bengtson, Peg Boulay, Tony Fasst, Maya Fuller, Norman Hesseldahl, Diane McClay, Tom Poplawski, and Mark Wilkening); and Whitney Tilt of the National Fish and Wildlife Foundation, who administered funds on behalf of the Marine Mammal Commission and made logistical arrangements for all stages of the consultation. John R. Twiss Jr. and Robert H. Mattlin, both former executive directors of the Marine Mammal Commission, are also acknowledged for their vision in helping to conceive the consultation and obtain the necessary funds.

Finally, the editors are grateful to the Johns Hopkins University Press, and particularly to Vincent Burke, acquisitions editor for biology and life sciences, for providing an outlet for this volume.

The editors believe strongly that, until scientists, managers, and legislators take a more anticipatory and proactive approach to marine mammal issues, it will be impossible to strike a balance between the public's desire to conserve marine mammals and their habitat on the one hand and our society's determination to pursue a wide range of activities in the marine environment, including resource use, on the other.

MARINE MAMMAL RESEARCH

JOHN E. REYNOLDS III

1

The Paradox of Marine Mammal Science and Conservation

I WAS LUCKY ENOUGH TO ENTER the field of marine mammal science at an exciting time, when giants such as Ken Norris, Dick Laws, Doug Chapman, Bill Schevill, Vic Scheffer, Masaharu Nishiwaki, Bill Evans, and David and Melba Caldwell (to name a few) stalked the scientific landscape. In those days, data were sparse but opportunities seemed endless, and there was a contagious spirit of conservation and hope. The thrill and the passion are still there for me to this day, albeit tempered by many years, miles, and debates.

I began my graduate research in 1974 at the University of Miami's Rosenstiel School of Marine and Atmospheric Science. There I eagerly and naively began my studies of endangered Florida manatees under the direction of Dan Odell, who would become both mentor and friend. Remarkable legislation—including the National Environmental Policy Act (NEPA 1969), the Marine Mammal Protection Act (MMPA 1972), and the Endangered Species Act (ESA 1973)—had been freshly forged by the U.S. Congress, and the world was going to save what my friend and colleague Paul Dayton sometimes calls "the poor whales." I was going to be a part of something special.

It has been more than 30 years since then, and it is time to assess our progress. As noted in "Future Directions in Marine Mammal Research" at the end of this volume, (the royal) we have had some success, including removal of the eastern stock of the gray whale from the Endangered Species list, increases in the number of Florida manatees, and growth of the Bering-Chukchi-Beaufort seas stock of bowhead whales. Nonetheless, several marine mammal populations hover on the brink of extinction, and the future of several others may also be bleak. What went wrong?

Perhaps our single biggest failure has been our inability to develop a disciplined, forward-looking approach to conservation and the science that supports it. Although we still profess commitment to our conservation goals, research and management efforts to achieve those goals tend to be driven more by economic interests and crises than by a well-integrated, forward-looking plan. Consider the species of marine mammals about which scientists possess some understanding. Their research and management generally reflect the fact that (a) people stand to make or lose money because of them (e.g., species that are commercially harvested; manatees slowing development in coastal waters of Florida; species held in public display facilities); (b) knowledge about them has implications for human health (e.g., marine mammals' ability to stay warm in cold waters) or safety (concerns regarding the extent to which beaked whale and other marine mammal conservation efforts may affect military readiness); or (c) they are poised to disappear. In other words, bursts of marine mammal science often follow money that appears as a reaction to human activities or crises.

The result is a reactive approach, punctuated by frenetic, lawsuit-driven responses to crises. The same approach is applied to other fields of natural resource study and conservation and appears to reflect a norm about which society and science have not been terribly bothered, certainly not to the point where the status quo has changed dramatically. This shortsighted approach is not working and is surely not what people follow when planning ahead in their personal lives.

People handle their private affairs so very differently from the way that they permit publicly owned resources to be handled. To the extent that their finances permit, reasonable people do all that they can to avoid crises and crisis-driven reactions. They use preventative medicine to avert life-threatening crises. They purchase life and health insurance to lessen the economic and other pain when crises do inevitably occur. And they put money into retirement accounts or other investments to ensure long-term quality of life for themselves and their families.

These and other actions are intended to avoid crises rather than to cope with them once they occur. When crises occur and people are unprepared, the number of options for responding may be limited, incomplete, expensive, and often unsuccessful. Those options may be painful, economically, physically, or emotionally. In contrast, a measured and anticipatory approach generates multiple options with various costs and benefits that can be weighed carefully before selection of those with the best outlook for long-term economic and physical health.

The same kinds of behavior should apply to natural resource research and management. Often they are not, and the costs tend to be borne by everyone who pays taxes, including you and me. In other words, how responsible agencies deal with resource-driven crises should concern all of us at a very personal level—and we should demand better.

An important milestone for me occurred in 1991 when I was appointed by President George H. W. Bush to chair the U.S. Marine Mammal Commission. That appointment provided me with an opportunity for professional growth and allowed me to learn daily from people far wiser than I, particularly John Twiss and Bob Hofman, who served as executive director and scientific program director, respectively, for the agency. Among the most valuable lessons I learned was that vision and foresight matter a great deal. People (and agencies) that plan ahead, anticipate crises, and take steps to avert those crises are simply better at achieving their goals. Nobody I have ever known is better at that than John Twiss.

In the mid- to late 1990s, in my testimony before congressional committees and in conversations with members of Congress, I urged a more forward-looking approach for agencies responsible for marine mammal research and management. In 2002, Congress provided the Commission with funds to hold a consultation on future directions for research needed to address critical issues affecting marine mammals in the coming decades. This was an exciting opportunity to look deliberately into the future in search of better ways to conserve marine mammals and marine ecosystems. This volume constitutes an important result of the congressionally supported consultation. Readers will find in its pages many truisms, such as the utility of planning ahead; the strength of a multidisciplinary team approach to problem solving; the power of good science to clarify cause-and-effect relationships rather than simply documenting correlations; and the need to avoid the seduction of technology as a generic fix. Although the contents were promoted by the U.S. Congress to provide guidance for the United States, several of the authors are from other countries, scientists from a half-dozen countries participated in the consultation, and many of the recommendations are, we believe, useful and relevant at the international level.

In the 30 years since I trundled into marine mammal science, the players have changed, the field has grown, considerable scientific progress has been made, and people still hope that their science may enhance conservation. It is at least as exciting to be a marine mammal scientist in 2005 as it was in 1974. Bright young people are enthusiastic and offer hope that research may provide both scientific insights and conservation solutions.

The fact remains, however, that some of Dayton's "poor whales" have not been saved. Several fundamental flaws in our science and conservation approach that existed 30 years ago remain today. Despite having made so much progress in so many ways, we still lack answers to a number of fundamental questions, answers that are necessary if "the poor whales" are going to make it. Where can we find them?

Ultimately, the real answer lies outside the realm of science. Science, as I noted earlier, reflects societal values. Imagine how science and conservation would be affected if society placed as much or more value on long-term environmental well-being as on immediate or short-term human needs. For example, just as human attitudes toward the harvest and consumption of marine mammals and various other species have shifted in the United States in my own lifetime, people can and should expect attitudinal and value shifts regarding the environment and natural resources in the future.

As with so many things in life, solutions revolve around the golden mean: we must develop, and our society must reflect, values that balance short- and long-term human needs, one of which is a healthy and productive environment. It is also important to bear in mind that normative values in the "lower forty-eight" states of the United States should not be considered as representative of values in the rest of the world. In fact, the values of many people in the lower forty-eight regarding whales and seals may be in opposition to those of subsistence hunters and others in Alaska. If conservation of marine mammals around the world is an important goal (and we believe it is!), then it is vital to look beyond U.S.-centric attitudes and to consider, with respect, global perspectives and values. In such a balanced setting, science would not dictate rules or provide answers. Rather, it would inform decision makers about the consequences of their choices. My co-editors and I hope that the information provided herein will initiate some important steps in that regard and promote a transition to an era in which people do a better job of understanding and conserving the world's living natural resources.

ANDREW J. READ

Bycatch and Depredation

DIRECT INTERACTIONS BETWEEN MARINE MAMMALS and fisheries pose some of the most serious and immediate threats to the animals, particularly to cetaceans. In January 2002 a group of experts on marine mammal bycatch concluded that "incidental capture in fishing operations is the major threat to whales, dolphins and porpoises worldwide. Several species and many populations will be lost in the next few decades if nothing is done. Urgent national and international action is needed" (Read and Rosenberg 2002). As we look to the future, it is clear that direct interactions with fisheries pose the single greatest threat to many populations of marine mammals, in the United States and elsewhere around the globe.

In 1985 Ray Beverton provided a useful classification of the various types of interactions between marine mammals and fisheries. He noted that some interactions are *operational,* or direct, in which marine mammals interact physically with fishing gear. These types can result in the mortality or serious injury of an animal that is "captured" but discarded, a process termed *bycatch* (Alverson et al. 1994; Fig. 2.1). In many areas of the world, marine mammals may be captured unintentionally but are retained for consumption or sale as *nontarget catch* (Hall 1996). As noted just a few years ago by the Scientific Committee of the International Whaling Commission, both bycatch and nontarget catch may have important demographic consequences for marine mammal populations (International Whaling Commission 2001). In some fisheries, marine mammals are first taken as bycatch, then retained as nontarget catch because of their value as food or bait, and finally they themselves become the target of the fishery (Read et al. 1988, Leatherwood and Reeves 1989, Dolar et al. 1994).

Fig. 2.1. Vaquita (*Phocoena sinus*) taken as bycatch in the gillnet fishery in the Gulf of California. (Photo courtesy of Bill Eppridge and Minden Pictures.)

Marine mammals may be captured, injured, or killed in discarded fishing gear or other marine debris (Laist et al. 1999).

In another form of operational or direct interaction, marine mammals may remove or damage fish captured in the gear, resulting in a reduction in the value of catches (Donoghue et al. 2003). This behavior, known as *depredation* (Fig. 2.2), may cause fishermen to take retaliatory measures against the mammals.

Interactions may also be *ecological,* in which marine mammals and fisheries interact through trophic pathways. The two major types of interactions (operational and ecological) require different conservation and management approaches (Northridge and Hofman 1999). In this chapter, I review direct or operational interactions, with a focus on bycatches; ecological interactions are covered elsewhere.

Conflicts between marine mammal populations and commercial fisheries are increasing in frequency and intensity, and this trend is likely to continue into the foreseeable future (DeMaster et al. 2001). In large part, this is because of continued human population growth, particularly in coastal regions. The increasing demand for marine protein results in a cycle of intensive harvesting and the serial depletion of fish stocks (Pauly et al. 2002). The global ocean has already lost more than 90% of its large, predatory fish (Myers and Worm 2003). The collapse of fish stocks results in an intensification and displacement of fishing effort, which increase conflicts with marine mammals. Such conflicts may be exacerbated in areas where marine mammal populations are recovering from previous directed harvests.

Fig. 2.2. Bottlenose dolphin (*Tursiops truncatus*) depredation on Spanish mackerel (*Scomberomorus maculates*) in North Carolina. Dolphin depredation is typically characterized by fish with missing heads, fish with distinct tooth marks, or fish with only the head remaining. (Photo courtesy of Danielle Waples.)

Despite recognition of the importance of this problem in the United States (National Oceanic and Atmospheric Administration 2003a), our knowledge of the global extent, nature, and effects of direct interactions between marine mammals and fisheries is fragmentary at best. We know very little about current levels of bycatches or depredation in most of the world's fisheries, although experience suggests that they are likely to be widespread, if not universal. Moreover, we have almost no information about the history of such interactions prior to the early 1970s (Reeves et al. 2003). It is likely that direct interactions with fishing operations caused significant adverse impacts on populations of marine mammals before scientists were in any position to take notice (Jackson et al. 2001).

Direct fisheries interactions are not unique to marine mammals; bycatches are an important conservation problem for many other marine organisms as well. Several species of albatrosses and sea turtles face extinction as a result of unsustainable bycatches in fisheries (National Research Council 1990, Tasker et al. 2000). It is possible to draw generalizations from a knowledge of interactions between these other species and fisheries and, conversely, to inform conservation efforts on behalf of those species from experience with marine mammals (Melvin and Parrish 2001). Other long-lived but less charismatic species are also in serious decline because of bycatch. For example, many long-lived elasmobranchs have suffered precipitous declines from bycatch (Casey and Myers 1998, Baum et al. 2003). Fisheries bycatch of these species is seldom monitored or regulated, so effects on their populations may go unnoticed until the species disappear.

The world's fisheries employ an enormous variety of gear types that vary greatly in their likelihood of interacting with marine mammals. It is not possible to review fully the many types and categories of fisheries here, but it is important to draw distinctions among a few categories of fisheries and several types of gear. A more complete description of the types of fishing gear and their potential for interactions with marine mammals may be found in the work of Northridge (1984, 1991a). Most of what is known concerning direct interactions between marine mammals and fisheries comes from commercial fishing operations, in which catches are landed and sold. Such fisheries range from large, industrial operations employing vessels with sophisticated electronic equipment to small-scale, artisanal fisheries that involve small boats and relatively unsophisticated gear. In some regions, catches are not sold but retained for consumption by the fishermen, their families, and local communities. Finally, in some areas, recreational fisheries are an important component of the total fishing effort, often using the same types of gear employed by commercial fisheries. In general, our knowledge of the nature and extent of interactions between marine mammals and fisheries varies directly with the scale of the fishery. We know much more about bycatches and depredation in industrial than artisanal fisheries, for example, because such large-scale operations are typically overseen by a management infrastructure of some kind in which catches (and occasionally bycatches) are monitored and reported. In most areas of the world, scientific knowledge of artisanal and aboriginal fisheries is scant or nonexistent.

It is important to draw a distinction between gear that is fished passively, such as gillnets and longlines, and gear that is fished actively, such as trawls and purse seines. Clearly, there is greater potential to modify the likelihood of marine mammal encounters with active fishing gear that is under the direct control of a fisherman rather than with gear that is left to fish and be retrieved later. The reduction in pelagic dolphin mortality in the eastern tropical Pacific purse seine fishery for yellowfin tuna is a prime example of fishermen's ability to modify the way active fishing gear is deployed to achieve a certain result (Joseph 1994, Hall 1998). As noted later in this chapter, however, direct interactions between marine mammals and fisheries occur much more frequently with passive fishing gear, particularly gillnets, than with active gear.

CURRENT STATUS OF GLOBAL FISHERIES

The Food and Agriculture Organization (FAO) of the United Nations has reported global fisheries landings since 1950. Catches increased dramatically in the 1950s and 1960s as a result of an expansion of fishing effort. The FAO and many national fisheries organizations promoted this expansion and for several decades the rate of increase of fisheries landings outgrew the rate of human population growth. The expansion of fishing capacity was fueled by government subsidies to the fishing industry in the form of low-interest loans to build new vessels, discounts for fuel, and support for marketing and infrastructure (Weber 2002). Despite this expansion, however, and since the FAO statistics were corrected for misreporting, landings have declined by approximately 0.7 million metric tons annually since the mid-1980s (Pauly et al. 2002). Consequently, an increasing number of fish stocks are overexploited and experiencing decreasing catches. In the United States, for example, approximately one-third of the stocks of known status are overfished (National Oceanic and Atmospheric Administration 2002).

The expansion of global fishing effort during the middle of the last century caused widespread depletion of large, predatory species (Myers and Worm 2003) and the collapse of many populations from direct overexploitation or unsustainable bycatches. The situation with elasmobranchs parallels that of marine mammals; both groups are long-lived with low fecundity and at severe risk of depletion from bycatch in fisheries that target more fecund species (Casey and Myers 1998, Baum et al. 2003). As a result of these depletions, fisheries shifted their efforts to lower trophic level species, essentially "fishing down marine food webs" and reducing ecosystem complexity (Pauly et al. 1998). The changes in biodiversity and ecosystem function brought about by fisheries are likely to have been most severe (but least documented) in the tropical and subtropical seas of the developing world.

It is impossible to forecast future trends in fishing effort and landings, but with the global human population predicted to increase until at least the middle of this century, it is likely that the intensity of fishing will increase in coastal waters and spread farther into the pelagic realm. This increase will bring more marine mammal populations into direct conflict with fisheries and intensify existing interactions. The declines in global landings and the increase in the number of overexploited stocks, combined with continued human population growth, are particularly worrisome signals for the future intensity and extent of direct interactions between fisheries and marine mammals.

CURRENT KNOWLEDGE OF DIRECT INTERACTIONS

Bycatch

Our understanding of the extent and magnitude of marine mammal bycatch is hindered by the almost complete lack of reporting on a global scale; very few countries have any formal reporting system for bycatches of *any* species. Neither is there any centralized global data repository that holds information on marine mammal bycatches. Some data on cetacean bycatches are reported annually to the Scientific Committee of the International Whaling Commission (IWC), but these reports are limited primarily to member states of the IWC and are acknowledged to be far from complete.

The United States is unique in that it has a formal observation and reporting system for marine mammal bycatches. This system was established under the 1994 amendments to the Marine Mammal Protection Act (MMPA). Under these amendments, the Secretary of Commerce is required to categorize fisheries according to their likelihood of taking marine mammals during the course of their operations. Vessels in fisheries deemed to have frequent (Category I) or occasional (Category II) bycatches are required to register to obtain authority to make such bycatches. Fishermen are required to report all bycatches but seldom do so, and in practice the only reliable data come from observer programs (Cornish et al. 1998). Therefore, to assess the magnitude of bycatches, the National Marine Fisheries Service places observers aboard commercial fishing vessels to estimate the bycatch rate. This observed rate is then applied to a measure of total fishing effort to estimate total bycatch for the fishery. Vessels in Category I and II fisheries are required by law to carry an observer if requested.

The Secretaries of Commerce and the Interior are also responsible for the preparation of assessment reports for each stock of marine mammals in the United States. These reports must contain information on stock structure, abundance, trends, sources, and magnitude of anthropogenic mortality, and an evaluation of whether this mortality exceeds threshold levels specified by the Act. The reports undergo external peer review by regional scientific review groups and are then published as a formal assessment of the status of each marine mammal stock. The assessment reports, which are updated regularly and are available on the agency's Web site (National Oceanic and Atmospheric Administration Fisheries 2003b), provide a tremendous amount of new, valuable, and largely untapped information on marine mammal stocks and their status (Read and Wade 2000).

In a recent paper (Read et al. in review), my colleagues and I compiled estimates of the magnitude of bycatches for all stocks of marine mammals in U.S. waters from the most recent stock assessment reports. We gathered information on annual bycatches for each stock from 1990 to 1999 and combined bycatches into three categories of fisheries: gillnets, trawls, and other (including, e.g., longlines, purse seines, and traps). The mean annual bycatch of marine mammals in U.S. fisheries between 1990 and 1999 was 6,215 (SD 1,415) (Tables 2.1, 2.2). Bycatches of cetaceans (3,029 ± 1,000) and pinnipeds (3,187 ± 1,078) occurred in similar numbers. The vast majority of both cetacean (84%) and pinniped (98%) bycatches occurred in gillnet fisheries. Almost all coastal marine mammals experience some level of bycatch in fishing operations, but some pelagic cetaceans, including sperm whales (*Physeter macrocephalus*) and beaked whales (family Ziphiidae), are also taken as bycatch in pelagic driftnet fisheries. Bycatches occur frequently in the Atlantic, Pacific, and Alaska regions, but seldom in the Gulf of Mexico or Hawaii, where gillnet fisheries are rare or nonexistent (Read and Wade 2000). These data do not include mortality and serious injury caused by bycatches in recreational fishing gear, which have not been quantified but are probably insignificant in comparison with the mortality in commercial fisheries.

Our estimates of the magnitude of marine mammal bycatches in U.S. fisheries are negatively biased. Potential bycatches in many fisheries, particularly those in Alaska, have yet to be properly assessed. Existing information for these and some other fisheries rely on anecdotal reports or logbook accounts, which do not reflect the true magnitude of marine mammal bycatches. As noted previously, few U.S. fishermen report such bycatches voluntarily although they are required to do so by the MMPA. For example, in 1990 fishermen reported 74 harbor porpoise (*Phocoena phocoena*) bycatches in the Gulf of Maine, but the total bycatch extrapolated from an observer program was more than 1,200 (Bisack and DiNardo 1992, Weber 2002:159). It is now widely accepted that accurate estimation of bycatch rates in any fishery requires an independent observer scheme (Northridge 1991a).

We can use two simple methods to extrapolate from the U.S. dataset to a global total. First, we use the ratio of U.S. to global landings to expand the annual U.S. estimates to global totals. We obtained data on U.S. and global landings of marine fishes from the FAOSTAT fisheries database (FAOSTAT 2002). We can also use the number of fishing vessels as the expansion metric, rather than fisheries landings. We stratified the U.S. marine mammal bycatch by fish-

Table 2.1 Cetacean bycatches in U.S. fisheries from 1990 to 1999

	1990	1991	1992	1993	1994	1995	1996	1997	1998	1999
Gillnet	4,902	3,154	2,373	2,489	2,928	2,261	2,624	2,095	1,481	1,051
Trawl	195	297	232	133	199	195	999	436	116	332
Other	3	9	256	60	388	475	114	11	70	408
Total	5,100	3,460	2,861	2,682	3,515	2,931	3,737	2,543	1,668	1,791

Source: Data taken from NOAA Stock Assessment Reports using methods described in Read et al. (2003a).

ery type and used ratios of the number of U.S.-to-global vessels in each fishery category to expand the U.S. bycatch estimates to a global total. We obtained data on the number of fishing vessels from the FAO FIGIS database of global fishing fleets (FIGIS 2003).

U.S. marine fisheries landings comprised approximately 5% of the global total from 1990 to 1999. Bycatch reduction measures were employed in the late 1990s in the United States but in few other parts of the world; therefore, we used data from 1990 to 1994 to generate estimates of the annual global bycatches of marine mammals (132,724 ± 18,964), cetaceans (64,120 ± 15,130), and pinnipeds (68,605 ± 26,236).

The U.S. fleet constituted 1.1%, 5.9%, and 4.2% of the global registry of gillnet, trawl, and other vessels, respectively, in the FIGIS database between 1990 and 1994. Using these ratios to extrapolate to global bycatches during this period yielded an annual estimate of 653,365 ± 108,851 marine mammals (307,753 ± 98,303 cetaceans and 345,611 ± 140,441 pinnipeds). The preponderance of gillnet vessels in the global fleet, coupled with the known high bycatch rates of marine mammals in gillnet fisheries in the United States and elsewhere, suggests that the vast majority of U.S. marine mammal bycatches occur in gillnet fisheries. At present, it is unclear to what extent this observation can be extended to the fleets of other nations, but initial efforts to document the magnitude of bycatches should perhaps focus on gillnet fisheries. However, this does not imply that significant bycatches do not occur in other gear types, such as some pelagic trawl fisheries. There is one important exception to the dominance of gillnets as a source of bycatch mortality. Baleen whales are taken frequently in gillnets, but in contrast to other marine mammals, important bycatches also occur in other types of fisheries, especially those that use vertical lines to mark traps, pots, or other demersal gear. The true magnitude of large-whale bycatches in different gear types is unknown, but, at least in U.S. fisheries, both gillnet and trap/pot fisheries are responsible for a significant portion of large-whale bycatches.

Clearly these global estimates are very crude. For example, bycatches of pinnipeds in many tropical countries are likely to be nonexistent. Furthermore, U.S. fisheries and their associated bycatch rates are unlikely to be representative of fisheries throughout the world. In addition, the registry of fishing vessels in the FIGIS database is both incomplete and of unknown accuracy. These data are contributed to the FAO voluntarily, but many states fail to report or do not specify the composition of their fleets by vessel type. Despite these limitations, however, it seems reasonable to conclude that the global bycatch of marine mammals numbers in the hundreds of thousands each year. This rough estimate is consistent with what we know about large, documented bycatches of marine mammals in some areas (Table 2.3).

Given our poor knowledge of the magnitude of marine mammal bycatches, it is perhaps not surprising that our understanding of their impact is very limited. We have some quantitative information from the United States from the stock assessment reports, but information from other parts of the world is scattered and incomplete.

Most stocks of marine mammals in the United States experience some level of bycatch in fishing operations. For only a few stocks, however, does bycatch exceed sustainable removal thresholds or potential biological removal (PBR) levels (Wade 1998). As depicted in Table 2.4, only 19 of 153 stocks experienced takes (which include all anthropogenic mortality and serious injury) that exceeded their PBR levels in 1997 (Read and Wade 2000). Pinniped stocks are taken as bycatch more frequently than cetaceans, perhaps because of

Table 2.2 Pinniped bycatches in U.S. fisheries from 1990 to 1999

	1990	1991	1992	1993	1994	1995	1996	1997	1998	1999
Gillnet	1,921	3,312	5,626	3,573	3,540	3,136	2,472	2,873	2,323	2,344
Trawl	19	36	34	10	29	3	15	17	14	11
Other	151	149	148	10	29	30	6	20	15	0
Total	2,091	3,497	5,808	3,593	3,598	3,169	2,493	2,910	2,352	2,355

Source: Data taken from NOAA Stock Assessment Reports using methods described in Read et al. (2003a).

Table 2.3 Large documented bycatches of marine mammals in the world's fisheries between 1990 and 1999

Species	Country	Fishery	Year	Estimate	Source
Dall's porpoise (*Phocoenoides dalli*)	Japan	Driftnet	1991	3,207	IWC (2002)
Harbor porpoise (*Phocoena phocoena*)	Denmark	Gillnet	1994–1998	6,785	Vinther (1999)
Harp seal (*Pagophilus groenlandicus*)	Canada	Gillnet	1994	36,000	Walsh, D., et al. (2001)
Pelagic dolphins	IATTC[a]	Purse seine	1990	47,448	MMC (2002)

[a]Inter-American Tropical Tuna Commission.

their coastal habitats, but few pinniped stocks experience takes greater than the calculated PBR level. Conversely, fewer cetacean stocks are impacted by bycatches, but a larger number of cetacean stocks experience bycatches greater than the PBR level. Of particular concern is the entanglement of North Atlantic right whales (*Eubalaena glacialis*) in fixed gear fisheries in New England, which is contributing to the decline of this small population (Caswell et al. 1999). Bycatch may also be contributing to the decline of California sea otters (*Enhydra lutris*) (Marine Mammal Commission 2000). Past threats include the bycatch of harbor porpoises in sink gillnets in the Gulf of Maine and bycatches of pelagic cetaceans in driftnet fisheries on both coasts. These conservation problems have been resolved by mitigation strategies derived by the take reduction team process (see later in this chapter) or by other fisheries management practices (such as the elimination of the Atlantic pelagic driftnet fishery for swordfish).

It is more difficult to draw conclusions regarding the global effects of bycatch on marine mammal biodiversity because the requisite information does not exist. It is clear, however, that such direct fisheries interactions are a major threat to many populations and some species of marine mammals. For example, more than half of the fifty-seven initiatives recommended in the recent IUCN–The World Conservation Union's Species Survival Commission Conservation Action Plan for the World's Cetaceans (Reeves et al. 2003) deal with bycatch. Bycatch is the primary threat to several endangered species, including the vaquita (*Phocoena sinus*) and Hector's dolphin (*Cephalorhynchus hectori*), and contributes to the dire conservation status of the baiji (*Lipotes vexillifer*), the Mediterranean monk seal (*Monachus monachus*), and the North Atlantic right whale. With a global marine mammal bycatch of at least several hundred thousand animals a year, it is likely that there are many other serious conservation problems that have not yet been identified. It is important to improve on the simple methods and incomplete data described here so that we can better understand the effects of bycatches on these populations.

Depredation appears to be common in some fixed gillnet and longline fisheries, particularly in coastal areas, but there is very little quantitative information on its extent, impact, or economic cost. Most reports of depredation are anecdotal, and few researchers have attempted to study this behavior or its ramifications for marine mammal populations. Nonetheless, marine mammal depredation has been demonstrated to have serious economic costs for some fisheries (Yano and Dahlheim 1995, Ashford et al. 1996, Secchi and Vaske 1998, Reeves et al. 2001), which derive from both a reduction in the value of the catch when fish are removed from the gear and damage to the gear itself.

Furthermore, depredation often results in fishermen taking retaliatory or precautionary actions that may have adverse consequences for marine mammal populations. For example, fishermen frequently shoot at pinnipeds engaging in depredation off the Pacific coast of the United States

Table 2.4 The status of U.S. marine mammal stocks during 1997

Taxon	Stocks	Stocks with takes	Stocks with takes greater than PBR level	Strategic stocks
Mysticetes	23	12 (0.52)	3 (0.13)	17 (0.74)
Odontocetes	100	60 (0.60)	12 (0.12)	22 (0.22)
Pinnipeds	23	22 (0.96)	2 (0.09)	5 (0.22)
Other	7	7 (1.00)	2 (0.29)	3 (0.43)
Total	153	101 (0.66)	19 (0.12)	47 (0.31)

Source: Data taken from Read and Wade (2000).

Note: The proportion of stocks in each category is presented in parentheses. "Other" taxa include two stocks of manatees (*Trichechus manatus*), three stocks of sea otters (*Enhydra lutris*), and two stocks of polar bears (*Ursus maritimus*).

(Mate and Harvey 1987) and in Alaska (Hoover 1988). In fact, such activities are considered a plausible hypothesis to account for the decline of the Steller sea lion (*Eumetopias jubatus*) in western Alaska (National Research Council 2003a). Fishermen also shoot killer whales (*Orcinus orca*) engaging in depredation in Alaska (Dahlheim 1988, Matkin and Saulitis 1994) and bottlenose dolphins (*Tursiops truncatus*) taking king mackerel off charter boat lines in Florida (Zollett and Read, unpub. obs.).

Depredation is no longer confined to coastal ecosystems. The decline in coastal fish stocks and the 1994 United Nations ban on large-scale high-seas driftnet fishing has led to a rapid expansion in the use of pelagic longline fisheries (Donoghue et al. 2003). In addition, demersal longline fisheries have expanded rapidly, particularly in the Southern Ocean, where effort has been directed at Patagonian toothfish and other deep-water species. The FAO considers such deep-water ecosystems as "the Final Frontier" of commercial fisheries expansion (Food and Agriculture Organization of the United Nations 2003). This expansion of industrial longline fisheries into pelagic and deep-water environments has brought new conflicts with pelagic marine mammals that have learned to engage in depredation. The problem is now documented in all the oceans (Donoghue et al. 2003).

We have no estimates of the total costs of depredation to fisheries in the United States, or to those of any other country. In addition, because of the nature of both the retaliatory actions taken by fishermen and the regulatory regime that protects marine mammals in the United States, we have no estimates of the numbers of marine mammals killed or seriously injured in this manner. It is clear, however, that with increased levels of fishing effort in coastal areas and the proliferation of fisheries in deep-water and pelagic habitats, such interactions are likely to increase in the future. Furthermore, such behavior will probably lead to the use of potentially harmful mitigation strategies, such as acoustic deterrent devices (Reeves et al. 2001, Donoghue et al. 2003). The ramifications of such interactions for marine mammal populations are far from clear. Some marine mammals may benefit from enhanced foraging success as a result of this behavior but might face an increased risk of entanglement in fishing gear or retaliation by fishermen. The overall effect on a marine mammal population will depend on the benefits and costs to individuals associated with this behavior and the frequency with which individuals engage in depredation. These costs and benefits have not been estimated rigorously for any interaction involving depredation.

Current Approaches to Mitigating Bycatches

In general, where attempts have been made to mitigate the problem of marine mammal bycatches and adequate resources are available, solutions have been found. Such solutions require an accurate assessment of the magnitude of the interaction prior to intervention, a clear management goal, resources to test and evaluate potential solutions, and the active participation of fishermen. With the amendments to the MMPA in 1994, the United States adopted a formal system to address the bycatch of marine mammals in commercial fisheries that included all these requirements. The stock assessment process determines whether bycatches exceed the PBR level for a stock of marine mammals. If so, the MMPA requires that a take reduction plan be developed. These plans must include measures that reduce anthropogenic mortality and serious injury to below the PBR threshold within 6 months of implementation. The take reduction plans are developed by groups of stakeholders, including representatives from federal agencies, academic and scientific organizations, environmental groups, and fishing groups. These stakeholders work through a process of negotiated rule making, assisted by a federally appointed mediator, to develop the plan.

The take reduction process is the crucible in which bycatch mitigation strategies are developed in the United States. The intense pressure under which these teams work forces fishermen, scientists, managers, and representatives of environmental groups to work together to find ways to reduce marine mammal bycatches in commercial fisheries. If the team cannot reach consensus, the Secretary of Commerce is required to develop a plan to reduce takes below the PBR level. To date, six take reduction teams have been convened; all of them have submitted draft plans to the Secretary of Commerce, and four of the teams were able to reach consensus on their entire plan. Detailed reviews of this process are available elsewhere (Bache 2001, Young 2001).

The take reduction process has been successful in reducing large bycatches of marine mammals to below PBR levels. It has not been successful in reducing very small bycatches (such as entanglements of North Atlantic right whales) to zero. Nevertheless, as a result of this process, total bycatches of marine mammals in the United States have declined over the past decade (Fig. 2.3). This reduction is more pronounced for cetaceans (Read et al. in review), perhaps, not surprisingly, because all six take reduction teams addressed cetaceans bycatches. Average annual marine mammal bycatch was significantly lower in 1995–1999 ($5{,}189 \pm 1{,}053$) than prior to the MMPA amendments in 1994 ($7{,}241 \pm 876$). These declines were not the result of a reduction in fishing effort, as landings from gillnet fisheries in the United States rose throughout this period (Read et al. in review).

Considerable progress has been made in the development, testing, and implementation of measures designed to reduce the incidental mortality of cetaceans in some U.S. fisheries. In general, however, progress in addressing the bycatch problem has been slow or nonexistent throughout much of the rest of the world, particularly in developing nations (Read and Rosenberg 2002). One exception is in the North Sea, where the Agreement for the Conservation of Small Cetaceans of the Baltic and North Seas (ASCOBANS) has adopted guidelines that define unacceptable levels of incidental mortality of small cetaceans in commercial fisheries (Agreement for the Conservation of Small Cetaceans of the Baltic and North Seas 2000). These guidelines have

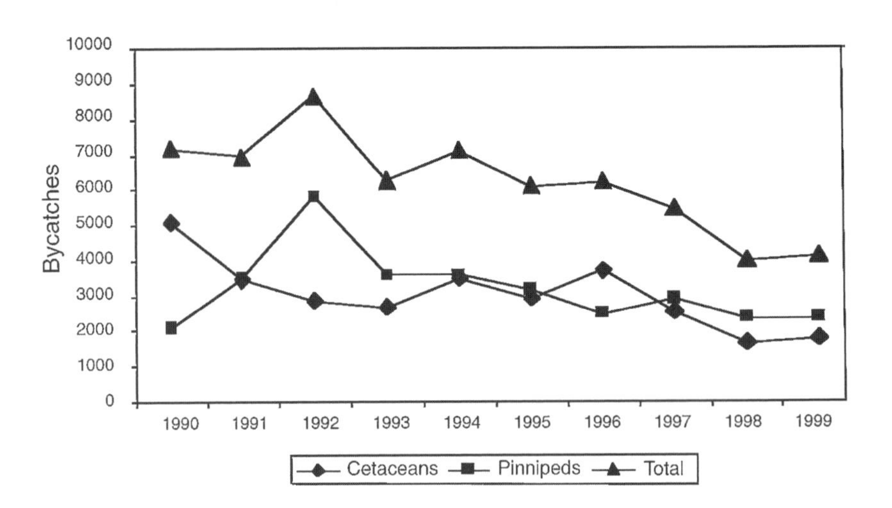

Fig. 2.3. Bycatches of cetaceans and pinnipeds in U.S. fisheries from 1990 to 1999. (From Read et al. in review.)

spurred some parties to ASCOBANS to evaluate the magnitude of bycatches in their fisheries and to test potential conservation strategies to reduce bycatches to below agreed thresholds. For example, acoustic alarms have been tested and found to be effective in reducing the bycatch of harbor porpoises in the Danish wreck gillnet fishery for cod (Larsen et al. 2002). These alarms are now required in this fishery (Vinther 1999, Larsen et al. 2002). Research is also being conducted to find ways to reduce the bycatch of small cetaceans in U.K. pelagic trawl fisheries (Northridge 2003). Similar programs exist in New Zealand (West et al. 1999).

As described by Hall et al. (2000), there are several general approaches to reducing bycatches. Total bycatch is the product of fishing effort and bycatch rate and, in principle, it is possible to reduce bycatches by reducing either effort or bycatch rate; in practice fishermen are understandably reluctant to accept the first alternative. In some cases, a threshold of bycatch mortality can be used to trigger the closure of a fishery (Hall 1998), which is essentially a dramatic and immediate reduction in effort. Similarly, the elimination of certain fisheries with high bycatch rates is essentially a drastic effort reduction measure.

Addressing the bycatch rate requires changes in fishing gear or practices that reduce the number of marine mammals killed per unit of fishing effort. This can be done by reducing the encounter rate, the entanglement rate, or post-entanglement mortality. Each approach has its merits and has been used successfully. It is possible to reduce the encounter rate by devising controls over when and where nets are fished, such as the time-area closures implemented in the Gulf of Maine (Murray et al. 2000). For such measures to work effectively, consideration needs to be given to the effects of displaced fishing effort away from the restricted area. Reducing the entanglement rate can be achieved by several means. It is sometimes possible to adjust the way gear is fished, such as the reduction in fishing depth employed by the California-Oregon driftnet fishery (Pacific Offshore Cetaceans Take Reduction Team 2003). Entanglement rate can also be reduced by displacing animals away from nets through the use of aversive stimuli or acoustic warnings (e.g., Kraus et al. 1997) or by modifying fishing gear so that entanglement is less likely after an encounter. Finally, it may be possible to reduce mortality after entanglement or entrapment by handling animals to ensure their safe release, by employing escape grids (Northridge 2003), or incorporating breakaway components in the gear (Atlantic Large Whale Take Reduction Team 2003).

One particular aspect of reducing mortality after capture deserves special mention here: the disentanglement network established by the Center for Coastal Studies on the U.S. East Coast. This network attempts to disentangle large whales that are caught in fixed gear in situ or are free-swimming and carrying fragments of gear. The network has been very successful in disentangling humpback whales (*Megaptera novaeangliae*) in both situations but less successful in disentangling right whales (Center for Coastal Studies 2003). It is widely recognized by the members of this network that disentanglement of large whales is a last resort and that efforts should be made to avoid entanglement in the first place. Thus, disentanglement should not be viewed as an appropriate long-term mitigation strategy.

Experience with the six take reduction plans indicates that each direct interaction between marine mammals and a fishery requires a solution specific to that combination of animals and gear. The behavior of marine mammals varies widely from species to species, as do the methods employed by fishermen. It is seldom possible to generalize from one bycatch problem to another; most interactions require a solution that reflects the unique combination of animal and fishery. In addition, it is noteworthy that none of the take reduction strategies developed in the United States relied on a single mitigation measure. In all cases, stakeholders decided to employ multiple strategies to address these problems, perhaps because of the uncertainty associated with any particular individual measure. Implicit in this approach (but explicit in the take reduction paradigm) is the ability to monitor the effectiveness of each strategy after implementation and to use data from observer programs as feedback to modify take reduction strategies to meet management goals.

Current Approaches to Mitigating Depredation

Surprisingly little research has been conducted on ways to reduce the extent or impact of marine mammal depredation of fish catches (Wickens 1995). This may be due to the effective, but typically destructive, methods used by fishermen to keep marine mammals away from their nets. For example, Northridge et al. (2003) reported that Greek fishermen used guns, dynamite, and gasoline-soaked rags attached to their nets to reduce depredation by bottlenose dolphins. A variety of crude explosive devices is used to deter sperm whales and other depredating cetaceans in the South Pacific (Donoghue et al. 2003).

There has been considerable interest in the use of sound to keep marine mammals away from fishing gear. Early attempts were largely unsuccessful (Mate and Harvey 1987), but there has been a recent resurgence of interest in the use of acoustic stimuli as a deterrence. In the Mediterranean, for example, bottlenose dolphins have been identified as causing economic losses to fisheries in several areas, and this has led to experiments testing the efficacy of acoustic alarms and acoustic harassment devices in mitigating these interactions (Reeves et al. 2001). In a recent study, Northridge et al. (2003) tested a commercial acoustic device that is marketed as "saving dolphins, saving nets, and ensuring a good catch" (Save Wave 2003). The use of these alarms (signals of 30–160 kHz with a source level of 155 dB re 1 μPa at 1 m) reduced damage (net holes) but did not adversely affect fish catches, leading the authors to suggest that such devices might work, at least in the short term. The mechanism by which these devices function is not known; nor is there any information on the potential for animals to habituate or sensitize to these sounds.

Economic Costs of Mitigation

There has been surprisingly little evaluation of the economic costs of the U.S. system of assessing and mitigating marine mammal bycatches. In a recent analysis, however, Griffin (2003) attempted to estimate the cost of take reduction measures associated with the Pacific Cetacean Take Reduction Plan, the Atlantic Large Whale Take Reduction Plan, and the Gulf of Maine Harbor Porpoise Take Reduction Plan. Griffin obtained economic data from the environmental assessment and regulatory impact review that accompanies each plan. These assessments include an evaluation of the plan's economic costs and potential alternatives. Griffin concluded that the cost of cetacean bycatch reduction represented 0.7–13.3% of the ex-vessel values of the fisheries impacted by these plans. The highest costs were incurred by portions of the Gulf of Maine sink gillnet fishery required to use acoustic alarms and implement time and area fishery closures. These costs are borne by the participants in the fisheries themselves.

Griffin's estimates did not include the cost of the marine mammal assessment program required to determine whether bycatches exceed PBR levels or of the monitoring program required to evaluate the success or failure of each mitigation measure. These costs are substantial and include surveys to estimate abundance and observer programs to generate estimates of bycatch rates.

Griffin noted that U.S. fishermen are at a competitive disadvantage with fishermen from similar fisheries in countries where marine mammal bycatch mitigation programs are not required. Most fisheries products now come into a global marketplace in which U.S. fishermen compete with those of many other countries. For example, foreign fisheries supply one-third of the Atlantic cod sold in the United States; half of this foreign market is controlled by Iceland. Icelandic gillnet fishermen take large numbers of harbor porpoises as bycatch or nontarget catch but do not employ mitigation measures to reduce this mortality. In contrast, gillnet fishermen in the Gulf of Maine must pay for acoustic alarms and bear the cost of fishing effort displaced by time and area closures. At some point, U.S. fishermen are likely to call for a leveling of this competitive field. This could occur through tariffs or embargoes, which are unlikely to survive challenges to the World Trade Organization (Joyner and Tyler 2000) or through consumer choice (see later in this chapter).

WHAT MORE DO WE NEED TO KNOW?

Bycatches

There is a clear need to improve our understanding of the magnitude and impact of marine mammal bycatches in the world's fisheries, particularly in the developing world. In addition, we have not yet completed our assessment of bycatches in the United States—many fisheries remain unmonitored (Dayton et al. 2002).

The World Wide Fund for Nature (WWF) Workshop on Cetacean Bycatches held in 2002 concluded that any assessment of bycatch should include collection and analysis of data describing:

- The composition of fishing fleets, including the size of the fleet (number of vessels), fishing methods, fishing areas, and measures of fishing effort.
- Status of marine mammal populations in the fishing areas, including estimates of abundance, population structure, trends, and potential rates of increase, where known.
- Estimates of bycatch rate (e.g., per haul or per trip), derived from independent observer programs where possible.
- Estimates of the total annual incidental catch, combining measures of bycatch rate with appropriate estimates of fishing effort.
- Estimates of the sustainable level of incidental mortality for each population or stock, using empirical data or proxies from comparable species or populations.

A lack of comprehensive information on the magnitude of marine mammal bycatches or the size and biology of particular marine mammal populations should not be used as a reason for inaction. Default or generalized models of marine mammal population dynamics and relative bycatch rates can and should be used to initiate action.

Obviously it will be difficult, if not impossible, to collect much of this information in many areas of the world that lack the infrastructure and expertise to conduct such research and monitoring. It is imperative, therefore, that capacity building be an integral component of any strategy designed to address the problem of marine mammal bycatches on a global scale. The transfer of technology and information is a critical prerequisite to any such strategy. The WWF Cetacean Bycatch Action Network and its Virtual Resource Center are examples of initiatives intended to meet this need (Worldwide Fund for Nature 2003). In these initiatives, scientists from the developed world have agreed to provide assistance to colleagues in developing nations who undertake such work.

We also need to know much more about the behavior of marine mammals around fishing gear if we are to understand both bycatches and depredation. There is great utility in making direct observations of interactions between marine mammals and commercial fisheries. We need to disseminate these findings widely throughout the fishing, management, and conservation communities. Very few studies have made detailed, fine-scale observations of the behavior of marine mammals around fishing gear. Research is required on the motivation of animals around nets and on the means by which they detect the gear. Most direct interactions between marine mammals and fisheries are not amenable to such observational research, but a few do offer opportunities (e.g., Read et al. 2003). In addition, studies in controlled, captive settings can offer a great deal of information on the process of detection and entanglement (e.g., Nachtigall et al. 1995).

There is much work to be done in developing, evaluating, and field-testing potential mitigation strategies. Direct involvement of the fishing industry is essential for developing solutions to bycatch (Hall et al. 2000). Fishermen are enormously creative, and incentives should be used to exploit this creativity. Here I highlight only one general area of research: the development of new conservation technology.

Some of the most successful solutions to the bycatch of seabirds, sea turtles, and fish have come from the development and use of innovative fishing gear. These innovations include the turtle excluder devices and Nordmore grates used in shrimp trawls and circle hooks and bait-casting devices employed in longline fisheries (Bergin 1997). New technological approaches, such as the use of acoustic alarms, or pingers, have been developed and implemented to reduce the bycatch of marine mammals. Acoustic alarms have now been shown to reduce the bycatches of several species of small cetaceans in a variety of gillnet fisheries (Kraus et al. 1997, International Whaling Commission 2000, Bordino et al. 2002, Barlow and Cameron 2003) and have also been used successfully to reduce seabird bycatches (Melvin et al. 1999). There is tremendous promise in the development of other technological innovations to reduce marine mammal bycatches, particularly with relatively simple concepts, such as stiff nets or breakaway gear. A modest investment in the development and testing of such approaches is likely to pay handsome dividends. Several existing programs, such as the National Whale Conservation Fund, already support such research, and others, such as the Saltonstall-Kennedy program, could be tailored to better encourage new technological approaches.

Depredation

Given our lack of information on the extent and cost of marine mammal depredation on fisheries, almost any standardized data collection program would be welcome. In particular, quantitative information is needed on the frequency of depredation by marine mammals (and other predators), the economic costs of the loss of catches and damage to gear, and links between depredation and the characteristics of fishing vessels and operations to understand why different vessels often experience markedly different levels of depredation (Donoghue et al. 2003). In addition, it is important to understand how prevalent this behavior is within populations of marine mammals—whether, for example, only a few individuals are responsible—and how the behavior is transmitted from individual to individual.

A stepwise, sequential approach to research and monitoring is desirable, with experiments and field trials being undertaken after documentation of the extent and costs of depredation. The Istituto Centrale per la Ricerca Applicata al Mare (ICRAM) workshop (Reeves et al. 2001) recommended such an iterative approach:

- Characterize the nature of depredatory interactions in a quantitative manner.
- If a problem exists, consult widely and locally for potential solutions.
- Undertake experimental testing of potential solutions while carefully assessing their adverse effects.
- If an experiment is successful, expand the approach and transfer information to equivalent fisheries.
- Establish a long-term monitoring program to ensure continued efficacy and document unforeseen impacts.

Participants in a recent workshop on depredation on longlines by cetaceans in the South Pacific encouraged research and development of acoustic and other approaches to mitigation (Donoghue et al. 2003). These individuals also emphasized the need for rigorous scientific trials to demonstrate effectiveness before broad-scale adoption of any particular mitigation device or procedure.

NEW AND PROMISING APPROACHES

In this section I identify several new approaches to management and research that hold promise for addressing the problem of direct interactions between marine mammals and fisheries. The approaches are presented in no particular order.

Stakeholder Involvement: Building on the Take Reduction Team Experience

The take reduction team process has facilitated the use of a science-based approach to the management of marine [illegible] ctly involves a wide range of [illegible] ing 2001). The direct involve- [illegible] mal, negotiated rule-making [illegible] f formulating strategies to re- [illegible] ammals in commercial fish- [illegible] ogical to involve fisheries par- [illegible] king; fishermen, after all, are [illegible] tices and the consequences of [illegible] ycatch problems may already [illegible] nt fishing practices—bringing [illegible] hese ideas to surface, to be re- [illegible] milar vein, it is both fair and [illegible] of environmental and animal [illegible] ded in the development of by- [illegible]

[illegible] participants were surveyed to [illegible] he negotiation phase of this [illegible] respondents indicated that the process was effective (86%) and fair (78%), although many participants (60%) were not satisfied with the outcome of the negotiations. Thus, although team members may not have approved of the outcome of their negotiations, they were willing to abide by them in efforts to reduce bycatches. In addition, participants believed that they were more likely to achieve a successful outcome by participating in the process than by opting out and allowing the government to draft a plan. Young (2001) concluded that several factors are essential for the success of this process, including the presence of an effective, neutral facilitator; the commitment of participants, including representatives of federal agencies, to work in good faith toward a consensus plan; sufficient time for negotiation and review of proposed regulations; accurate and accessible scientific information; and timely implementation of the plan by the relevant management agency.

The take reduction team approach should be maintained as the primary mechanism for dealing with marine mammal bycatches in U.S. commercial fishing operations. In fact, the Pew Oceans Commission (2003) has recommended that this process be used as a model for the development of plans to deal with bycatches of other species in the United States. It remains to be seen how useful this model will be in other countries, especially those in which stakeholders are not routinely involved in the management of fisheries.

International Action: The Role of the FAO

Few other countries are likely to implement a formal system to assess and manage direct interactions between marine mammals and fisheries in the foreseeable future. This is particularly true in the developing world, where such assessments are urgently needed. The FAO can play a critical role in encouraging member states to consider interactions with marine mammals as an essential component of fishery management plans. The FAO is the only body with the resources, political standing, and expertise required to facilitate such action on a broad scale.

There are several means by which the FAO can encourage member states to deal with marine mammal/fisheries interactions. One possible mechanism is an International Plan of Action (IPOA), which outlines a series of measures that should be undertaken by countries to deal with a particular issue. An IPOA for reducing incidental catch of seabirds in longline fisheries serves as a useful model for an IPOA dealing with interactions between marine mammals and fisheries (e.g., Read and Rosenberg 2002). In and of itself, the IPOA provides no guarantee that action will be taken, but it does serve to heighten awareness among national fisheries managers that this issue should be considered. Other possible means of facilitating international action through the FAO include expert consultations or technical workshops that bring together managers from various states to consider a particular issue.

The United States, which has the experience of the take reduction team process to draw on, should take a leadership role at the FAO's Committee on Fisheries to promote international action to deal with direct marine mammal/fisheries interactions to reduce bycatches. Moreover, as U.S. fishermen are placed at a competitive disadvantage because of the take reduction plan measures, it is in its own enlightened self-interest for the United States to promote marine mammal bycatch mitigation abroad. The National Oceanic and Atmospheric Administration (NOAA) has an International Bycatch Reduction Task Force that is examining existing international agreements for their effectiveness in resolving bycatch problems (National Oceanic and Atmospheric Administration 2003a). This task force should work to promote bycatch reduction initiatives in the FAO and through appropriate regional fishery management organizations. Such action would cost little but potentially reap tremendous dividends for the global conservation of marine mammals.

The Role of the Consumer: Seafood Labeling and Consumer Choice

Conservation advocacy groups have successfully influenced consumers to make certain choices regarding seafood purchases. The "Dolphin-Safe" label is now an integral component of the fisheries for trade in tuna products and has profoundly influenced interactions between pelagic dolphins and the purse seine fishery for yellowfin tuna in the eastern tropical Pacific (Joseph 1994, Gosliner 1999). The recent "Give Swordfish a Break" campaign raised awareness of the plight of swordfish and helped provide the momentum required for development of a more effective management plan for this species in the United States (Lee 2000). Several other initiatives, including seafood wallet cards distributed by the National Audubon Society's Living Oceans Program and other groups, help to inform consumers about the sustainability of seafood products and fishing practices (Brownstein 2002).

Such campaigns are not limited to the United States. The Marine Stewardship Council (MSC) has developed an international environmental standard for sustainable fisheries that uses a product label to reward responsible fishery practices (Marine Stewardship Council 2003). Consumers concerned about the environmental consequences of fishing practices are able to choose seafood products that have been assessed against the MSC standard and labeled to identify them as environmentally friendly products. The MSC uses several criteria to determine sustainability, including a statement that "the fishery is conducted in a manner that does not threaten biological diversity at the genetic, species or population levels and avoids or minimises mortality of, or injuries to endangered, threatened or protected species."

Such campaigns to influence consumer choice are likely to proliferate in the future as conservation groups become more sophisticated in influencing the marketplace and fishing organizations respond to such pressures. The potential to influence fisheries practices, including those that pertain to interactions between marine mammals and fisheries, is almost untapped. It is important that such campaigns be based on sound, unbiased science so that consumers can make informed choices about the sustainability of seafood products. Conservation organizations and the MSC have struggled to establish clear and universal criteria by which to judge fishing practices, including the effects of bycatch. This is an area where science can help to make consumer choice a sharper and more effective tool.

Toward Zero Mortality?

The MMPA, as amended in 1994, set forth an immediate goal of reducing marine mammal bycatches to below the PBR level and an ultimate goal of a "zero mortality rate." The NOAA recently proposed a definition of the zero mortality rate goal as 10% of the PBR level for each stock. Takes of this magnitude are assumed to pose an insignificant risk to any stock of marine mammals. Fishermen are understandably nervous about meeting such a goal, given the arduous negotiations required to meet PBR levels under the take reduction team process. Achieving this goal will require changes in the practices of many fisheries, including those that take only a sustainable number of marine mammals as bycatch. It is unclear how such changes in fishing practices will be negotiated and implemented, although the take reduction team process provides an attractive mechanism. Nevertheless, there is widespread support within the environmental and scientific communities for reducing bycatches to levels approaching zero (e.g., Pew Oceans Commission 2003).

Paying the Bills

For the last decade, federal spending on ocean science has averaged approximately $750 million (Pew Oceans Commission 2003). It is clear that this level of funding is insufficient to meet existing domestic mandates regarding direct interactions between marine mammals and fisheries. There are no additional funds available to expand these responsibilities or to support international conservation measures. Instead, it may be useful to look for other models of funding to support the needed research and management. One promising approach is in place in New Zealand, where since 1995 the government has implemented a scheme to recover funds from the fishing industry for the investigation and mitigation of impacts of fishing activities on protected species of marine wildlife (West et al. 1999). Conservation services levies are set annually following consultation between the relevant government agencies and stakeholder groups, including both the fishing industry and environmental groups. Levies have been used to pay for observer programs, to monitor species affected by bycatches, and to fund the development of mitigation strategies. Once a bycatch problem has been resolved, levies are no longer charged, thus providing an economic incentive for fishermen to cooperate with mitigation programs. In at least one instance, a bycatch of marine mammals was eliminated by changes in fishing practices (West et al. 1999).

The New Zealand system is a fair and equitable means of funding conservation measures designed to assess and mitigate marine mammal interactions with fishing operations, particularly in a system in which fishermen exploit a common resource. Such a program could be instituted with either commercial or recreational fisheries. In some cases, the costs of supporting such initiatives may exceed the economic capacity of a particular fishery; in such instances, consideration should be given to closing the fishery because of its unsustainable environmental and economic costs.

There are a few examples of industry-funded observer programs in the United States, but these are the exception rather than the rule. Adoption of this type of system in the United States, and elsewhere, would provide a significant new source of funds with which to assess and mitigate bycatches and, at the same time, provide a compelling economic incentive for fishermen to cooperate with such conservation measures.

New Paradigms in Fisheries Management

Finally, it is important to note that some fundamental changes in the management of fisheries are required to fully mitigate the effects of interactions with marine mammals. Foremost among these changes is the need to reverse the burden of proof in fisheries management (Dayton 1998). At the present time, the burden of proof lies with those individuals or organizations (generally environmental groups) that perceive that a fishery has caused some adverse environmental effect, such as an unsustainable bycatch. Management and regulatory agencies must then be persuaded, often through litigation, to take action to assess or mitigate these problems. This seems odd given that fisheries are profiting from common resources that are owned, in a general sense, by society at large. A more rational approach would be for the individuals profiting from fisheries to demonstrate that their activities have no adverse environmental effects. Participants in a new fishery, for example, might pay for the cost of an observer program to demonstrate that they are not taking marine mammals or other protected species as bycatch. In general, an assessment of the potential magnitude and impact of bycatches should be a prerequisite to approval for fishing. As noted previously, fishery management needs to move to a more holistic view of marine ecosystems, in which the need to protect certain components, such as marine mammals, is weighed carefully against the desire to exploit others for economic gain. The passage of the Sustainable Fisheries Act in 1996 represented only a very small step toward a required change in the burden of proof (Weber 2002). In cases where information is uncertain or inconclusive, the need to protect, maintain, and restore ecosystem structure and function should outweigh social or economic needs of the fishery (Pew Oceans Commission 2003).

CONCLUSIONS

Fisheries bycatch presents a real threat to many populations and some species of marine mammals. Other direct interactions, such as depredation, cause economic loss and other hardship to fishermen. In the United States, we have an effective system of assessing the magnitude and impact of marine mammal bycatches in commercial fisheries, although our work is far from complete in this regard. We have only begun to examine the effects of depredation on fisheries and marine mammal populations. Elsewhere in the world, the situation is dire—many large and unsustainable bycatches likely occur undetected, and depredation appears to be frequent and perhaps increasing in extent and intensity. There is still much to learn but also much work to do that does not require additional information. The United States has led by example in the assessment and mitigation of direct interactions between marine mammals and commercial fisheries. Our assessment and mitigation schemes work effectively and serve as models for other countries. We need to build on the success of these domestic programs and translate our experiences and expertise to other parts of the world, especially to developing countries, where such resources are lacking. This is our responsibility as the leading first-world country, but such actions will also benefit U.S. fishermen and consumers, who rely increasingly on a global marketplace for fisheries products. There is much to do.

ACKNOWLEDGMENTS

I thank my colleagues Phebe Drinker, Simon Northridge, and Paul Wade for allowing me to borrow liberally from papers that they have co-authored. Colleagues at the IWC Scientific Committee, especially Phil Clapham and Finn Larsen, helped to refine some of these ideas. Tim Ragen and John Reynolds initially asked me to prepare this chapter and waited with great patience during its preparation. Jess Maher helped considerably with preparation of the literature cited list. Bill Perrin and Randy Reeves provided useful comments on an earlier draft of the manuscript.

ÉVA E. PLAGÁNYI AND
DOUG S. BUTTERWORTH

3

Indirect Fishery Interactions

INDIRECT FISHERY INTERACTIONS ARE ECOLOGICAL interactions between marine mammals and fisheries, as opposed to directed marine mammal takes and operational (also termed technical) interactions in which marine mammals damage or become entangled in fishing gear with negative consequences for both the fishery and the animals (see Read this volume). Ecological and operational conflicts are sometimes difficult to separate because, for example, animals may damage fishing gear (an operational conflict) in the process of removing fish from the gear (a form of competition). There are many more published studies pertaining to operational interactions than to so-called biological or ecological interactions, no doubt because the former are more obvious and easier to quantify (see Northridge 1991b for a comprehensive review of both forms of interaction).

In this review we distinguish further between feeding-related and other indirect interactions. The former include exploitative competition (in ecologists' terminology), which involves reduction by consumption or utilization of a limited resource without any direct interaction among the competing species (Clapham and Brownell 1996): for example, a marine mammal eating a fish that might otherwise have been caught by a fisherman or, even more indirect competition, where the competitors target different resources that are linked through a food web effect (e.g., when a marine mammal consumes a fish that is an important prey species of a commercially desirable fish species). Another form of indirect interaction is interference competition, in which species actively disrupt the activities of others (Clapham and Brownell 1996). This would include, for example, disruption of the feeding activities of marine mammals as a result of disturbances such as noise from fishing operations. A final set of indirect ecological interactions does not involve any direct interaction between marine mammals and fisheries but considers the effects on one party of, for example, habitat destruction by the other. The focus in this chapter is on feeding-related interactions, but we briefly discuss other indirect interactions as well.

In almost all cases the available evidence for feeding-related competition between marine mammals and fisheries is currently inconclusive (Plagányi and Butterworth 2002). However, perceived conflicts between the two in pursuit of common sources of food have come increasingly to the fore in recent years (e.g., DeMaster et al. 2001, Bjørge et al. 2002). Escalating pressures on shared resources are expected in the future because of both the growth of a few marine mammal populations and the needs of an increasing human population. Reductions in direct takes in response to recognition that many populations of marine mammals were heavily overexploited in the nineteenth and early part of the twentieth century, as well as a widespread change in people's perceptions of whether marine mammals should still be regarded as renewable resources available for harvest, have meant that some marine mammal populations are currently increasing, sometimes by as much as 5–12% per annum (e.g., Shaughnessy et al. 2000, Bowen et al. 2003). In contrast, long-term (20- to 30-year) declines in the populations of several pinnipeds have been attributed to a combination of environmental factors and the depletion of prey species by commercial fisheries (Wade 2002), but the relative importance of these factors remains unknown.

The Food and Agriculture Organization of the United Nations (FAO) has estimated that more than 1 billion people worldwide currently rely on fish and shellfish for more than 30% of their animal protein (Food and Agriculture Organization 2002). The predicted increase in the global annual per capita consumption of fish is of concern given that marine fisheries are unlikely to appreciably exceed the present level of global landings (Food and Agriculture Organization 2002). It is expected that the global shortfall between supply and demand over the next few decades will be ameliorated to some extent by growth in aquaculture (Food and Agriculture Organization 2002). Nevertheless, if the world's human population continues to increase faster than the total food-fish supply from aquaculture, pressure to increase fisheries production will mount.

It is therefore imperative that scientists focus their efforts to achieve a better understanding of the nature and implications of indirect interactions. Concerns about the consequences for fisheries of an increasing marine mammal population have already been expressed in southern Africa, for example, where in 1990 South African fur seals (*Arctocephalus pusillus pusillus*) were estimated to consume some 2 million tons of food a year (Butterworth et al. 1995). Considering that this amount just about equals the annual human catch of fish in the region and that the fur seal population was then expected to increase further (Butterworth et al. 1995), the reasons for concern and potential for conflict are obvious.

Another example is found in the Pacific Ocean, where marine mammals may consume about 150 million tons of food per year, which is some three times the current annual human fish harvest (Trites et al. 1997). Tamura (2003) similarly deduced that, on a worldwide scale, fish consumption by cetaceans is approximately equivalent to commercial fisheries catches. He argues that competition between cetaceans and commercial fisheries is particularly severe in the North Pacific and North Atlantic oceans.

In Newfoundland, following the collapse of the cod fisheries in the late 1980s and early 1990s, the possible impact of an increasing harp seal (*Pagophilus groenlandicus*) population became a focal topic (Stenson et al. 1997, 2002). Given the difficult nature of this problem and consequent lack of clear scientific guidelines, lobbying and political pandering have tended to dominate, as exemplified by the confident declarations of Canada's then–fisheries minister Brian Tobin in 1995. Without acknowledging the associated scientific uncertainties, he announced that the total allowable catch for northwest Atlantic harp seals was being increased from 186,000 to 250,000 seals because it was "intellectually dishonest to claim that seals have not had a significant impact on recovering fish stocks." He then added that "there is only one major player fishing that stock and his first name is Harp and his second name is Seal," and he concluded that the seals "must . . . be harvested . . . in the context of building the groundfish stocks."

In this chapter we first review the competition for food and fishery resources between marine mammals and fisheries and then present a brief summary of some specific examples that address the question of whether marine mammal populations have had a negative impact on potential yields from fisheries through competition. These examples generally concern perceived competitive interactions, as in nearly all cases the evidence is inconclusive. We then summarize some examples pertinent to the opposite situation—whether fisheries have had a negative impact on marine mammals. As pressures mount to provide scientifically defensible answers to questions regarding marine mammal / fishery interactions, marine resource dynamics modelers are responding by including marine mammals in their schemes or constructing new models. Such models are appropriate tools for this task in that they effectively summarize the situation in terms of data availability and understanding of how systems operate; therefore, the models themselves are reviewed.

Moving one step further into a realm where little is known, we provide a brief overview of additional indirect interactions between fisheries and marine mammals that do not involve competition for resources. Specifically, potential instances are noted where fishing has affected (or is likely to directly affect) marine mammals by damaging critical habitats, by altering the structure of ecosystems, or by otherwise altering marine mammal population dynamics and/or population parameters. We also review the possible impact of mammal-borne parasites on fisheries. As empirical data and knowledge of ecosystem structure and functioning continue to improve, scientists, conservationists, and other interested and affected parties will increasingly bring these concerns to the fore.

An initial attempt is made to rank the various indirect interactions in terms of their overall importance. Focusing on the most important factors, we identify criteria to assist in comparing these various indirect fisheries effects in terms of their potential impact. Finally we identify and prioritize research needs.

FEEDING-RELATED INTERACTIONS

A summary of the range of possible indirect interactions between marine mammals and fisheries is given in Figure 3.1a–h, of which a–f refer to feeding-related interactions. The types of interactions may range from a direct marine mammal/fisheries competition for a target prey species, predator species, or both (Fig. 3.1a–c) to an indirect food web effect whereby a fish species targeted by a fishery is, in turn, the predator or prey of another species that is an important component of the diet of a marine mammal (Fig. 3.1d,f). The various interactions may be further complicated if there is a marked degree of cannibalism (of the target species in particular; Fig. 3.1f). Additional interactions include scenarios in which (a) a fishery of one species has a negative impact on the habitat of a marine mammal prey species (Fig. 3.1g) and (b) marine mammals indirectly reduce the value of the catch landed by a fishery through the transmission of parasites (Fig. 3.1h).

Table 3.1 lists marine mammals for which potential feeding-related interactions with fisheries have been identified. We included six mysticete species, fourteen odontocetes, sixteen pinnipeds, and two marine otters. Note that the information presented in the table is a first step only and should be interpreted broadly. We do not claim comprehensive capture of every record of small-scale marine mammal/fishery interactions.

The following general features are evident from Table 3.1 and the associated literature:

- In almost all cases, perceived interactions are difficult to establish conclusively.
- Reports in the literature may be biased one way or the other, depending on whether the concern arises from fishing interests (wanting to justify reducing marine mammals to improve fishery yields) or conservationist groups (wanting to ensure that adequate food supplies remain for threatened or potentially threatened marine mammal species). However, whereas more than twenty of the species listed have at one time or another been killed (legitimately or otherwise) in the belief that reducing marine mammals will improve fishery yields, concerns that overfishing is negatively impacting local marine mammal populations have been documented for only a few species.
- Fisheries that allegedly suffer from competition from marine mammals include the following: (a) Fisheries operating near pinniped breeding colonies. (b) Fisheries (both commercial and small-scale) operating in nearshore coastal areas (that overlap with restricted cetacean distributions such as that of the Indo-Pacific humpbacked dolphin, *Sousa chinensis*). (c) Fisheries reliant on seasonal pulses in the abundance of a target species (e.g., pelagic schooling fish such as anchovy and migratory fish species such as tuna and yellowtail); marine mammals generally congregate to take advantage of these pulses, with the consequent increased visibility further fueling suspicions that fewer marine mammals translate into more fish for the fisheries. (d) Fisheries reliant on species that are also important in the marine mammal diet, particularly where the midtrophic layers are dominated by only a few species. (e) Fisheries for species where estimates of consumption of those species by marine mammals in the fishery area are substantial compared with fishery takes.
- Most of the actual, perceived, or predicted marine mammal/fisheries interactions take the form of direct competition for a shared prey resource, even though the actual sizes or year classes of fish targeted may differ. However, there are several documented examples of indirect interactions whereby, for example, a fishery impacts a predator or prey species, which is in turn the prey and predator, respectively, of another species fed upon by a marine mammal (e.g., Punt and Butterworth 1995, Crespo et al. 1997). With the exception of the dugong (*Dugong dugon;* Mukerjee 1998, United Nations Environment Programme 2002), West Indian manatee (*Trichechus manatus;* Anon. 1999), and walrus (*Odobenus rosmarus*), there are very few documented cases (even speculations) that damage to a habitat by a fishery is indirectly responsible for reducing food supplies to a marine mammal.

INSTANCES OF POTENTIALLY DETRIMENTAL EFFECTS OF MARINE MAMMALS ON FISHERIES

Pinnipeds (Seals, Sea Lions, and Walruses)

In the early 1990s, major collapses occurred in the Atlantic cod (*Gadus morhua*) fisheries on the east coast of Canada and several explanatory hypotheses were posited (Bundy 2001). Harp seal populations off Newfoundland and Labrador have been increasing at an estimated rate of 5% a year since the mid-1980s and are known to consume a substantial tonnage of juvenile cod (Stenson et al. 1997). Using a bio-

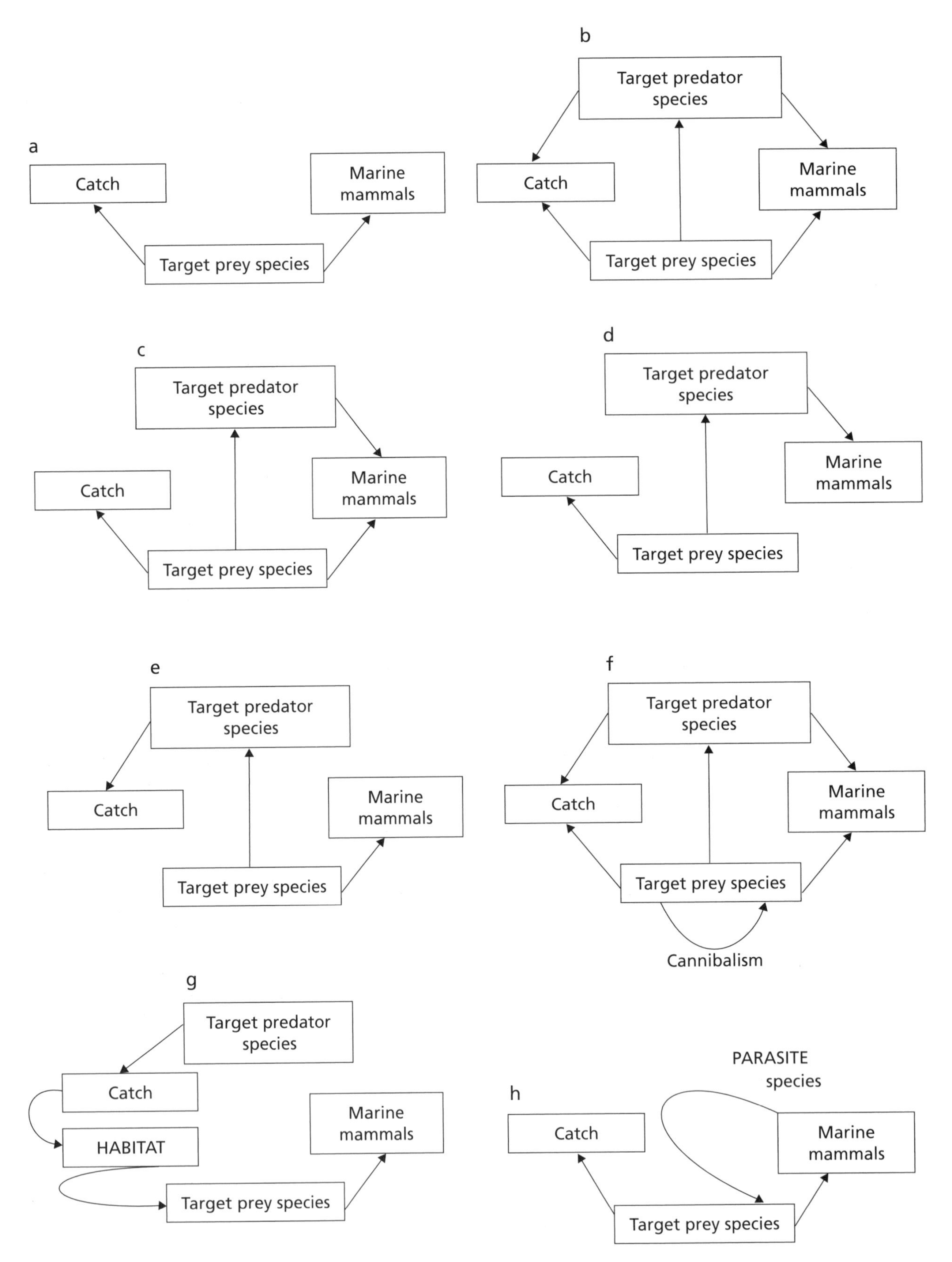

Fig. 3.1. Schematic summary of range of possible indirect interactions between marine mammals and fisheries. The type of interaction may be either that of direct competition between fishermen and marine mammals for a target prey species, predator species, or both (a, b, c) or through an indirect food web effect, whereby a fish species targeted by a fishery is, in turn, the predator or prey of another species that is an important component of the diet of a marine mammal (d, e). The various interactions may be further complicated if cannibalism (of the target species in particular) occurs to a marked degree (f). Note that these simplified figures depict only predation/fishery interactions, but these various interactions may be mediated to various degrees by environmental, spatial, and age-structure effects. Additional interaction representations include scenarios in which a fishery has a negative impact on the habitat of a prey species (example shown in g) and scenarios in which marine mammals indirectly reduce the value of the catch landed by a fishery because they transmit parasites (example shown in h).

Table 3.1 Identified possible and considered potentially possible marine mammal/fisheries feeding-related interactions

Taxonomic group	Common and scientific name	Form of possible competition/interaction	Geographic location	Notes
Order Mysticeti	Baleen whales			
Family Balaenopteridae	Minke whale (incl. dwarf minke subspecies) *Balaenoptera acutorostrata*	*Direct/indirect	Northern Hemisphere Dwarf minke whale subspecies—Southern Hemisphere	Perceived competition because targets cod, herring, capelin, and sandlance, e.g., ostensible direct competition for cod with fisheries off Iceland and Norway (Stefansson et al. 1997, Schweder et al. 2000); considered a "key species" competing for fishery resources in the western North Pacific (Government of Japan 2002, Tamura and Fujise 2002).
	Antarctic minke whale *Balaenoptera bonaerensis*	*Direct/indirect	Circumpolar in Southern Hemisphere	(a) Possible direct competition particularly if krill fisheries were to expand. (b) Indirect interaction: krill surplus hypothesis (see text) (Laws 1977).
	Bryde's whale *Balaenoptera edeni*	*Direct/indirect	Worldwide in tropical to temperate waters	Resident populations, e.g., off the southern African coast rely on pelagic fish populations also targeted by commercial fishery. In the western North Pacific ostensibly competes directly with fisheries for Japanese anchovy (*Engraulis japonicus*) and indirectly because both whales and skipjack tuna (*Katsuwonus pelamis*) (an important commercial species) target anchovy (Tamura and Fujise 2002).
	Fin whale *Balaenoptera physalus*	Direct	Cosmopolitan	Southern Hemisphere—possible direct competition with fisheries for krill. Northern Hemisphere—possible direct competition for herring, capelin, and sand-lance. Possible competition for cod off Iceland and Norway (Stefansson et al. 1997). Hypothesized indirect interaction caused by collapse of Canadian capelin fishery (Whitehead and Carscadden 1985) (see text).
	Blue whale *Balaenoptera musculus*	Direct/indirect	Cosmopolitan	(a) Expanding krill fisheries may impede recovery. Their high feeding costs limit their dive time and hence they may be particularly reliant on the availability of dense prey aggregations (Acevedo-Gutierrez et al. 2002). (b) Indirect interaction: krill surplus hypothesis (Laws 1977) (see text).
	Humpback whale *Megaptera novaeangliae*	*Direct/indirect	Cosmopolitan	Southern Hemisphere—possible direct competition with fisheries for krill. Northern Hemisphere—possible direct competition for herring, capelin, and sand-lance. Possible competition for cod off Iceland and Norway (Stefansson et al. 1997). Hypothesized indirect interaction caused by collapse of Canadian capelin fishery (Whitehead and Carscadden 1985) (see text).
Order Odontoceti	Whales and dolphins			
Family Physeteridae	Sperm whale *Physeter macrocephalus*	Direct	Cosmopolitan	Unlikely unless major expansion of deepwater squid fisheries. Ostensibly an important competitor in the western North Pacific (Government of Japan 2002); In the Southern Hemisphere competition very likely for deep-sea species such as orange roughy (*Hoplosthethus atlanticus*) and Patagonian toothfish (*Dissostichus eleginoides*) (e.g., off Macquarie Island) (Ashford et al. 1996, Goldsworthy et al. 2001).
Family Phocoenidae	Dall's porpoise *Phocoenoides dalli*	Direct	North Pacific Ocean	Possible because diet includes several commercially fished species such as hake, herring, mackerel, capelin. Conflicts off Japan (Kishiro and Kasuya 1993).
	Harbor porpoise *Phocoena phocoena*	*Direct/indirect	Coastal Northern Hemisphere	Diet includes commercially fished species such as herring, mackerel, and anchovy; coastal distribution renders interactions likely (e.g., MacKenzie et al. 2002, Notarbartolo di Sciara et al. 2002).
	Finless porpoise *Neophocaena phocaenoides*	Direct/indirect	Asian coastal waters	Diet includes small fish, prawns, squid, cuttlefish; may be vulnerable because of living near large population centers in Asia that have depleted local fish stocks (e.g., Yu 2002).

continued

Table 3.1 continued

Taxonomic group	Common and scientific name	Form of possible competition/interaction	Geographic location	Notes
Family Dephinidae	Indo-Pacific humpback dolphin *Sousa chinensis*	Direct	Indian Ocean and western Pacific Ocean	Possible competition with nearshore fisheries because their restricted distribution often overlaps with heavily fished areas (Peddemors 1999).
	Atlantic white-sided dolphin *Lagenorhynchus acutus*	Direct/indirect	North Atlantic	(a) Direct competition unlikely, although there is some overlap with commercially fished species. (b) Indirect: decline in herring in the Gulf of Maine (possibly as a result of overfishing) resulted in an increase in sandlances, thought to be a major prey item of these dolphins (Kenney et al. 1996).
	Pacific white-sided dolphin *Lagenorhynchus obliquidens*	Direct	North Pacific	Perceived competition with fisheries, e.g., culling operations off Japan during the 1970s (Kishiro and Kasuya 1993). Overlaps with fisheries in inshore waters for small schooling fishes.
	Striped dolphin *Stenella coeruleoalba*	Direct	Worldwide in tropical to warm temperate waters	Unlikely although possible threats from overfishing in the Mediterranean (Notarbartolo di Sciara et al. 2002, Bearzi et al. 2003).
	Short-beaked common dolphin *Delphinus delphis*	*Direct	Tropical to warm temperate waters worldwide	In Black Sea, e.g., expanding anchovy and sprat fisheries thought likely to negatively affect local dolphin populations (Birkun 2002). Perceived as a threat to the purse-seine and trawl fisheries in the Mediterranean, and reduced prey availability considered a factor in a population decline in this region (Bearzi et al. 2004).
	Common bottlenose dolphin *Tursiops truncatus*	*Direct	Cosmopolitan	Perceived competition with fisheries, especially in nearshore areas (e.g., Kasuya 1985). Overfishing in the Mediterranean may be a threat to local populations, with some descriptions of individuals being emaciated (Politi et al. 2000, cited in Bearzi et al. 2003).
	Risso's dolphin *Grampus griseus*	Direct	Tropical and temperate waters worldwide	Alleged competition with commercial fisheries (e.g., killed in Japan) (Kasuya 1985, Kishiro and Kasuya 1993). Generally unlikely barring intensive development of squid fisheries (Northridge 1991b).
	False killer whale *Pseudorca crassidens*	Direct	Worldwide in tropical and warm temperate	Perceived competition with, e.g., yellowtail fishery off Japan (Kasuya 1985).
	Killer whale *Orcinus orca*	*Direct/indirect	Cosmopolitan	(a) Direct: depends on locality but, e.g., possible competition with herring fisheries in coastal Norway; salmon (*Oncorhynchus* spp.) stock declines may negatively affect killer whale pods resident off western North America (Ford et al. 1998). (b) Indirect: possible altered diet because of overfishing (Estes et al. 1998) (see text for details).
Subfamily Orcaellinae	Irrawaddy dolphin *Orcaella brevirostris*	Direct	Coastal Indo-west Pacific	In coastal and fresh waters, reduced food supplies from overfishing may be a problem (Bannister et al. 1996).
Order Pinnipedia Superfamily Phocoidea	Seals and sea lions			
Family Phocidae ("true" seals)	Hawaiian monk seal *Monachus schauinslandi*	*Direct	Hawaii	Population decline at French Frigate Shoals possibly linked to commercial overfishing of lobsters and associated bycatch species that are important in the diet of these seals (Marine Mammal Commission 2001).

	Mediterranean monk seal *Monachus monachus*	Direct	Mediterannean and Northwest African coast	Threatened species because of several other factors but cannot exclude possibility of reduced food supply contributing to pup mortality (González et al. 1997, Gucu et al. 2004). Deliberately killed in the past because perceived to be competitor for fish and squid.
	Southern elephant seal *Mirounga leonina*	Direct	Southern Hemisphere	Concern that the ever-increasing volume of squid caught by commercial fishermen may be negatively impacting Península Valdés population. Possible overlaps if deepwater fisheries expanded. Off Macquarie Island, e.g., fisheries for Patagonian toothfish considered a threat to the elephant seals, who depend heavily on this deep-sea species. Population declines have been linked to food availability (Hindell 1991, Pistorius et al. 1999), although not necessarily to fishing.
	Crabeater seal *Lobodon carcinophaga*	Direct/indirect	Southern Ocean	(a) Direct: possible direct competition for krill, its primary prey, if commercial krill harvests increase. (b) Indirect: krill surplus hypothesis (Laws 1977).
	Hooded seal *Cystophora cristata*	Direct	North Atlantic Ocean	Suspected direct competition for Greenland halibut and redfish in the Gulf of St. Lawrence, although seals primarily target prerecruits to the commercial fishery (Hammill et al. 1997).
	Gray seal *Halichoerus grypus*	*Direct	North Atlantic Ocean	Large annual losses to Canadian fisheries claimed (but not substantiated) in the past based on allegations of substantial dietary overlaps with commercial demersal species (Northridge 1991b). Harwood and Croxall (1988) found little evidence for competition with fisheries. On the eastern Scotian Shelf seals implicated in causing the decline and preventing the early recovery of the cod stock, but model results suggest this is unlikely (Mohn and Bowen 1996).
	Harbor seal *Phoca vitulina*	*Direct	Northern Hemisphere	Potential for competition, e.g., in the Moray Firth, Scotland, seals showed evidence of reduced body condition when clupeids absent from inshore waters (Thompson et al. 1997). Decline in western Gulf of Alaska population correlated with an an increase in commercial trawl fisheries, for shrimp in particular (Pitcher 1989, 1990). Killed both historically and currently (e.g., in Norway and Canada) because of perceived competition with commercial fishermen.
	Harp seal *Pagophilus groenlandicus*	*Direct/indirect	North Atlantic and White Sea	(a) Interactions likely as the harp seal is the most abundant Northern Hemisphere pinniped. Implicated as possibly retarding recovery of cod stocks off Newfoundland and Labrador. Direct competition considered a distinct possibility if fisheries for polar cod (*Boreogadus saida*) are expanded further in the Barents Sea, particularly given the link between harp seal condition and the collapse of the Barents Sea capelin stock in the 1980s (Nilssen et al. 1997, 2000, Frie et al. 2003). Considered a major predator on capelin in the Barents Sea ecosystem (Nilssen et al. 2000). (b) Indirect interactions possible as Atlantic cod (a commercially sought after species) feed on polar cod, also eaten by the seals. Harp seals feed on juvenile Atlantic cod, which are not recruited into the commercial fishery (Stenson et al. 1997).
Superfamily Otariidae Subfamily Otariinae	Steller sea lion *Eumetopias jubatus*	*Direct/indirect	North Pacific	(a) Direct: Population decline possibly linked to decline in food supply partly attributable to commercial fishing (Committee on the Alaska Groundfish Fishery and Steller Sea Lions 2003). (b) Indirect: possible competitive release effect re pollock (Rosen and Trites 2000) (see text for details).
	California sea lion *Zalophus californianus*	Direct	California	Potential for competition with fisheries because of overlap of feeding and fishing habitats (García-Rodríguez and Aurioles-Gamboa 2004).

continued

Table 3.1 continued

Taxonomic group	Common and scientific name	Form of possible competition/interaction	Geographic location	Notes
	Southern sea lion *Otaria flavescens*	Direct	South America	Possible competition with expanding commercial fisheries for squid and nearshore fish species. Off Patagonia possible conflicts with commercial hake (*Merluccius hubbsi*), squid (*Illex argentinus*), and anchovy (*Engraulis anchoita*) fisheries as these constitute the main common prey of sea lions (Koen-Alonso et al. 2000, Koen-Alonso and Yodzis 2005). Calls for reductions as a result of perceived competition, e.g., off Uruguay. Possibly only minor impact on harvests of jack mackerel (*Trachurus symmetricus*) off central Chile (George-Nascimento et al. 1985, Hückstädt and Antezana 2003).
Subfamily Arctocephalinae	New Zealand fur seal *Arctocephalus forsteri*	Direct	New Zealand and Australia	Some conflicts because diet includes some commercial finfish species such as hoki (*Macruronus novaezelandiae*) (Carey 1992), but apparently eat predominantly noncommercial species such as lanternfish (Myctophidae), anchovies, octopus (Harcourt et al. 2002).
	Antarctic fur seal *Arctocephalus gazella*	*Direct/indirect	Antarctic	(a) Direct: commercial trawl fisheries pose a threat if allowed to expand near breeding colonies such as those at South Georgia and the Antarctic Peninsula (Barlow et al. 2002, Boyd 2002). (b) Indirect: krill surplus hypothesis (Laws 1977).
	2 subspp. (a) Cape fur seal *Arctocephalus pusillus pusillus* (b) Australian fur seal *A. p. doriferus*	*Direct	Southern Africa and Australia	South Africa—perceived competition with hake fishery in particular (see text) and possible competition with other fisheries as well (Punt and Butterworth 1995). Australia—more than half the diet consists of commercially targeted fish and squid species; hence concerns about conflicts with fisheries (Southern Squid Jig Fishery in particular) given that the seal population is rapidly recovering from past hunting (Arnould 2002, Hume et al. 2004).
	Northern fur seal *Callorhinus ursinus*	Direct/indirect	North Pacific ocean, Bering Sea, Sea of Okhotsk	Posited competition with Alaska pollock fishery—interaction is indirect because seals prey on smaller pollock than those targeted by the fishery, but these smaller fish are in turn cannibalized by the larger fish (Trites et al. 1999) (see text). Fisheries management actions posited to affect the foraging patterns of the seals (Robson et al. 2004).
Family Odobenidae	Walrus *Odobenus rosmarus*	Direct/indirect	Arctic	(a) Direct: potential competition with shellfish fisheries (Born et al. 2003). (b) Indirect: habitat destruction from bottom trawling.
Order Carnivora Family Mustelidae	Otters Sea otter *Enhydra lutris*	*Direct/indirect	North Pacific nearshore waters	(a) Direct: perceived competition with abalone fishery off California (Fanshawe et al. 2003). (b) Indirect: overfishing claimed responsible for pinniped decreases with the subsequent reduction in prey availability to killer whales resulting in their switching to preying on sea otters (Estes et al. 1998).
	Marine otter *Lontra felina*	Direct	West coast of South America	Possible direct competition with fishermen for nearshore marine fish and shellfish (Medina-Vogel et al. 2004).

Note: This list includes marine mammals for which potential feeding-related interactions with fisheries have been identified in the literature, as well as examples where a case can be made that there is potential for a problematic interaction. Inferences presented here are preliminary and largely speculative because of the lack of information and the difficulties in conclusively demonstrating linkages. Taxonomic and distribution information is based mostly on Perrin et al. (2002) and Reeves et al. (2002a). The form of possible competition is described as direct if the marine mammal species and fishery potentially compete for a common food source (see Fig. 3.1a,b,c,f) and as indirect if a fishery potentially affects a predator or prey species that is in turn the prey or predator, respectively, of another species fed upon by the marine mammal species (see Fig. 3.1d,e). Species thought to be appreciably affected (currently or historically) or particularly vulnerable to indirect fishery interactions are marked with an asterisk in the third column.

energetics model, Stenson et al. (1997) estimated that in the Gulf of St. Lawrence alone harp seals annually consume some 445,000 tons of capelin (*Mallotus villosus*), 20,000 tons of polar cod (*Boreogadus saida*), and 54,000 tons of Atlantic cod. There is some overlap between the sizes of capelin taken by the seals and by the commercial capelin fishery, but the juvenile Atlantic cod age classes targeted by the seals are smaller than those recruited into the commercial fishery. Nonetheless, there is an obvious temptation to argue a causal relationship between the failure of the cod population to recover as rapidly as expected after its protection and the increase in harp seal abundance, especially considering that the socioeconomic implications of the collapse of the cod fishery were substantial, as some 40,000 fishermen were put out of work. Although the results of at least one ecosystem modeling study (Trites et al. 1997) support the hypothesis that the recovery of these cod populations is being retarded to some extent by the increased biomass of harp seals, ecosystem models generally have poor predictive reliability, largely because of data limitations.

Off the west coast of South Africa, Cape fur seals are estimated to consume almost as much hake as is taken by the commercial fishery (Punt and Butterworth 1995; Fig. 3.2). There are however, two species of commercially valuable hake: a shallow-water (*Merluccius capensis*) and a deep-water species (*M. paradoxus*), with the larger of the former eating the smaller individuals of the latter. The results of multispecies models suggest that the net effect of a seal reduction would likely be fewer hake overall because fewer seals (which are thought to prey primarily on shallow-water hake) would mean more shallow-water hake and hence more predation on small deep-water hake (Punt and Butterworth 1995). This relatively simple example highlights the complexity of predation, food-fish, and fishery interactions and hence the difficulties of demonstrating conclusively that marine mammals are in direct competition with humans for food fish, as may superficially appear to be the case.

Whales

Numerous multispecies modeling studies have investigated the direct and indirect effects of common minke whales (*Balaenoptera acutorostrata*) on the cod, herring (*Clupea harengus*), and capelin fisheries in the Greater Barents Sea (e.g., Schweder et al. 2000). Minke whales are abundant in this region and prey on all three species, prompting the question of whether fishermen could expect larger catches if the populations of minke whales were reduced. These studies indicate that there is competition between the whales and fisheries in this region and that the effect on the fisheries would correspond roughly linearly to changes in whale abundance. They estimate that each minke whale reduces the potential annual sustainable catches of both cod and herring by some 5 tons (Schweder et al. 2000). Similarly, studies off Iceland suggest that the piscivorous minke, humpback (*Megaptera novaeangliae*), and fin whales (*Balaenoptera physalus*) may have a considerable impact on the region's cod stock (Stefansson et al. 1997). The cod fishery is of key importance to the Icelandic economy, and the rebuilding of the cod population and catches is recognized as an important economic consideration. It is therefore not surprising that there have been arguments in favor of reducing whale populations so that commercial fisheries can increase.

It is noteworthy that the main focus of the Japanese Whale Research Program in the North Pacific (JARPN II) has shifted from a study of the stock structure of minke whales to feeding ecology and ecosystems research (Government of Japan 2002). Minke whales show seasonal and geographical changes in diet, and there might be direct competition between minke whales in the western North Pacific and the fishery for Pacific saury (*Cololabis saira*) in particular (Tamura and Fujise 2002). As part of the JARPN II program, an ECOPATH with ECOSIM (Walters et al. 2000) modeling approach has been applied to the western

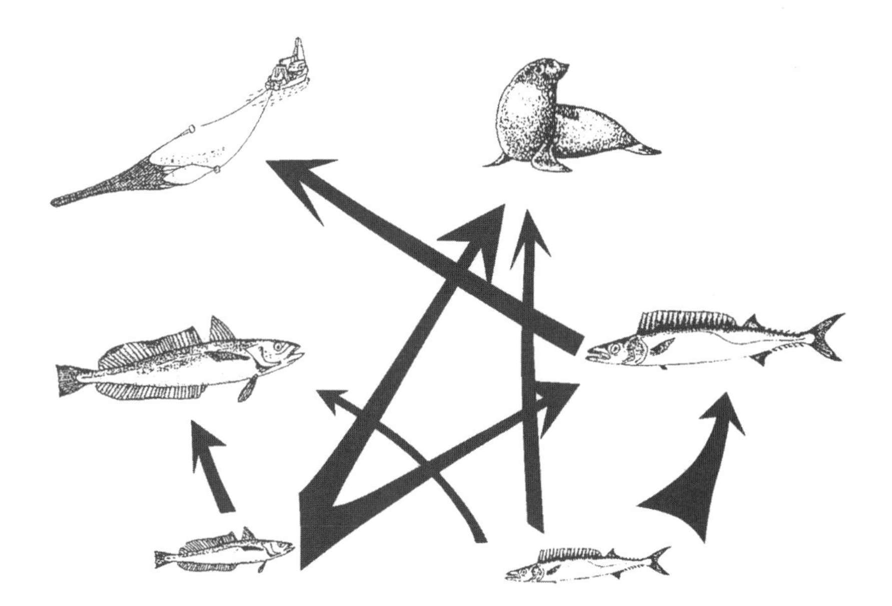

Fig. 3.2. Schematic illustration of the complexities of predation, food-fish, and fishery interactions as summarized in a minimal realistic model of South African fur seals and Cape hake interactions off South Africa. (From Punt 1994.)

North Pacific, with results suggesting that removal of both minke and sperm (*Physeter macrocephalus*) whales would result in increases in the biomass of prey species (Government of Japan 2002). Bryde's whales (*Balaenoptera edeni*) are also an important species for consideration in the Japanese study, given that they are abundant and reportedly play an important role in the western North Pacific ecosystem. Work is also in progress to develop a MULTSPEC-type model (Bogstad et al. 1997) incorporating minke whales, Pacific saury, Japanese anchovy (*Engraulis japonicus*), and krill (*Euphausia superba*) (Government of Japan 2002). However, questions have been raised as to whether the modeling approaches to be applied are overly simplistic as they fail to take enough species into account and do not incorporate adequate feedback mechanisms and second-order effects (International Whaling Commission 2005).

Marine mammals have been identified as potentially serious competitors, for example, off the northeastern United States, a region that includes important fishery areas such as the Gulf of Maine and Georges Bank (Overholtz et al. 1991, Kenney et al. 1996). The latter area exemplifies the conflicts that can arise between fishery management plans tasked with rebuilding prey populations and requirements, in this case by the U.S. Marine Mammal Protection Act, to facilitate an increase in the abundance of marine mammal predators.

Small Cetaceans

In many areas of the world, coastal fishermen consider dolphins to be serious competitors, and fisheries have embarked on culls or control programs. In some instances a bounty has been offered for culling dolphins (e.g., Kishiro and Kasuya 1993). In the Mediterranean, both short-beaked common dolphins (*Delphinus delphis*) and common bottlenose dolphins (*Tursiops truncatus*) were culled in the past because of inter alia their reputation as competitors with fisheries (Bearzi et al. 2003).

One of the largest culls on record occurred at Iki Island, Japan, in response to declines in catches of yellowtail (*Seriola dorsalis;* Kasuya 1985). From 1976 to 1986, drive-fishing was responsible for the deaths of thousands of bottlenose dolphins and hundreds of Pacific white-sided dolphins (*Lagenorhynchus obliquidens*), false killer whales (*Pseudorca crassidens*), and Risso's dolphins (*Grampus griseus*) (Kasuya 1985).

Sea Otters

Sea otters prey on a variety of marine invertebrates such as urchins and abalone (*Haliotis* spp.). Off southern California, southern sea otters (*Enhydra lutris nereis*) have been blamed by some for the decline of the red abalone (*Haliotis rufescens*) fishery (Fanshawe et al. 2003), but there is little direct evidence to support this notion. The area's commercial abalone fishery was closed in late 1997, and factors such as commercial fishing, disease, and changing environmental conditions are all thought to have contributed to the decline of these commercially valuable shellfish (Tegner et al. 2001). Although the southern sea otter population is not increasing, there are perceptions that the level of competition between otters and commercial abalone fishermen has increased because of the recent southward movements of otters, which has enlarged the overlap of otter feeding grounds and abalone fishing areas (Fanshawe et al. 2003).

INSTANCES OF POTENTIALLY DETRIMENTAL EFFECTS OF FISHERIES ON MARINE MAMMAL POPULATIONS

Pinnipeds

The western Alaska population stock of Steller sea lions (*Eumetopias jubatus*) has shown a continuous decline since the 1970s (Loughlin et al. 1992, Marine Mammal Commission 2004) and was listed in 1997 as an endangered species under the U.S. Endangered Species Act. Several groups have argued that this decline is due in part to the large fishery harvests of walleye pollock (*Theragra chalcogramma*) and Atka mackerel (*Pleurogrammus monopterygius*), which are both key sources of food for sea lions and the basis for commercially valuable U.S. fisheries (Committee on the Alaska Groundfish Fishery and Steller Sea Lions 2003). This argument is supported by research demonstrating considerable overlap in the sizes of these species taken by the western stock of sea lions and the commercial trawl fishery (Zeppelin et al. 2004). Measures to reduce the perceived competition between sea lions and fisheries for groundfish stocks include the establishment of "buffer" (no-trawl) areas to include important locations where the sea lions breed, feed, and rest, as well as requiring the pollock harvest to be more evenly distributed over the remaining areas and spread throughout the year (Marchant 2001).

However, modeling studies (Trites et al. 1999) suggest that the observed decline in the sea lion population is unlikely to be explained solely through trophic interactions and is more likely linked to shifts in environmental conditions that have led to changes in the favored complex of species. Moreover, Trites et al. (1999) and others highlight the difficulties of predicting the direction and magnitude of a change in an ecosystem arising from a reduction in predation or fishing pressure. They posit that, paradoxically, Steller sea lion and northern fur seal (*Callorhinus ursinus*) populations might realize greater benefits if adult pollock and large flatfish were more heavily fished. This competitive release effect may occur because, for example, pollock are cannibalistic and hence decreased adult pollock abundance as a result of heavier fishing may result in increased numbers of juvenile pollock available to marine mammals.

An alternative hypothesis, the so-called "junk-food" hypothesis (Rosen and Trites 2000), links the decline of Steller sea lions in the Gulf of Alaska and the Aleutian Islands to the dominance of pollock (a gadoid fish) in their diet instead of higher-quality (in terms of energy content) fishes such as Pacific herring (*Clupea pallasi*) and sandlance (*Ammodytes hexapterus*). In a similar vein, Loughlin and York (2000) conclude that nutritional stress cannot be excluded as a possible contributory source of Steller sea lion mortality. Furthermore, Winship and Trites (2003) recently estimated that in southeastern Alaska, seasonal changes in the energy density of the diet of Steller sea lions (the threatened eastern stock, in this case) result in the animals requiring approximately 45–60% more food a day in early spring than in late summer. Regional differences in the energy density of the diet similarly account for substantial differences (up to 24% based on summer diets) in food requirements among the southeastern and western Steller sea lion populations (Winship and Trites 2003). Thus it has been suggested that it may be important to consider not only the quantity but also the distribution and quality of food available to marine mammal populations (Trites and Donnelly 2003).

The role of commercial fisheries (targeting pollock and herring) compared to the environment (e.g., through a shift in ocean climate in the late 1970s) in altering the forage fish abundance and community structure of this region remains unclear. However, the Endangered Species Act listing of the eastern stock of Steller sea lions has drawn attention to the need for analyses of indirect fisheries/marine mammal interactions. The Committee on the Alaska Groundfish Fishery and Steller Sea Lions (2003) concluded that the reasons for the observed decline in the western stock of Steller sea lions are still open to debate and that there is currently insufficient evidence to exclude fisheries as a contributory factor.

A second example of a probable fisheries-generated lack of food concerns the mass migration of harp seals south along the coast of Norway in 1987 in response to the collapse of the capelin stock in the mid-1980s (Nilssen et al. 1997, Livingston and Tjelmeland 2000). Heavy fishing played an important role in that collapse (Gjøsæter 1998), as well as in earlier declines in the cod and herring stocks, with the associated sequence of events further highlighting the complexities of cod, capelin, herring, and fisheries interactions in the Barents Sea. The effects of fishing should generally be easier to detect in relatively simple ecosystems such as the Barents Sea, where most of the energy is channeled through a few species positioned at an intermediate level in the food web (Gislason 2003).

Whales

Competition effects are difficult to quantify, but Whitehead and Carscadden (1985) proposed that the collapse of the eastern Canadian capelin fishery in the 1970s had a negative effect on fin whales. They suggested that a shortage of capelin might have allowed humpback whales to out-compete fin whales because the latter rely principally on capelin as a prey source.

A competitive predation between a marine mammal and a fishery implies that the marine mammal population is limited by the food available and hence that it should be possible to demonstrate a response of some vital population parameters to a change in that availability. Recent probable population increases of several krill-eating marine mammals, such as Antarctic minke whales (*Balaenoptera bonaerensis*), crabeater seals (*Lobodon carcinophagus*), and Antarctic fur seals (*Arctocephalus gazella*), have been attributed by some investigators to a probable large increase in the availability of krill in Antarctic waters. Following the substantial reduction through overexploitation of large whale populations in the early twentieth century, some researchers argued that about 150 million tons of "surplus" krill became available to other predators annually. However, this "krill surplus" theory, enunciated by Laws (1977), has yet to be universally accepted. Supporting evidence has been sought (and argued to have been found) by addressing questions such as: (a) whether the mean age at maturity of minke whales (Thomson et al. 1999) and crabeater seals (Bengtson and Laws 1985) has dropped in more recent years as a response, which might result from an increase in food availability, and (b) whether minke whale assessments based upon catch-at-age data indicate population increases (Butterworth et al. 1999). Clapham and Brownell (1996) stress several difficulties in conclusively linking purported changes in the abundance and reproduction of baleen whales generally to interspecific competition following large-scale commercial exploitation, including insufficient data on levels of prey biomass and predator consumption, the unconfirmed validity of many of the purported changes, a wide range of alternative explanations, and a lack of sufficient data to discriminate among the foregoing.

Small Cetaceans

Dolphin populations that have localized coastal distributions, such as the Pacific humpback dolphins off KwaZulu-Natal, South Africa, may be vulnerable to commercial fishery expansions because of increased competition with fisheries for limited food resources (Peddemors 1999).

In the Mediterranean and the Black seas, prey depletion is considered the primary or secondary cause of habitat degradation and the loss of at least four marine mammal species: bottlenose dolphins, short-beaked common dolphins, striped dolphins (*Stenella coeruleoalba*), and harbor porpoises (*Phocoena phocoena*) (Notarbartolo di Sciara et al. 2002). Declines in short-beaked common dolphins have also been attributed to a number of human-induced threats, including both direct and incidental catches by the fishery and contamination by xenobiotics, as well as to global climate change (Bearzi et al. 2003). Of all of these, prey depletion is the most likely proximate cause of the observed trends (Politi and Bearzi 2004). Other marine mammals in the Mediterranean, such as bottlenose dolphins, also show signs of malnutrition, presumably as a response to overfishing of their prey stocks and intensive trawling (Politi et al. 2000, cited in Bearzi et al. 2003). An additional complication in assessing the impacts of prey depletion concerns its role in compromising animal health, as was possibly the case in the large die-off of Mediterranean striped dolphins (as the result of an epizootic) in 1990–1992 (Aguilar 2000).

The decline in sardines and anchovies, previously preferentially targeted by fishermen in some areas of the Mediterranean, has caused a shift in focus to other small pelagic fish [e.g., round sardinella (*Sardinella aurita*) and garpike (*Belone belone*)] that are targeted by common dolphins and also needed by the growing aquaculture industry (Bearzi et al. 2003).

Sea Otters

Dramatic declines in northern sea otter (*Enhydra lutris kenyoni*) populations have been indirectly linked to competition with fisheries. As noted previously, overfishing is argued to be one of the factors in the decline of pinniped populations (harbor seals [*Phoca vitulina*] and Steller sea lions) in Alaska's Aleutian Islands (Loughlin and York 2000). Killer whales (*Orcinus orca*) in this region may feed preferentially on pinnipeds (e.g., Heise et al. 2003), but the pinniped decline may have caused them to switch to sea otters as prey, as described in Estes et al. (1998). That work argues that overfishing populations of high-caloric and nutritive value that serve as pinniped fish prey may have an impact not only on the latter populations but also indirectly on killer whales and sea otters.

Sirenians (Dugongs and Manatees)

Dugongs and manatees feed primarily on vascular aquatic plants so that there is no direct competition with humans for a shared food resource. Rather there may be indirect interactions that are due to the degradation and loss of their habitat, which is partly attributable to fishing activities such as trawling (United Nations Environment Programme 2002).

FOOD WEB INTERACTIONS

Trites et al. (1997) and Kaschner et al. (2001) estimate the degree of competition between fisheries and marine mammals for prey and primary production in the Pacific and North Atlantic oceans, respectively. In both instances, marine mammals are estimated to consume about three times the amount of food that humans harvest, but mostly of species not of current commercial interest. Kaschner (2004) presents similar assessments based on a global analysis of food consumption and catch by marine mammals and fisheries. Trites et al. (1997) estimate that the greatest overlaps in catches and prey in the Pacific occur for pinnipeds (60%) and dolphins and porpoises (50%), but such estimates are likely subject to appreciable uncertainty. Kaschner (2004) estimates that in the Northern Hemisphere resource overlaps are greatest for these same groups, whereas in the Southern Hemisphere overlaps for baleen whales and large toothed whales are the most substantial.

Trites et al. (1997) argue that although direct competition for prey between fisheries and marine mammals appears limited, indirect competition for primary production may have an appreciable effect. Such so-called food web competition may occur if there is overlap between the trophic flows supporting the two groups (see Fig. 3.3). Evidence in support of marine mammal/fisheries food web competition is provided by a negative correlation between the estimate of the primary production required to support fisheries

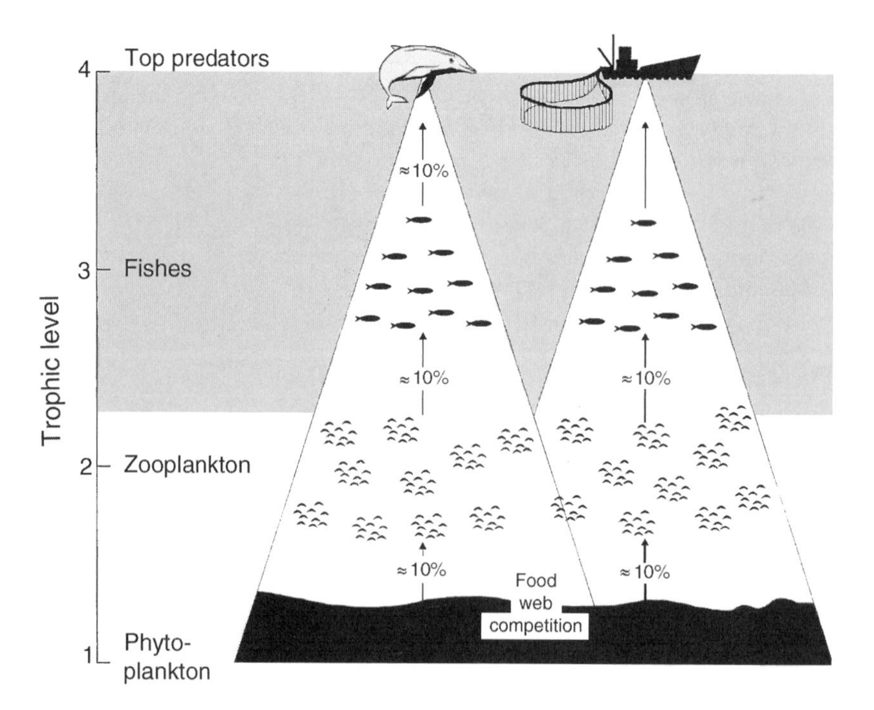

Fig. 3.3. Schematic example of indirect competition for food by marine mammals and fisheries. The representation shows how top predators, such as marine mammals, may be affected by fisheries because of limits on the primary productivity available to support the two groups. Thus, even though the mammals' prey and species taken by fisheries may not overlap, so-called food web competition occurs at the base of the food pyramids. (From Trites et al. 1997.)

catches and that to support the number of marine mammals thought to be present in the different FAO statistical areas that span the Pacific Ocean (Trites et al. 1997). Nonetheless, the conclusions to be drawn here depend on the extent to which these estimates are considered reliable.

METHODS TO ASSESS FEEDING-RELATED INTERACTIONS AND THEIR APPLICABILITY

Whereas commercial fishing interests in many parts of the world perceive marine mammals as serious competitors for scarce resources, many others argue that marine mammals are being used as scapegoats for management policies in failed fisheries (Butterworth 1999). Scientific evidence is therefore increasingly being sought to settle these disputes, but there is growing appreciation that the scientific methodologies required to address them are complex, time-consuming, and beset with difficulties. Rice (2000) notes that even though most modeling frameworks give competition a strong role in structuring marine communities, this is questionable considering that, after decades of research in terrestrial ecosystems, the role of competition there remains unresolved.

SIMPLE APPROACHES

Initial approaches (e.g., Beverton 1985) to quantify the impact of consumption by marine mammals on fish catches took account of the fact that, particularly for pinnipeds, the fish that are eaten tend to be smaller than those taken in commercial fisheries. There is thus not an exact correspondence between a ton of a commercially desired fish species eaten by seals, say, and a ton caught by a fisherman. This follows because, although a fish eaten by a seal would have grown larger by the time it became vulnerable to fishing, it might also have died before reaching that size as a result of other sources of natural mortality, such as consumption by other predators.

Such computations, which essentially treated marine mammals as the equivalent of another fishing fleet, are now generally considered to be inadequate because of oversimplification, particularly as they disregard potentially important secondary (indirect) interactions. To perform more realistic computations while still accepting that both data and computing power limitations necessarily restrict the degree of complexity that is viable for multispecies models, the following three complicating factors need to be addressed:

- How many of the large number of interacting species in any ecosystem have to be considered and what are the functional forms of these interactions?
- Do age-structure effects have to be taken into account? This can become important when, for example, one species that preys on the small juveniles of a second finds itself the prey of the larger adults of that same species.
- Is the assumption that species interactions occur homogeneously over space adequate or an oversimplification? Accurate representation of spatial overlaps in prey availability and marine mammal feeding distributions are complicated by the fact that for pinnipeds, for example, the distribution of breeding and resting sites does not necessarily reflect their feeding distributions (see, e.g., Bradshaw et al. 2003).

More recent attempts to quantify marine mammal/fishery interactions have taken account of some or all of the above complicating factors although studies have not yet reached the stage where all three are satisfactorily considered simultaneously. The scientific methodologies required to address these aspects are complex, time-consuming, and beset with difficulties as is illustrated through selected examples below.

MORE COMPLEX MODELING APPROACHES

In discussing the methods currently being used to provide a more reliable basis for the scientific evaluation of the competitive effects at issue, it is useful to classify models as either "efficient predator" or "hungry predator" models (Butterworth and Plagányi 2004). In the former, the predator is assumed to always get its daily ration (e.g., MSVPA, MULTSPEC) although the species composition of this ration may change over time with varying prey abundances. In contrast, in the latter, a predator is assumed to compete with others of the same (and possibly other) species for limited vulnerable proportions of prey (e.g., "foraging arena," based models applied in approaches such as ECOSIM; Walters et al. 1997).

A second useful classification groups models as minimum realistic models (MRM) on the one hand and "ecosystem" models on the other. An MRM seeks to include only those species considered likely to have important interactions with the species of primary interest. The MRM group includes MSVPA, MULTSPEC, BORMICON/GADGET, Seastar, Scenario Barents Sea, and the original seal-hake MRM of Punt and Butterworth (1995). Shared characteristics of these models include the following (North Atlantic Marine Mammal Commission 2002):

- They are system-specific.
- Only a small, selected component of the ecosystem is modeled.
- Lower trophic levels and primary production are modeled as constant or varying stochastically.

In contrast, the ECOPATH/ECOSIM models, for example, are generic and capable of explicitly including most ecosystem components as well as incorporating lower trophic levels and primary production, although naturally they can also be applied in a simplified form closer to the MRM concept.

A general problem in reviewing the suite of models currently available to explore indirect marine mammal/fisheries interactions is that many of these models have essentially been constructed using a generic template tailored to fish species. Fish generally show a much greater range of values of their basic life history parameters than do marine mammals (Plagányi and Butterworth 2004). Moreover, in species such as large mammals, which have low reproductive rates and long life spans, most density-dependent changes in vital rates occur at high population levels (Fowler 1981) so that the levels providing the maximum sustainable yield tend to be those closer to carrying capacity. In contrast, density-dependent changes occur at low population levels for species with life history strategies typical of most fishes.

ECOPATH with ECOSIM (EwE)

Because the ECOPATH (Polovina 1984, Christensen and Pauly 1992), ECOSIM (Walters et al. 1997, Christensen et al. 2000), and ECOSPACE (Walters et al. 1999) suites currently dominate attempts worldwide to provide information on how ecosystems are likely to respond to changes in fishery management practices, it is critical to review the applicability of these approaches to answering questions in the context of marine mammals and indirect interactions (Aydin and Friday 2001, Aydin 2004, Plagányi and Butterworth 2004). ECOPATH is a static mass balance approach based on equations originally proposed by Polovina (1984) for estimating biomass and food consumption of species (or groups of species) within an aquatic ecosystem. The production/consumption relationships for each functional group i in an ecosystem are described by

$$B_i(P/B)_iEE_i = Y_i + \sum_j B_j(Q/B)_jDC_{ji} + BA_i \qquad (1)$$

where B_i and B_j are the biomasses of i and the consumers (j) of i, respectively; $(P/B)_i$ is the production/biomass ratio for i; EE_i is the fraction of production of i that is consumed within, or caught from the system (the balance being assumed to contribute to detritus); Y_i is the fisheries catch ($Y = FB$; F is the proportion fished); $(Q/B)_j$ is the food consumption per unit biomass of j; DC_{ji} is the fractional contribution by mass of i to the diet of j; and BA_i is a biomass accumulation term that describes a change in biomass over the time period studied and/or net immigration (Christensen 1995).

The ECOSIM model (in its simplest form) converts these steady-state trophic flows into dynamic, time-dependent predictions by assuming that the trophic flows Q_{ij} between prey pools i and predator pools j are determined as follows:

$$Q_{ij} = a_{ij}v_{ij}B_iB_j/(v_{ij} + v'_{ij} + a_{ij}B_j) \qquad (2)$$

where a_{ij} is the rate of effective search for prey type i by predator j, and v_{ij}, v'_{ij} are prey vulnerability parameters, with $v_{ij} = v'_{ij}$ as the default setting (Walters et al. 2000). As in the classic Lotka-Volterra formulation ($Q_{ij} = a_{ij}B_iB_j$), flows are determined by both prey and predator biomasses, but equation (2) incorporates an important modification in that it encompasses a framework for limiting the vulnerability of a prey species to a predator, thereby including the concept of prey refugia and also tending to dampen the unrealistically large population oscillations usually predicted by the Lotka-Volterra formulation (see, e.g., Mori and Butterworth 2004).

The recommended default vulnerability value ($v^{\star}_{ij} = 0.3$, where $v^{\star}$ is a rescaled value of v; see Plagányi and Butterworth 2004) implies that the recent predator population level input corresponds to the "half-saturation" point on the consumption curve as shown in Figure 3.4. Specifying such a default for all predators irrespective of their various prior exploitation histories seems open to question.

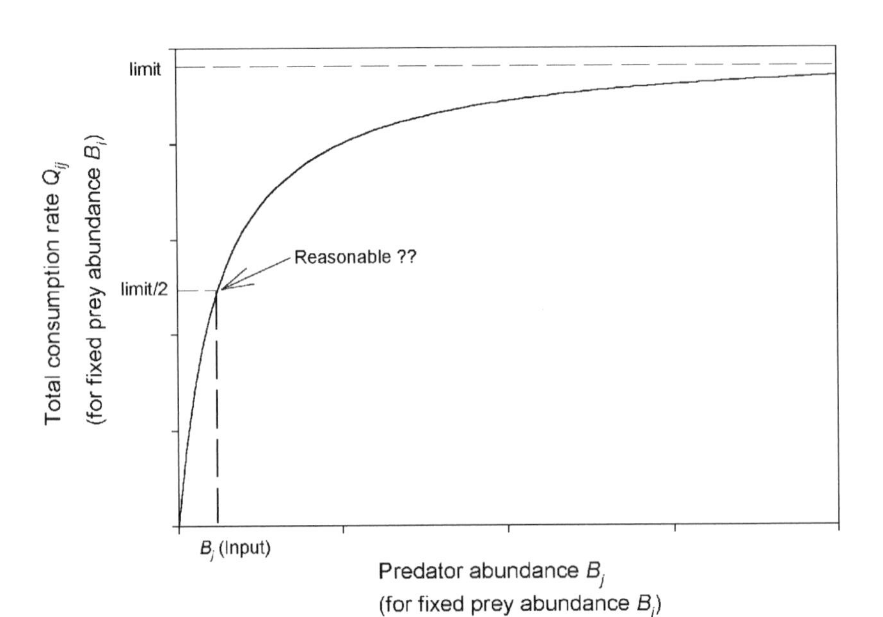

Fig. 3.4. A closer look at the ECOSIM foraging arena representation and its implications, when using a default vulnerability value of 0.3, for conclusions related to the predicted effects of marine mammal reductions on fishery yields (see text for details). The central question depicted in the figure is whether it is reasonable to assume that the point as shown is the present (input) situation irrespective of prior exploitation history (see text).

Recent versions of ECOSIM allow a "split pool" representation of juvenile and adult biomasses to allow representation of trophic ontogeny (Christensen and Walters 2004). ECOSPACE (Walters et al. 1999) is a dynamic version of ECOPATH that allows for spatial structure and also incorporates all the key elements of ECOSIM.

Plagányi and Butterworth (2004) list the following as some of the strengths of the ECOPATH with ECOSIM approach: (a) the structured parameterization framework; (b) the inclusion of a well-balanced level of conceptual realism; (c) a potentially improved representation of predator-prey interaction terms; and (d) the inclusion of a semi-Bayesian resampling routine (ECORANGER) to take account of the estimation uncertainty associated with model inputs. Some weaknesses noted include: (a) the constraining nature of the mass-balance assumption (of ECOPATH), which is used to provide initial values for ECOSIM model variables; (b) the questionable handling of some life history responses; and (c) the paucity of systematic and stepwise investigations into model behavior and properties. In particular, attention is drawn to the following limitations, which have to be borne in mind when interpreting any outputs of EwE pertaining to the question of the extent and nature of indirect interactions between marine mammals and fisheries.

BIOENERGETIC CONSIDERATIONS. In the interests of simplicity, EwE does not make allowances for detailed bioenergetic considerations (Aydin and Friday 2001, Aydin 2004), and unfortunately some of this information is critical in the current context. For example, alternative prey types are treated as energetically equivalent in EwE (note, e.g., in equation (1) that (P/B), with increasing food supply and a consequent increase in per capita predator consumption in terms of equation (2), animals may either, depending on user inputs (Walters et al. 2000): (a) allocate surplus to reproduction rather than somatic growth (food allocation hypothesis); or (b) spend less time foraging so as to decrease time at risk to predation so that the density-dependent response to population reduction would be in the form of a drop in natural mortality) (predation hypothesis).

Plagányi and Butterworth (2004) suggest that, for marine mammals in particular, there is a problem with both of these ECOSIM limits. Under changing conditions, natural mortality is conventionally considered one of the least labile population parameters for these animals. Furthermore, physiological limitations mean that above a certain threshold level there is generally no appreciable response of population variables to increasing food availability. One danger of inadvertently modeling marine mammal populations as overly compensatory (by treating them in the same way as fishes) is the possibility that this could lead to overly optimistic predictions of the likely recovery of populations following their reduction or of likely increases in response to increased food availability.

IMPLICATIONS OF THE ECOSIM INTERACTION REPRESENTATION. Given current pressures to accurately quantify the predicted effects of marine mammal reductions on fisheries or, conversely, the effects of increases/decreases in fishery catches on marine mammal populations, it is imperative that models employed for this purpose be closely scrutinized to understand the extent to which underlying model assumptions predetermine or have implications for the results obtained. This is particularly important in the case of "blackbox" types of models such as ECOSIM, so this aspect is addressed briefly here.

The ECOSIM "foraging arena" concept (Walters and Kitchell 2001), as embodied in equation (2), is a novel functional response representation that, according to these authors, is supported to some extent by studies of fish populations. However, EwE cannot straightforwardly depict instances where the foraging arena *V*'s (vulnerability pools) are used simultaneously by multiple predators. This may be important in instances in which a fish predator targets prey similar to marine mammal prey or where there are overlaps in the vulnerability pools available to marine mammals and to fisheries. The EwE as presently configured implicitly assumes that direct interference among predator species (which it ignores) is inherently different from within-species interference [explicitly modeled by equation (2)].

Plagányi and Butterworth (2004) argue that with default model settings, ECOSIM cannot yield pure-replacement results when predicting the effects of a "predator" (a fishing fleet, say, that acts exactly the same way in terms of prey selection) in supplanting marine mammals. This can be explained by considering, for example, the fundamental ECOSIM relationship between total consumption Q_{ij} of prey pool i by a marine mammal predator j and the abundance of predators B_j as shown in Figure 3.4. As is evident from this plot, if marine mammals (as predators) are present in high numbers (toward the right-hand side of the plot) and these numbers are subsequently halved (say, e.g., via a cull), when ECOSIM's default vulnerability values are used the total consumption Q_{ij} hardly drops because the per capita consumption (Q_{ij}/B_j) nearly doubles. The default parameter value selections for the model thus effectively hardwire it to such an extent that they are likely to effectively swamp other signals pertinent to predicting the effects of a marine mammal reduction. One's belief in the ECOSIM predictions would in this case hinge on the extent to which the ECOSIM interaction representations are considered overcompensatory. The assertion by Walters and Kitchell (2001, based on unpublished data) to support the foraging arena model that "predators with full stomachs are not a common field observation" is one that is still open to debate. Furthermore, digestion time constraints likely put a cap on consumption rates of marine mammals [contrary to the implications of equation (2)], as suggested by work such as that of Rosen and Trites (2000).

Arguments offered in Cooke (2002) reinforce these concerns with the ECOSIM defaults by demonstrating through the use of a simple model that whether or not a reduction in cetaceans results in higher fishery yields than would otherwise, other things being equal, be obtained depends critically on the assumed vulnerability of the fish to the cetaceans. It is only under scenarios assuming this vulnerability to be high that fishery yields are predicted to be sensitive to the abundance of cetaceans. Cooke's (2002) results highlight the importance of first exploring the probable relationship between cetacean abundance and consumption of a target fish species because a priori assumptions in this regard strongly influence model outcomes in terms of whether or not they yield pure-replacement results.

By virtue of being packaged in a form that is readily accepted by as many people as possible, the EwE model loses some transparency and the code is rendered less accessible. Further dissections of this model have been conducted by, inter alia, Aydin (2004) and by Mackinson et al. (2003), who explore the consequences of alternative feeding interactions between cetaceans and their main fish prey. The latter authors, for example, demonstrate very different predicted model outcomes depending on the type of functional response formulation being implemented.

Multispecies Virtual Population Analysis

Multispecies Virtual Population Analysis (MSVPA) is a technique that uses commercial fisheries catch-at-age and fish stomach-content data to estimate both past fishing and predation mortalities of some of the chief fish species of interest (see, e.g., Sparre 1991, Magnússon 1995). Unlike VPA (Virtual Population Analysis), which assumes that the natural mortality rate remains the same over time and usually also age, here natural mortality is split into two components: (a) predation that is due to predators explicitly included in the model (M2), which depends on time and age because of variations in predator abundance; and (b) residual mortality (M1) caused by all additional factors, which are customarily taken to be constant. Based on the resulting estimates of M2, forward-looking simulations (MSFOR) are then used to determine the average long-term consequences of changing fishing patterns.

This approach requires substantial data pertaining to the predation ecology of the predators included in the model, to the extent that tens of thousands of stomachs were sampled in the North Sea in 1981 and 1991, the "Years of the Stomach," under the auspices of the International Council for the Exploration of the Sea (ICES). MSVPA applications have thus far focused primarily on the North Sea, with the considerable data requirements generally impeding its use in other areas, although similar approaches have been applied in the Bering, Baltic, and Barents seas as well as in the Gulf of Maine. Given the various difficulties associated with sampling marine mammal diets, it would be problematic to expand this approach to model indirect marine mammal / fisheries interactions. However, gray seals (*Halichoerus grypus*) were included in the 1991 North Sea model because scat analyses were available to estimate food suitability constants ("suitability" is an important input to MSVPA and specifies a predator's relative preferences should different prey species all be present in equal abundances).

A second potential problem with the MSVPA approach is that it concentrates on the impacts of predators on prey but ignores any potential effects that changing prey populations may have on the predators themselves (because of its constant ration assumption; see the next paragraph). Nonetheless, MSVPA has some utility in quantifying the relative losses in prey biomass attributable to other predatory fish, marine mammals, and commercial fisheries. Interestingly, MSVPA results for the North Sea support the results of other studies suggesting that losses that are due to marine mammals are small relative to losses attributable to predatory fish (Bax 1998, Bogstad et al. 2000, Furness 2002). However, given that the consequences of overlaps in consumption probably depend heavily on the associated temporal and spatial scales, standard MSVPA (as with most other models) falls short in not explicitly modeling migration into and out of the area under analysis.

The MSVPA approach differs fundamentally from ECOSIM in its assumption that the intake by a predator is time invariant—a so-called efficient predator model in which predators always attain their daily ration in contrast to the "hungry predator" assumption of ECOSIM's foraging arena model (Butterworth and Plagányi 2004). Once again it is critically important to recognize that this model assumption, in many circumstances, leads to predictions of a near-linear relationship between reductions in marine mammal biomass and increases in fishery yields (and hence results in different predictions than ECOSIM given that, as explained previously, ECOSIM with default settings cannot yield exact-replacement results). This stresses the need for improved understanding of feeding ecology and relationships before selections can be made among models that have different implications for the key question under review here. The MSVPA studies have at least made a start (e.g., Rice et al. 1991, Rindorf et al. 1998) in trying to determine the extent to which the consumption of a given prey is a simple linear function of its relative abundance in an ecosystem (the constant suitability assumption). Put simply, such models have problems accounting for scenarios in which predators keep preferentially eating a given food item despite its increasing scarcity, as has been demonstrated, for example, for fishes such as cod and whiting (*Micromesistius poutassou;* Rindorf et al. 1998).

Marine mammals as a group presumably vary substantially in terms of the extent to which their preferences for different prey remain stable over time. At least a few species may prey-switch (e.g., Estes et al. 1998). Harbitz and Lindstrøm (2001) demonstrate the use of a stochastic spatial analysis framework to derive relationships between expected proportions of prey biomass in the sea and in the diet of minke whales. They illustrate, for example, that whereas minke whales in the Barents Sea appear to actively select capelin in preference to other species, this is no longer the case once their preference for foraging in the upper water layers is also taken into account. Their methods, which highlight the importance of interpreting results at the correct spatial scale, could usefully be applied to other predator-foraging studies.

An MSVPA model is almost certainly not adequate on its own as a tool for investigating indirect interactions involving marine mammals. However, together with an ECOSIM model, it can at least make a first attempt at bounding the likely impact of a marine mammal reduction on a fishery as, based on the assumed forms of interaction, the former is likely to overestimate the effect and the latter to underestimate it (at least when using default or low-vulnerability settings). MSVPA is probably too data hungry to be widely applied, but that does not necessarily preclude the use of MSFOR to predict forward, provided the model is initialized using sensible assumptions based on at least some data (International Whaling Commission 2004). ECOSIM is also data hungry for dietary information but does not require the detailed catch-at-age information that is needed by MSVPA for all the major species considered.

Although there are often insufficient data to permit the application of a full MSVPA approach, a slightly simpler or even hybrid version could be used, such as Mohn and Bowen's (1996) approach when they modeled the impact of gray seal predation on Atlantic cod on the eastern Scotian Shelf. They first ran a standard VPA using commercial landings and research survey data, then added the consumption of cod by gray seals to the commercial landings, and finally repeated the VPA, which was retuned to take gray seal predation into account. They incorporated two alternative models of food consumption by seals (a constant ration predation model, in which the fraction of cod in the diet was assumed constant, and a proportional ration model, in which that fraction was assumed proportional to cod abundance), with the two predation models yielding substantially different estimates of the amount of cod consumed by gray seals.

Minimum Realistic Models: A Specific Example

It is estimated that the fur seal population off the west coast of South Africa consumes at least as much hake as is landed by the fishing industry. In response to substantial pressure as to whether the fishery for hake, the most valuable of the region's commercial fish species, would benefit from a seal reduction, Punt and Butterworth (1995) developed the first so-called minimum realistic model (MRM) following recommendations made at a preceding international workshop (Butterworth and Harwood 1991). The key feature of the MRM approach is the focus on only those species considered likely to have substantial interactions with the species of interest (typically the aim is to account for approximately 80% of the predation mortality). Model components in this case therefore included the two hake species, the fishery, seals, and an "other" predatory fish component (see Fig. 3.2). The level of detail taken into account for each component depended on what was considered necessary to capture the key aspects of its dynamics and was a function as well of the data that were available. Thus fully age-structured models were used for the two hake species (to capture cannibalism and interspecies predation effects), but the "other" predatory fish components were simply lumped into either a small- or large-fish category.

As detailed food web structure can make a big difference in predicting the ecosystem effects of fishing (Yodzis 1998), there is clearly a need for more work in order to systematically assess the effects of ignoring weak ecosystem links in models such as the one described previously. Yodzis (1988, 2000) also highlights the inherent indeterminacy of multispecies manipulations. Given the plethora of uncertainties, it is somewhat surprising that work to date has not concentrated more on calculating probability distributions for a response (Yodzis 1998).

The models described in the next section can also be categorized as of the MRM type. One advantage of the Punt and Butterworth (1995) model in the current context is that a realistic population dynamics model (Butterworth et al. 1995) was used to simulate the seal population, in contrast to the more usual practice of trying to adapt models originally constructed to simulate fish dynamics. The downside of such carefully tailored models is the considerable amount of work and expertise required, but this has to be weighed against the availability of suitable generalized modeling approaches that are sufficiently flexible to be adapted to local scenarios. As an example, models of marine mammal/fishery interactions can be expected to vary substantially between the Northern and Southern Hemispheres, given the substantially greater importance of zooplankton (e.g., krill) in the diet of the Southern Hemisphere animals.

MULTSPEC, BORMICON, and GADGET

These models (and others not mentioned in detail here such as Scenario Barents Sea (Schweder et al. 2000), Seastar (stock estimation with adjustable survey observation model and tag-return data; Tjelmeland and Lindstrøm 2005) and FLEXIBEST (International Whaling Commission 2004a) are all of Northern Hemisphere origin and have variously incorporated predation by marine mammals. A common feature is that they are area-disaggregated, which is a definite advantage given the migratory behavior of many marine mammals and the consequent importance of considering spatial-temporal overlaps among fisheries, marine mammals, and shared prey species. In brief, MULTSPEC (see Bogstad et al. 1997) is a length-, age-, and area-structured simulator for the Barents Sea, which includes cod, capelin, herring, polar cod, harp seals, and minke whales. Predation interactions are modeled only as one-way for marine mammals, which do not themselves react to changes in prey availability in the model. BORMICON (a boreal migration and consumption model) is another area-structured approach for the multispecies modeling of Arcto-boreal ecosystems (Stefansson and Palsson 1998). It refines MULTSPEC and generalizes it for use in other regions.

Given that work is not currently continuing on MULTSPEC and that BORMICON is being incorporated as a special implementation of GADGET, we focus here instead on a brief review of GADGET (globally applicable area-disaggregated general ecosystem toolbox) (see, e.g., web page www.hafro.is/gadget/; coordinator G. Stefánsson). Current case studies include the Barents Sea, the Celtic Sea, Icelandic waters, and the North Sea (Begley and Howell 2004). Although marine mammals have not been included in any of the case studies to date, this project is still being developed and shows promise for modeling these indirect interactions (and has been recommended for such—North Atlantic Marine Mammal Commission 2002) because of some of the following features (Begley 2004):

- Population trends can be split by species, size class, age group, area, and time step.
- The model platform is flexible in permitting the easy addition or substitution of alternative model components of biological processes such as growth and prey suitabilities.
- Change from age- to length-based predation selectivity models is considered an advance, given that length is generally a better predictor of prey vulnerability than age.
- Marked improvements in representing uncertainty are possible, given the inclusion of a range of likelihood functions that can be maximized to obtain parameter estimates and their confidence intervals when fitting to data. The ability to conduct global maximization of the likelihood is a definite advantage, as is the continuing work to derive improved statistical measures of uncertainty.
- GADGET incorporates a data warehouse that facilitates investigations using data at various levels of aggregation.
- Migration is implemented through movement matrices that can be used to capture broad seasonal patterns even without the finer details for marine mammals being known.

As with the other modeling approaches, a major impediment is the current lack of adequate data to describe feeding relationships, especially when considering situations far from the current level. There is in any case a general paucity of adequate data on cetacean diets (e.g., Stefansson et al. 1997).

Bioenergetic Models

A separate suite of models includes those based on allometric and bioenergetic reasoning. Yodzis (1998) used a twenty-nine-species food web model incorporating allometric reasoning to investigate the effects of a reduction of fur seals on fisheries in the Benguela ecosystem. However, the model structure was arguably too linear and lacked age, spatial, and seasonal structure. More recently, an improved bioenergetics model was constructed to describe interactions among squid (*Illex argentinus*), anchovy (*Engraulis anchoita*), hake (*Merluccius hubbsi*), and South American sea lions (*Otaria flavescens*) off the Patagonian shelf (Koen-Alonso and Yodzis 2005). These models reinforce the importance of correctly specifying the form of the functional response because of its critical influence on model behavior. They address many of the aspects outlined previously related to modeling feeding interactions, but at the cost of requiring much more detailed data.

Although likely to be too data intensive for most studies, the combined geographic information system (GIS) and energetics modeling approach of Bjørge et al. (2002) shows promise. They used radio-tracking data to construct an energetics simulation model of a population of harbor seals in Norway. By integrating their results into a GIS model, they were able to analyze the co-occurrence of fishing operations and seals. Interestingly, they concluded that harbor seal predation probably impacted some fisheries negatively but had a positive effect on shrimp catches because the seals removed benthic-feeding fishes.

Convention for the Conservation of Antarctic Marine Living Resources Predator-Prey Models

The adoption of the Convention for the Conservation of Antarctic Marine Living Resources (CCAMLR) and, particularly, Article II thereof (for a discussion of the implications, see, e.g., Butterworth 1986) was a crucial step forward in acknowledging the importance of maintaining the ecological relationships among harvested, dependent, and related populations of marine resources. Krill is the primary food source of a number of marine mammal species in the Antarctic, and concern has been expressed that the krill fishery, if it expanded rapidly, might retard the recovery of previously overexploited populations such as the large baleen whales of the Southern Hemisphere (e.g., Nicol and Robertson 2003).

In response to the CCAMLR mandate, modeling procedures have been developed to assess the impact of Antarctic krill harvesting on krill predator populations and to explore means of incorporating the needs of these predators into the models used for recommending annual krill catch levels. For example, Antarctic fur seals from Bird Island, South Georgia, are another example of a previously overexploited population that has recently grown rapidly so that its needs for krill as a food source may, in due course, conflict with the objectives of krill harvesters. Initial modeling procedures estimated the level of krill fishing intensity that would reduce krill availability and, hence, the population of a predator to a particular level (Butterworth and Thomson 1995, Thomson et al. 2000).

A particular concern in CCAMLR has been the potential negative effects of a concentration of krill fishing in the vicinity of land-based breeding colonies, where the foraging range of parents is necessarily restricted. Mangel and Switzer (1998) developed a model at the level of the foraging trip for the effects of a fishery on krill predators, using the Adelie penguin (*Pygoscelis adeliae*) as an example. Their approach of incorporating advection and diffusion processes in a spatial-temporal framework to model krill availability in relation to the location of breeding colonies can be extended and applied to seal populations. Given the large interannual fluctuations observed in krill biomass, these models might also have to include the capacity to incorporate physical forcing of prey dynamics (Constable 2001). Moreover, Constable's (2001) approach recognizes the need to consider potential competitive effects among a suite of predators.

IMPORTANCE OF CONSIDERING UNCERTAINTY ISSUES AND MANAGEMENT CONSTRAINTS

Punt and Butterworth's (1995) MRM approach incorporated a noteworthy feature that involved taking explicit account of uncertainty and management issues through a simulation framework that used the feedback control rules actually in place for setting annual hake fishery catch levels. Its purpose was to determine whether the management system applied to compute catch levels could take advantage of an increase in sustainable yields of hake if brought about by a seal reduction. Cooke (2002) stresses the importance of considering management constraints and issues of uncertainty as integral components of attempts to assess the effects of changing cetacean abundance on fishery yields. There is currently a dearth of studies incorporating these aspects.

ADDITIONAL INDIRECT INTERACTIONS

The previous section focused primarily on so-called feeding-related interactions (cf. Fig. 3.1a–f), such as perceived competition between fisheries and marine mammals for target prey species. This section describes some examples of other more indirect ecological effects (cf. Fig. 3.1g,h).

HABITAT DEGRADATION THROUGH FISHING ACTIVITIES

Effects of Trawling

Bottom trawling is undoubtedly the greatest fishing-caused disturbance in the ocean, with bottom trawls reportedly annually scraping across an area of the ocean equivalent to half the entirety of the world's continental shelves. Trawling nets often have heavy rollers, chains, and wooden "doors" attached to enable them to penetrate some sediments to a depth of as much as 6 cm. Hence they impose substantial noncatch and discard mortalities on benthic species in the trawl's path (see also Ragen this volume).

In a study of the long-term effects of trawling, McConnaughey et al. (2000) detected significant differences in the number and types of seafloor organisms between trawled and untrawled areas. However, further studies of the ecological implications of the short- and long-term impacts of trawling are necessary before any definitive conclusions can be drawn regarding effects propagated higher up in the food web to the marine mammals. That such effects are quite possible is highlighted by a study modeling the effects on the survival of juvenile cod of creating a range of marine protected areas (MPAs) of various sizes and configurations, where trawling and dredging are prohibited in portions of Stellwagen Bank National Marine Sanctuary off the coast of Massachusetts (Lindholm et al. 2001). There were dramatic increases in the survival rates of juvenile cod when the highly complex seafloor habitats in Stellwagen Bank were protected in a MPA. Other studies that link the survival of juvenile Atlantic cod with the type and quality of seafloor habitat where the juvenile cod settle also suggest that trawling may have important effects on the fish populations upon which both fisheries and marine mammals depend.

Examples of marine mammals that are potentially highly sensitive to the deleterious effects of trawling on bottom sediments include the gray whale (*Eschrichtius robustus*), which feeds primarily on species such as benthic amphipods, and the walrus, which feeds on molluscs and other invertebrates that live on the seafloor and in bottom sediments (Born et al. 2003). Fisheries-generated habitat destruction may be particularly serious when it affects habitats that are expected to exhibit slow recovery times and/or are disproportionately important ecosystem components, such as areas with bathymetric features that render them productivity "hot spots." One example concerns seamounts, which are important feeding areas supporting a number of marine mammals. These habitats are in many cases being severely impacted by destructive deep-sea fishing practices (Koslow et al. 2000).

Other Deleterious Fishing Practices

A number of other destructive fishing practices are on record, such as the use of explosives or cyanide in coral areas. These practices seriously damage the habitat with consequent repercussions for both fish communities and the marine mammal populations that rely on these habitats for their foraging. The mobile nature of cetaceans provides them some capacity to mitigate against such habitat destruction, provided that it is not overly widespread.

The bycatch of marine mammals in fisheries constitutes one of the major threats to these animals (see Read in this volume for a review of this form of direct interaction).

Noise Pollution

Although its impact is poorly known, noise pollution has been argued to be one of the major causes of concern for marine mammals because of the importance of sound in their adaptive fitness at all levels from the individual to the species (Roussel 2002, Hildebrand this volume). Noise pollution from fishing vessels is relatively minor compared to that from other sources. Nonetheless, noise from fishing vessels has been implicated in affecting marine mammals negatively in at least two ways:

- Noise from fishing operations may interfere with the sensitive echolocation systems of toothed cetaceans and thereby indirectly reduce their foraging efficiency. Perez et al. (in Roussel 2002) found a negative correlation between cetacean clicks and whistles and shipping noise (from commercial ships and fishing fleets), either because the cetaceans reduced their echolocation in response to shipping noise or, of equal concern, their sounds became masked by such noise.
- Noise from fishing operations may keep marine mammals away from or reduce the amount of time they spend in preferred (presumably more productive) feeding areas.

Echo sounders are increasingly being used onboard fishing vessels to assist in fish and depth detection, but their effect on marine mammals has not yet been documented (Roussel 2002). The use of acoustic deterrents (pingers) to prevent net entanglement, for example, has received slightly more attention but remains a controversial topic because of concerns related to their role in provoking cetaceans to reduce or stop echolocation (Tregenza in Roussel 2002, Read this volume). Moreover, acoustic harassment devices are implicated in displacing both target (e.g., harbor porpoises) and nontarget species (e.g., killer whales) from areas where noise has deliberately been introduced into their environment (Morton and Symonds 2002, Roussel 2002).

ROLE OF FISHING IN ALTERING THE STRUCTURE OF ECOSYSTEMS

One of the best-known effects of fishing is a change in the overall size distribution of fish. For example, the larger sizes and some larger species (e.g., halibut *Hippoglossus hippoglossus*) in the North Sea have become uncommon because of fishing (Pope 1991, Daan et al. 2005). This has two main implications for marine mammals. First, given that most marine mammals target somewhat smaller fish than commercial fisheries, a possible effect of an erosion of the larger sizes over time is to increase the fisheries/marine mammal competitive overlap. On the other hand, the impact of an increased proportion of small fish may be positive in the case of marine mammals with a preference for smaller fish. Secondly (but less likely given the first point noted previously), marine mammals that used to prey on relatively larger fish species or individuals would be forced to shift to preying on yet smaller fish, with consequent (largely unquantified) negative effects on their bioenergetics.

There is documentation regarding some sixteen cetacean species (Fertl and Leatherwood 1997) and several pinniped species (e.g., Shaughnessy and Payne 1979) feeding in association with trawling. Trawlers can either act as a mobile patch, concentrating food to the benefit of marine mammals (Fertl and Leatherwood 1997) or they can broaden feeding resources by disturbing and/or bringing prey species higher in the water column, where they are easier targets for marine mammals.

Bycatch and discards associated with trawling have to various extents played a role in altering benthic community structure (Alverson et al. 1994). Apart from habitat modifications resulting from the trawl gear itself, changes in bottom structure may result from factors such as reductions in oxygen saturation (affecting community structure) that are due to localized dumping of discards such as fish heads and increases in scavenger or decomposer species attracted to such areas (Alverson et al. 1994). Alverson et al. (1994) stress the fact that discards provide ready forage for several species (including dolphins and seals) and, hence, that the resulting population enhancements should be borne in mind when considering how to deal with discards. On the other hand, Fertl and Leatherwood (1997) caution that shifts in marine ecosystems as a result of trawling may seem beneficial to species such as dolphins initially but may prove detrimental in the long run, especially in cases where trawling ultimately results in severe declines in prey species. They note that although trawling often provides more easily captured food (e.g., the long-standing association between bottlenose dolphins and shrimp trawlers) and opens new feeding niches for marine mammals, it probably destroys other niches. It also exposes marine mammals to greater risk of injury or death associated with fishing operations.

FISHERIES-GENERATED ALTERATIONS IN FORAGING STRATEGY

Estes et al. (2003) highlight the important but hitherto mostly ignored role of individual dietary variation in population and community ecology, using sea otters as an example. In response to intraspecific competition, dietary diversity in sea otters decreases with decreasing population density (Estes et al. 1981 cited in Estes et al. 2003, Watt et al. 2000). That individual differences in dietary diversity may be promoted by strong intraspecific competition adds complexity to the indirect fisheries/marine mammal interactions: reducing marine mammals to increase fishery yields may thus be much less effective than predicted if the consequent relaxation in intraspecific competition translates into reduced dietary variation among the surviving animals as they switch to more extensive feeding on a preferred prey species (which happens to be of commercial interest). Note that this runs contrary to what the constant suitability assumption underlying MSVPA (see earlier section on MSVPA) would predict.

It is noteworthy that several marine mammals display individual variation in diet (on a variety of spatial scales) and actively teach foraging skills to their offspring (Heithaus and Dill 2002). If overfishing substantially reduces the abundance of one prey species and the principal effect of cultural transmission is one of inertia (because transmitted specializations act as local optima), then the associated danger exists of apparently "suboptimal" specializations being maintained over multiple generations (Estes et al. 2003). While still in their infancy, studies of interindividual and intraindividual variation in foraging tactics, and the circumstances in which they are employed, will improve the ability to predict marine mammal/fisheries interactions (Heithaus and Dill 2002). This could potentially further complicate the specification of functional responses in ecosystem models.

ROLE OF FISHING IN ALTERING PREDATION MORTALITY

As alluded to in the earlier section on the potentially detrimental effects of fisheries on marine mammals, fishing may play an indirect role in increasing the mortality rates of a marine mammal species as a result of predation by species such as killer whales. The interaction may be relatively complex as with the Steller sea lion/otter example or more straightforward as when killer whales switch to targeting other marine mammals in instances where there has been overfishing of their preferred fish stocks. Whether or not killer whales can, in fact, substantially and negatively impact other marine mammal species is currently open to debate. A perhaps more worrisome negative effect of fishing resides in the possibility that as fisheries deplete prey stocks upon which marine mammals depend, the latter become exposed to increased predation risk themselves by being forced to either increase their foraging time or to forage further afield in areas where they themselves are at higher predation risk.

Turning the tables, some shark fisheries may have a positive indirect effect on marine mammals because of the associated reduced predation on (immature in particular) animals. Sharks represent an important predatory threat to most pinnipeds and some cetaceans (Weller 2002).

MARINE MAMMALS AS HOSTS FOR PARASITES OF COMMERCIAL FISH SPECIES

One of the best-known examples of marine mammals indirectly influencing a fishery concerns anisakid nematodes, whose larvae use commercial fish and squid. Infected fish may be eaten by marine mammals with the larval nematodes then growing to maturity and continuing the cycle (Raga et al. 2002). Although the effects of the larvae scarcely go beyond rendering the fish unappealing to consumers, there are naturally serious economic (marketing) consequences for some major commercial species. For example, the parasite *Pseudoterranova decipiens* ("codworm") infected cod in eastern Canada to such an extent that the losses it caused in 1982 alone were estimated to be $Can 20 million (Raga et al. 2002).

Elsewhere, extensive investigations have been conducted concerning the dynamics of seal parasites, the second most common nematode being the "herring worm" *Anisakis simplex,* and their interactions with commercially important fish species (Podolska and Horbowy 2003). For example, Ólafsdóttir and Hauksson (1997) found geographical and seasonal differences in the anisakid infestations of Icelandic gray seals. Ultimately these studies may shed some light on mitigation methods to reduce infestation levels in commercial catches by coupling to models incorporating seasonal and geographic dynamics of parasite infections. Management measures to ameliorate the adverse effects of this indirect interaction are considered the most sensible way to proceed, as it seems highly unlikely that there are any simple linear relationships between marine mammal populations (acting as hosts) and parasite loads of fish that feed, for example, on planktonic crustaceans that have in turn fed on free-living parasite larvae. Any arguments put forward to reduce marine mammals on this basis would thus hardly have an unequivocal scientific basis.

ASSIGNING IMPORTANCE TO INDIRECT FISHERIES EFFECTS IN TERMS OF THEIR POTENTIAL IMPACTS

Summary and Ranking of Interaction Types

A number of different indirect fishery interactions have been described here. The potential importance of each depends to a large extent on the local context. Very generally we have attempted to rank the nine major interaction categories identified on a scale from 1 (very important) to 5 (least important) based on an overall impression of the following three factors:

- How widespread and likely the effects.
- The economic repercussions of the effects (on a wide scale).
- The extent of likely ecological repercussions.

A summary of the different interactions and provisional rankings is given in Table 3.2. The direct competition for food and fishery resources between marine mammals and fisheries emerges as the major potential impact (to either or both). Again the relative importance of this factor can be expected to vary substantially from region to region.

Considerations in Identifying Important Interactions

Based on insights gained in reviewing this topic, we make the initial proposal that the following represent important

Table 3.2 Initial summary and ranking of types of biological interactions between marine mammals and fisheries

Ranking	Type of biological interaction
1	Direct competition for commercial fish species
2	Indirect competition for commercial fish species
3	Habitat degradation owing to trawling and other fishing practices
	Marine mammals as hosts for parasites of commercial fish species
	Indirect competition for primary production
	Changes in shark/killer whale predation rates affecting marine mammals
4	Fishery-induced shifts in the size structure of ecosystems
	Alterations in foraging strategies in response to fishing effects
5	Effects of noise from fishing operations on marine mammal foraging activities and effectiveness

Note: Rankings are from 1 (most important) through 5 (least important). Rankings serve as a guide to the relative potential importance of each interaction (to either or both the fishery and the marine mammals) based on a global overview of the possible number and likelihood of occurrences and the associated economic and ecological implications.

considerations in assigning importance to indirect fisheries effects in terms of their potential impact:

1. High degrees of overlap (known or assumed) in the type and/or size classes of prey targeted by marine mammals and a fishery. This needs to be assessed within a spatiotemporal framework to account, for example, for the disproportionately large impact of fisheries in instances where they overlap spatially and/or temporally with increased seasonal or age-specific energetic demands of marine mammals.
2. Cases involving species with no easy access to alternative food sources—examples include species with restricted nearshore distributions (see, e.g., DeMaster et al. 2001) or limited mobility.
3. Spatial overlaps between commercial fishing grounds and the foraging areas for pinniped breeding colonies (a special case of 2).
4. Cases involving species with high degrees of specialization in prey species (stenophagous), particularly species with physiological or other constraints that restrict the range of prey that can be eaten.
5. Trawling in nearshore waters, and particularly in areas known to be inhabited by dugongs.
6. Cases involving marine mammals with very large and possibly localized population numbers because of the large total amounts of prey required by such populations.

These six points take account of the importance of considering the impacts in both directions. Examples of potentially important fishery effects based on these criteria would therefore include (but are not restricted to):

- Sea otter/killer whale/pinniped/fishery interactions off Alaska.
- Harp seal/fishery interactions in the North Atlantic.
- Gray seal/fishery interactions in the North Atlantic.
- Cape fur seal/fishery interactions off the coast of southern Africa.
- Krill/predator interactions in the Antarctic.
- Minke whale/fishery interactions in the Barents Sea and the North Pacific.
- Dugong/seagrass/trawling interactions.

Highlighting Species of Particular Concern

Our preliminary synopsis suggests that the following seventeen species (not ordered in priority) should be given priority (in terms of data acquisition and focused modeling studies) when assessing the effects of indirect fishery interactions (see Table 3.1 for more information):

1. Antarctic minke whale (*Balaenoptera bonaerensis*)
2. Common minke whale (*Balaenoptera acutorostrata*)
3. Bryde's whale (*Balaenoptera edeni*)
4. Humpback whale (*Megaptera novaeangliae*)
5. Harbor porpoise (*Phocoena phocoena*)
6. Common bottlenose dolphin (*Tursiops truncatus*)
7. Short-beaked common dolphin (*Delphinus delphis*)
8. Killer whale (*Orcinus orca*)
9. Gray seal (*Halichoerus grypus*)
10. Harbor seal (*Phoca vitulina*)
11. Harp seal (*Pagophilus groenlandicus*)
12. Hawaiian monk seal (*Monachus schauinslandi*)
13. Steller sea lion (*Eumetopias jubatus*)
14. Cape fur seal (*Arctocephalus pusillus pusillus*)
15. Antarctic fur seal (*Arctocephalus gazella*)
16. Sea otter (*Enhydra lutris*)
17. Dugong (*Dugong dugon*)

One additional consideration relates to the conservation status of a species. For example, sirenians face a multitude of threats worldwide, such as poaching and bycatch, which are potentially more serious than indirect fishery interactions (in the biological sense of trawling damaging their feeding grounds), and further consideration of these effects obviously has to have high priority to avoid exacerbating population declines. Similar examples of indirect fishery effects that are important to consider despite being contributory, rather than major, threats to species are interactions involving blue whales (*Balaenoptera musculus*) (with krill) and humpback dolphins (competing with fisheries for estuary-associated fish; Peddemors 1999).

SUGGESTIONS FOR FUTURE STUDIES

Feeding-Related Interactions

In analyzing the diet composition of marine mammals to assess overlaps with fisheries, the following factors should be borne in mind:

1. The possibilities for using nonlethal methods to determine diet composition and foraging strategies (e.g., Hooker et al. 2002).
2. The importance of distinguishing between prey size preference and prey availability. Floeter and Temming (2003) present a good example (albeit applied to North Sea cod) of methods for distinguishing the two. Furthermore, prey preferences determined at local levels by experiment may not be representative of inputs required for interaction models at the population level because of spatial-temporal variations in predator-prey overlap over the region of interest (Lindstrøm and Haug 2001).
3. The spatial-temporal overlap between marine mammals and their prey. This can be addressed, for example, by weighting prey ingestion estimates by local predator abundance (see, e.g., Floeter and Temming 2003). Moreover, focused research is needed as to when and where prey are ingested; this may need longitudinal studies of a sufficient sample size to provide a representative set from the population (Boyd pers. comm). It is particularly pertinent to take spatial-temporal overlaps into account because: (a) baleen whales, because of their migratory habits, are generally present on feeding grounds for only part of the year; and (b) some pinnipeds haul out to breed so that there is a gradient in their use and need for prey as one moves away from their breeding grounds (e.g., Harwood 2001).
4. Age, sex, and regional differences in the feeding behavior of marine mammals (e.g., Flinn et al. 2002). Females in particular are likely to be more sensitive to indirect fishery impacts because of their increased energetic requirements and because, for example, the foraging range of a female adult otariid is limited by her need to return regularly to the breeding colony to nurse her pup (Boyd 1998, Boyd et al. 1998). In some cases it may be important to account for intraspecies variation in diet (e.g., Estes et al. 2003).
5. The need to place investigations into an appropriate spatial and temporal framework (e.g., Bradshaw et al. 2002).
6. The logistical difficulties involved in working with large whales or with particularly cryptic or elusive animals such as the beaked whales.

Data Requirements

A framework (United Nations Environment Programme 1999) for assessing the data requirements for scientific evaluation of proposals to reduce marine mammal populations to benefit fisheries is given in Table 3.3, but these criteria cannot be met even in situations where relatively large datasets are available (Harwood and McLaren 2002). Data requirements for future investigations of fishery effects depend to a large degree on the local context so that a comprehensive overview here would be impractical. Nonetheless, the following general points should be borne in mind:

1. Data needs should be assessed in relation to modeling requirements, preferably with simulations conducted initially to highlight critical parameters.
2. Clarity concerning stock identity of cetaceans (International Whaling Commission 2004) is important because a localized effect is likely to have a much greater impact if the animals involved constitute a small, localized stock rather than being a component of a larger, rapidly mixing assemblage.
3. There is an overriding need to collect data (experimental and/or field based, and on foraging in particular) to assist in resolving issues related to the relationships between the rates of predation and the types and densities of available prey. An example is provided by Lindstrøm et al. (2002), who recently showed that the dietary importance of herring to minke whales in the Barents Sea varies with herring abundance in a nonproportional manner. In particular, data are required in the form of time series in contrast to the more conventional snapshot-type estimates.
4. Data collection efforts should focus on areas where there is the greatest chance of success, either because

Table 3.3 Data required to evaluate proposals to reduce marine mammal populations

Marine mammals
- Abundance, distribution, and migration
- Per capita food/energy consumption
- Diet composition, including methods of sampling and estimation
- Demographic parameters

Target fish species
- Abundance, distribution, and migration
- Demographic parameters (weight at age, age at spawning, etc., commercial catch per unit effort)
- Details of assessment models and results

Other predators and prey of the target species
- Abundance, amounts consumed, details of stock assessment if any

Other components of the ecosystem
- Two-way matrix of "who eats who," with estimated or guessed annual consumptions
- Estimated abundance by species

Source: Based on fishery impacts, from Table 2 of United Nations Environment Programme (1999).

of the structure of the ecosystem or because there are large signals in the data (e.g., the Antarctic).
5. Attention needs to be given to the consequences of data-related uncertainties and their effects on the bias and variance of quantities estimated from such data.

Modeling Studies

The North Atlantic Marine Mammal Commission (NAMMCO) has focused on marine mammal/fisheries interactions for a number of years. For example, workshops have been convened to investigate the role of minke whales, harp seals, and hooded seals (*Cystophora cristata*) in the North Atlantic (North Atlantic Marine Mammal Commission 1998), the economic aspects of marine mammal/fisheries interactions (North Atlantic Marine Mammal Commission 2001), and the main uncertainties in extrapolating from feeding behavior or stomach contents to annual consumption (North Atlantic Marine Mammal Commission 2002), as well as to model marine mammal/fisheries interactions in the North Atlantic (North Atlantic Marine Mammal Commission 2003). The conclusion of the first of these workshops, that marine mammals have substantial direct and/or indirect effects on commercial fisheries in the North Atlantic (North Atlantic Marine Mammal Commission 1998), focused attention on studies related to competition and the economic aspects of marine mammal/fisheries interactions (e.g., North Atlantic Marine Mammal Commission 2001).

In light of uncertainties in calculations of marine mammal consumption, concrete recommendations were sought with regard to estimating consumption in the North Atlantic (North Atlantic Marine Mammal Commission 2002). The next step was to review how available ecosystem models could be adapted to quantify marine mammal/fisheries interactions in the North Atlantic. The following requirements (North Atlantic Marine Mammal Commission 2003) are particularly relevant in identifying the desirable features of a multispecies modeling framework:

1. Flexibility of functions for prey selection.
2. Flexibility of age structuring (from fully age-structured to fully aggregated).
3. Accessible code and transparent (not "black-box") operation.
4. Able to be tailored to area and species of concern.
5. Includes interactions accounting for most of the natural mortality M for species of concern.
6. Spatial and temporal resolution able to be tailored for target species.
7. Uncertainty in data and model structure reflected in results.

We have coarsely assessed the main groups of models discussed here in terms of these seven requirements, as well as the additional requirement that marine mammals be explicitly included, rather than treated as exogenous components (Table 3.4). This last requirement is included because the models presented differ substantially in terms of whether they represent: (a) only the effects of marine mammals on a commercial prey species (e.g., MSVPA, BORMICON, and other models were originally constructed with the primary aim of assessing fish stocks); (b) only the effects of prey on marine mammal populations (e.g., CCAMLR models constructed with this aim in mind); or (c) effects operating in both directions (e.g., ECOSIM).

Although the finer details of the coarse assessment presented in Table 3.4 could be argued, our general impression is that GADGET and minimum realistic models (MRM), such as the approach suggested by Punt and Butterworth (1995), show the most promise as tools to assess indirect interactions between marine mammals and fisheries. Bioenergetic/allometric modeling approaches as, for example, that of Koen-Alonso and Yodzis (2005) also have a role to play in attempting to characterize the finer details of these interactions. Given that the Antarctic marine ecosystem could be viewed as a case on its own, further development of the suite of CCAMLR predator-prey models (essentially also MRM types) is considered the most appropriate approach for this region.

We cannot overemphasize the importance of applying different modeling approaches to the same system (provided that appropriate resources, in terms of both personpower and data, are available). This is particularly useful for qualitative cross-checking to determine whether different kinds of models give similar results, in order to gauge how much confidence one can have in their reliability. Furthermore, given the importance of comparing the outputs of various approaches, as well as the need to test model predictions against both simulations and reality, there needs to be an internationally coordinated effort to provide a framework for testing the models being suggested and applied (Boyd pers. comm.).

An appreciation of the need to understand the assumptions underlying each model emerged from both the NAMMCO meeting on modeling marine mammal/fisheries interactions (North Atlantic Marine Mammal Commission 2003) and the International Whaling Commission (IWC) meeting on cetacean/fishery competition (International Whaling Commission 2004). Both meetings stressed the need for:

- Careful consideration as to whether or not underlying model assumptions are appropriate for the case under investigation.
- Tests of the sensitivity of predictions to alternative assumptions, particularly regarding interaction terms (e.g., Mackinson et al. 2003).
- Addressing uncertainty, in particular by focusing research on discriminating between alternative assumptions that yield appreciably different predictions.

A further pragmatic recommendation that came out of the IWC meeting was that modeling efforts should focus on

Table 3.4 Summary of modeling methods employed to investigate indirect interactions between marine mammals and fisheries

Model	(1) Flexibility regarding prey selection?	(2) Flexibility regarding age structuring?	(3) Accessible code and transparent?[a]	(4) Can be tailored to other areas and species of concern?	(5) Accounts for most of M of species of concern?	(6) Spatial and temporal resolution can be tailored?	(7) Uncertainty adequately reflected ?	(8) Marine mammal dynamics explicitly modeled?
ECOPATH with ECOSIM (EwE)[b]	No/some	Some	Not in general but is possible	Partly	Yes	Maybe some (ECOSPACE)	No	Yes
MSVPA	No	Yes	Yes	Limited	Yes	Yes	No	No
MULTSPEC	No	Yes	Yes	No	No/sometimes	Yes	No	No
BORMICON	No	Yes	Yes	No	Yes	Yes	Yes	No
GADGET[c]	Yes	Yes	Yes	Yes	Yes	Yes	Yes	Yes
Minimum realistic model (MRM) such as Punt and Butterworth (1995)	Yes	Yes	Yes	Yes in general	Yes	Yes	Yes	Yes
Bioenergetic models (e.g., Yodzis 1998, Koen-Alonso and Yodzis 2005)	Yes	No	Yes	Data limited	Yes	Yes	Yes	Yes
CCAMLR predator-prey models (e.g., Thomson et al. 2000)	No/some	Yes	Yes	No	No	Yes	Yes	Yes

Note: The different approaches are here evaluated by the authors in terms of the extent to which they "meet" (have met or in principle could be applied to meet) the seven criteria identified in the Workshop Report of the North Atlantic Marine Mammal Commission (2003) as desirable features of a multispecies modeling approach (see text for further details), as well as the additional criterion of whether marine mammals are explicitly modeled or simply included as exogenous components of the model.

[a]Comments here refer at times to the mathematical specification of the code, rather than necessarily to the code itself.

[b]Note comments with respect to EwE mostly refer to "the modeling package" that is widely used, not to modified versions such as considered in Mackinson et al. (2003).

[c]Note comments with respect to GADGET refer more to the potential than demonstrated ability of this approach given that it is still under construction.

specific areas/systems where there is the greatest chance of success. Table 3.5 summarizes the key characteristics of such systems as identified at that workshop. One ideal ecosystem for such investigations is the Barents Sea, where there is evidence of relatively tight predator/prey coupling with only a few fish species (herring, cod, and capelin) playing key roles. Systems characterized by strong physical forcing (bottom-up control) are likely to show little or no response to the removal of predators. This is because even strong trophic interactions may be insufficient to increase the spatial and temporal variability in the abundance of a species in systems characterized by high residual variabilities as a result of such physical forcing (Benedetti-Cecchi 2000). Food web and interaction web models should accordingly take account of the relationship between the variance of a trophic interaction and the residual variability of a resource (Benedetti-Cecchi 2000).

The Antarctic has often been proposed as a suitable starting point for developing models because it has a relatively simple ecosystem that has suffered large impacts from overfishing (e.g., Mori and Butterworth 2004). However, as with other high-latitude regions with short links to high trophic levels, it is subject to great physical variability, which may have to be better understood before reliable conclusions can be drawn regarding trophic interactions.

In considering future investigations of marine mammal/fishery interactions, the following additional points are worth reiterating:

- The overriding importance of further investigations regarding the appropriate form for functional responses (the prey/predator interaction terms) and feeding selectivities/suitabilities.
- The need to consider operational (i.e., management) issues, such as how fisheries respond to variability in target stocks and whether perceived gains from a marine mammal reduction are actually realizable, given the various constraints on fishing operations (including the manner in which total allowable catches are computed; Cooke 2002).
- The need for more realistic modeling of marine mammal populations in the sense that they are not simply large fish, generally have more limited reproductive (pregnancy) rates, and may exhibit different functional responses.
- The importance of appropriately modeling fish/fish and fish/fishery dynamics before trying to draw conclusions regarding fish/fishery/marine mammal interactions.
- The need for further systematic investigation (presumably through simulation studies) of the numbers of links that have to be included in a nontrivial ecosystem model for reliable predictive ability.

THE ROLE OF UNCERTAINTY

Given the demands of an increasing human population for more food from the marine environment, coupled with the continued overexploitation of a sizable proportion of fisheries worldwide (Food and Agriculture Organization 2002), there is likely to be increased pressure on managers from both fisheries desiring greater yields and conservation groups concerned about the ecosystem effects of fishing and pressing for more cautious management. Because of the difficulties involved in providing definitive scientific advice on the questions that these divergent objectives engender, scientists often equivocate. It is virtually impossible to substantiate, incontrovertibly, claims that predation by marine mammals is adversely affecting a fishery or vice versa.

In the absence of conclusive evidence, managers are increasingly called upon to apply the "Precautionary Principle" (Principle 15 of the UNCED Rio Declaration [Agenda 21] of 1992), which requires that, "where there are threats of serious or irreversible damage, lack of full scientific certainty shall not be used as a reason for postponing cost-effective measures to prevent environmental degradation" (Food and Agriculture Organization 1995). However, this has been argued both ways in this context: Either that a marine mammal cull should not take place in the absence of

Table 3.5 Summary of system characteristics in relation to modeling feasibility

System property	More feasible to model	Less feasible to model
Data availability	High or reasonable	Low or nonexistent
Food web properties		
Number of species	Relatively low	Moderate or high
Number of species interactions (i.e., complexity of the food web)	Low	High
Species interaction strength	Strong	Weak or diffuse
Habitat properties		
System openness	Relatively closed	Wide open
System boundaries	Tight and obvious	Loose or merging
Depth	Shallow or moderate depth	Deep
Physics (e.g., environmental forcing)	Low—or else obvious	High or else unclear

Source: From Table 4 of the report of the IWC workshop on cetacean/fishery competition (International Whaling Commission 2004).

clear evidence that it will benefit fisheries or, alternatively, that a marine mammal population should be held at its existing level in the absence of clear evidence that the additional consumption of fish accompanying its continued increase will not possibly damage fisheries (Butterworth et al. 1988, Butterworth 1999).

SUMMARY AND CONCLUSIONS

We have identified marine mammal species that potentially interact with fisheries at present (or may do so in the future) (see Table 3.1) and highlighted some seventeen species that should be given priority in terms of data acquisition and modeling to assess indirect fishery interactions.

Although in most cases there is as yet insufficient evidence to support claims related to indirect fishery interactions, allegations that they negatively affect fishery yields have been made for about 20% of sixty-one species of marine mammals reviewed in the literature. Regarding the reverse concern, tentative links between overfishing and marine mammal population declines have been suggested for at least a dozen species (see Table 3.1). Two of the most striking arguments for fishing-generated food declines affecting marine mammals are illustrated by the Steller sea lion and harp seal examples discussed in this chapter. Until fairly recently, only claims of negative effects on fishery yields received much attention. However, the adoption of the CCAMLR Convention some three decades ago, with its stress on the needs of dependent and related species in determining harvest levels (Butterworth 1986, Constable 2001), has proved to be a watershed, mandating management action to ameliorate the potential effects of fisheries on marine mammals, among other taxa.

Nine different types of biological interactions between marine mammals and fisheries are identified in this chapter, and a broad attempt has been made to rank these on a global basis (see Table 3.2). The principal factor of concern is the perceived competition between marine mammals and fisheries, which means that there is an urgent need for scientific evaluation to confirm the existence of appreciable competition and to estimate its scale and extent in terms of its impact on either or both the mammals and the fisheries.

Despite the persistent assumption by nonscientists that there is a mass-for-mass equivalence in the prey of marine mammals and the yields available to fisheries, such equivalence is called into question even under the simplest form of analysis. In fact, the complexity of ecosystems could well be such that the response to a marine mammal reduction could be highly dispersed through the food web, involving many other species. In some cases, competition effects are moderated because, for example, one of the putative competitors in fact reduces the abundance of a predatory fish species, in turn affecting the abundance of the target prey species. Although marine mammals are the most obvious scapegoats for fisheries because of their visibility, there is typically greater competitive overlap in the feeding "niches" of fisheries and fish that eat other fish. In spite of these considerations, there appears to be a dichotomy in the current thinking of policy makers, with most parties arguing either that fewer marine mammals would translate into greater fishery yields or that such links are insubstantial and more consideration should be given to the impacts on marine mammals of prey reductions caused by fisheries.

Even though more and better information on marine mammal diets is becoming available, limited understanding of the feeding strategies of marine mammals remains a key uncertainty. There is a need to quantify not only present diets and their spatial-temporal variation, but also the associated functional responses, that is, how these diets will change as the abundances of the various species change. It is important also to bear in mind that some highly specialized marine mammals (or, similarly, highly specialized fisheries) are particularly vulnerable to competition effects because they cannot simply change their diet in response to overfishing of an important food source (Plagányi and Butterworth 2002). Rapidly expanding squid fisheries constitute one area in which problems might arise in the future. As yield from the world's major fisheries declines as a result of overexploitation and associated corrective management measures, attention is likely to shift to less-developed fisheries, such as those for cephalopods—the marine capture fisheries predicted by FAO to have the greatest growth in consumption (Food and Agriculture Organization 2002).

Given current pressures to quantify the predicted effects of marine mammal reductions on fisheries, or, conversely, the effects of increased/decreased fishery catches on marine mammal populations, it is imperative that the models employed be closely scrutinized to understand the extent to which their underlying assumptions predetermine or have implications for the results obtained. Having critically reviewed some of the main models that have or might have been applied in the current context, we conclude that, in general, GADGET and minimum realistic models (MRM), such as the approach of Punt and Butterworth (1995), show the most promise as tools to assess indirect effects between marine mammals and fisheries.

ACKNOWLEDGMENTS

We are grateful for constructive comments on earlier versions of the manuscript provided by Ian Boyd, William Perrin, André Punt, and Randall Reeves.

FRANCES M. D. GULLAND
AND AILSA J. HALL

The Role of Infectious Disease in Influencing Status and Trends

THE MOST COMMON DEFINITION OF DISEASE is "any departure from normality" (Rothman 1986). This description may seem simple enough, and it may indeed be straightforward to diagnose disease among individual animals in close contact with humans. Yet establishing its occurrence, frequency, and particularly its impact on wildlife populations presents a challenge that has been confronted only in recent years.

For a disease to have an effect on a population, it must first affect an individual. Its consequent effect at the population level then depends on what is happening to the population's life history (i.e., its fecundity, survival, and dispersal). Little is known about the ecological significance of disease in marine mammal populations because work to date has focused mostly on individual animal health. For the most part, effort has been directed toward infectious disease, the category that we emphasize here. However, it should be remembered that many other diseases could concurrently have a negative impact on a particular species or population. As they are difficult to diagnose and are perhaps relatively rare in marine mammals, they may be easily overlooked. Such diseases are likely to belong to the other major categories: malignant, congenital, degenerative, deficiency, functional, toxic, and metabolic.

Theoretical studies suggest that infectious diseases regulate host abundance by exerting density-dependent effects on reproduction or survival (Anderson 1979). However, there have been few empirical studies on free-living animals to determine whether the effects of disease on populations are indeed density dependent. This is partially because: (a) density-dependent effects are hard to detect when populations are at equilibrium, and (b) most disease investigations in wildlife have focused on determining the proximate rather than predisposing causes of large die-offs.

A few studies of terrestrial wildlife have shown that interactions between disease and factors such as host nutrition, behavior, genetics, and climate can influence life history parameters and population dynamics (reviewed in Hudson et al. 2002). Such investigations depend upon concurrent monitoring of both infectious disease and host population dynamics, which has rarely happened in marine mammal disease research. To date, most research has focused on the causes of morbidity and mortality in stranded and captive animals and on the health of hunted animals. These studies have shown that there are a number of diseases in free-living marine mammal populations that can cause mortality or decrease growth and reproduction both as primary or secondary factors, interacting with other factors discussed later in the chapter.

Epidemiology (the study of the occurrence of disease) is a relatively new discipline with a number of specifically defined terms that can be confused with each other, but which, in fact, mean quite different things. In order to avoid such misunderstanding, some definitions of common epidemiological terms are given in Table 4.1.

Table 4.1 Definitions of terms commonly used in epidemiological studies

Term	Definition
Prevalence	This is a static measure; the proportion of a defined group having a condition at one point in time, often expressed as a percentage. A period prevalence is the proportion having a condition at any time within a stated period. P = number with disease / total number sampled.
Incidence	This is a kinetic measure; the proportion of a defined group developing a condition within a stated period. The term is often used in reference to a disease frequency but epidemiologically it is a proportion, that is, a number of cases related to a defined population and a stated period of time. I = number of new cases / total number individuals × time sampled.
Intensity	This is usually used in reference to parasites either: (i) the mean number of parasites within infected members of the host population, or (ii) the mean parasite burden of the entire population. It is important to distinguish between these two, as unless the prevalence is 100%, the population parasite burden will be less than the mean number within the infected members of the host population.
Rate	Rates constitute the underlying principal concept in epidemiology as the basis for comparison among population groups and most rates take the form: Frequency of observed state or event / Total number in which this state or event *might* occur.
Microparasite	Parasites that undergo direct multiplication within their definitive hosts (e.g., viruses, rickettsia, bacteria, fungi, and protozoa).
Macroparasite	Parasites that in general do not multiply within their definitive hosts, but instead produce transmission stages (eggs and larvae) that pass into the external environment or to vectors.
Endemic	A disease whose prevalence or occurrence does not exhibit wide fluctuations through time in a defined location or population.
Epidemic	A sudden, rapid spread or increase in the prevalence or intensity of a parasite or disease. An epidemic is often the result of a change in circumstances that favors pathogen transmission.

THE INFLUENCE OF DISEASE ON STATUS AND TRENDS OF MARINE MAMMAL POPULATIONS

Disease and Marine Mammal Mortality

EPIDEMIC DISEASES. Epidemic diseases (sometimes referred to as epizootics when dealing with animals rather than humans) are those that occur at a time or in a place where they are not usually found or with a greater frequency than expected in a certain period. The most dramatic effect of disease on marine mammal populations is the increase in mortality during an epidemic. Severe epidemics might reduce host population density to such an extent that stochastic events or previously unimportant ecological factors may further reduce the host population size (Harwood and Hall 1990). For example, canine distemper dramatically reduced black-footed ferret (*Mustela nigripes*) populations in Wyoming, bringing them to extinction in the wild (Thorne and Williams 1988), and avian malaria reduced native Hawaiian honeycreeper (*Hemignathus parvus*) populations to such small numbers that many were finally eliminated by predation or habitat loss (Warner 1968). The importance of infectious disease epidemics in causing declines in marine mammal populations as a result of increased mortality is unclear, as few marine mammal die-offs have been sufficiently investigated to determine their cause and because it is often difficult to determine pre-epidemic host population size accurately.

Recent epidemics among marine mammals have caused dramatic mortality, but the effects on host population size have varied. For example, approximately 18,000 harbor seals (*Phoca vitulina*) (70% of the population) died in the phocine distemper virus (PDV) epidemic in Europe in 1988, and age and sex distributions were skewed for several years after the epidemic (Heide-Jørgensen et al. 1992a). Ten years later, however, the population in most of Europe had recovered to pre-epidemic numbers (Reijnders et al. 1997). The origin of this morbillivirus epidemic is unclear, but it may have been introduced to harbor seals by the southward dispersal of harp seals (*Phoca groenlandica*) from the Barents and Greenland seas, where the virus is believed to be endemic

(Stuen et al. 1994). This introduction of a virus into a naïve population with large numbers of susceptible animals without antibodies allowed rapid transmission above a threshold level, resulting in an epidemic.

A mathematical model developed in 1992 to investigate the infection dynamics of this disease predicted that reintroduction of the virus resulting in large-scale mortality would not occur again for at least 10 years, after which time the number of susceptible (nonimmune) harbor seals would again be sufficient to maintain an epidemic (Grenfell et al. 1992). A new outbreak of PDV occurred in the North Sea in 2002 (Jensen et al. 2002), killing more than 20,000 harbor seals (cwss.www.de/news/publications/Wsnl/Wsnl02-2/articles/1-seal-epidemic.pdf; and www.waddensea-secretariat.org/news/news/Seals/01-seal-news.html). Interestingly, no influx of harp seals was reported prior to this outbreak. The questions about triggers and disease vectors of the 2002 epidemic still have to be answered.

The impacts of morbillivirus epidemics on other marine mammal populations are less well documented. Outbreaks of canine distemper (CDV) killed 5,000–10,000 Baikal seals (*Pusa sibirica*) in 1987–1988 (Grachev et al. 1989) and 10,000 Caspian seals (*P. caspica*) in 2000 (Kennedy et al. 2000), and may also have been responsible for the deaths of 2,500 crabeater seals (*Lobodon carcinophagus*) in the Antarctic in 1955 (Laws and Taylor 1957, Bengtson et al. 1991). A morbillivirus was isolated from Mediterranean monk seals (*Monachus monachus*) that died during an epidemic, but its importance relative to biotoxins in causing mortality remains controversial (Hernandez et al. 1998). Dolphin morbillivirus (DMV) caused the deaths of several thousand striped dolphins (*Stenella coeruleoalba*) in the Mediterranean Sea in 1990–1992 (Domingo et al. 1990) and then spread to short-beaked common dolphins (*Delphinus delphis*) in the Black Sea in 1994 (Birkun et al. 1999). It was also found in tissues of bottlenose dolphins (*Tursiops truncatus*) from a die-off of approximately 750 animals along the eastern United States in 1987–1988 (Lipscomb et al. 1994). As population estimates and stock structure for this latter species are unclear, the percentage of the dolphin population affected is unknown but may have been as much as 50% of one stock (National Oceanic and Atmospheric Administration 2000). A fourth morbillivirus, porpoise morbillivirus (PMV), was detected in dead bottlenose dolphins in 1993 in the northern Gulf of Mexico and in a die-off of harbor porpoises (*Phocoena phocoena*) in the Irish Sea in 1994 (Kennedy et al. 1988, Taubenberger et al. 1996). The PMV and DMV viruses are antigenically similar and may represent different strains of a "cetacean" morbillivirus referred to as CMV (Blixenkrone-Møller et al. 1996). Although CMV infection has been documented to have caused die-offs of cetaceans, the numerical effect on cetacean populations is still uncertain, mostly because of a lack of knowledge regarding mortality at times other than during epidemics and changes in population size and structure over time.

A smaller epidemic in harbor seals occurred along the New England coast of the United States in 1979–1980 and was associated with influenza A infection (Geraci et al. 1982). Unusually warm conditions, crowding, and a concurrent respiratory infection with mycoplasma may have triggered this epidemic, which killed at least 450 seals (3–5% of the local population) (Payne and Schneider 1984).

Bacteria can also cause epidemic mortality in marine mammals. *Leptospira pomona* was first isolated from California sea lions (*Zalophus californianus*) in 1970 and has since caused periodic die-offs of sea lions along the northern California coast about every 4 years (Vedros et al. 1971, Gulland et al. 1996). Despite hundreds of animals dying in each outbreak, the impact of the mortality has not prevented the growth of the California sea lion population, currently estimated at about 10% a year (Forney et al. 2000). Similarly, a gram-negative pleiomorphic bacterium that was believed responsible for an epidemic in the endangered Hooker's sea lion (*Phocarctos hookeri*) population in New Zealand in 1998 caused mortality of 60% of that year's pup production (Baker 1999), followed by a 5% decrease in pup production the following year. No long-term impact on the population dynamics of these sea lions has been detected (I. Wilkinson, pers. comm.).

ENDEMIC DISEASES. Endemic diseases (or enzootic diseases in animals) are those that occur in a population at a regular, predictable rate. Endemic diseases, which can also cause widespread mortality, do not exhibit wide spatial or temporal fluctuations in occurrence although sudden changes in host susceptibility or pathogen transmission can cause an endemic disease to become epidemic. A number of microparasites that cause endemic mortality have been identified in stranded marine mammals. These include at least ten families of viruses (reviewed by Visser et al. 1991, Van Bressem et al. 1999), thirty genera of bacteria (reviewed by Dunn et al. 2001), and several fungi (reviewed by Reidarson et al. 2001) that have been associated with disease in individual marine mammals. Their effects on populations are currently unknown as their dynamics are not understood.

Macroparasites (helminths and arthropods) are rarely associated with epidemics because their transmission dynamics are different than those of the microparasites (viruses and bacteria). They do not reproduce in the host, do not elicit long-lasting immunity, and have an aggregated distribution within the host population (Shaw and Dobson 1995). This aggregation has important implications for effects on the host; as a rule, the stability of the host-parasite interaction is enhanced by parasite aggregation. The interaction among aggregation, parasite virulence, transmission efficiency, and the host's population growth rate in the absence of parasites determines whether the host-parasite interaction is stable or exhibits cyclic or chaotic dynamics (Wilson et al. 2002). As macroparasites tend to have dose-dependent effects on host morbidity and mortality, effects are more likely in individuals at the "tail" of the distribution. Small sample sizes are thus likely to miss affected hosts.

The effects of macroparasite-induced mortality on marine mammal host populations are thus difficult to document, and few studies have attempted to assess them. An exception is a study of cranial lesions caused by the nematode *Crassicauda* sp. in spotted dolphins (*Stenella attenuata*) killed accidentally in the eastern tropical Pacific tuna fishery (Perrin and Powers 1980). Changes in the prevalence of irreversible lesions with age suggested that *Crassicauda* had caused 11–14% of the natural mortality, although biases such as dolphins with lesions being caught preferentially in the fishery could have affected the study results. Lambertson (1986) similarly examined the prevalence of *Crassicauda boopis* in North Atlantic fin whales (*Balaenoptera physalus*) and estimated parasite-associated mortality to be 4.4–4.9% of the total. He also suggested that infection of blue whales (*Balaenoptera musculus*) with *Crassicauda boopis* could limit recruitment in this species as mortality attributed to infection occurred at about the same age (1 year) as the inferred peak in natural mortality estimated from population models of this species (Lambertson 1992).

Disease and Marine Mammal Reproduction

Infectious diseases alter reproductive rates in terrestrial mammals and birds by decreasing fertility and causing abortion, premature parturition, and neonatal mortality (Scott 1988, Feore et al. 1997). These effects may be destabilizing and can regulate host population dynamics in some cases—for example, the nematode *Trichostrongylus tenuis* causes cycles in red grouse (*Lagopus lagopus*) populations by decreasing reproduction (Hudson 1986). Many infectious agents that could have similar effects in marine systems have only recently been isolated from marine mammals, and their prevalence and impact on free-ranging marine mammal populations are unknown (Dailey 2001, Dunn et al. 2001).

Caliciviruses and *Leptospira interrogans var. pomona* have been isolated from cases of abortion and premature pupping in California sea lions with high levels of organochlorines (Gilmartin et al. 1976); herpesviruses have been detected in aborted harbor seal fetuses and dead pups (Osterhaus et al. 1985, Gulland et al. 1997); *Coxiella burnetti* was detected in a case of placentitis in a harbor seal (LaPointe et al. 1999); *Toxoplasma gondii* was detected in a fetus of a dead Risso's dolphin (*Grampus griseus;* Resendes et al. 2002); and a *Brucella* sp. has been isolated from an aborted bottlenose dolphin (Miller et al. 1999). Papillomaviruses have been detected in cases of genital warts in Burmeister's porpoises (*Phocoena spinipinnis*) in Peru, where the lesions may be sufficiently severe to impede copulation (Van Bressem et al. 1999).

Although such pathogens have rarely been isolated from marine mammals and to date have only been detected in a few individuals, serological surveys suggest that exposure to these diseases is widespread. Antibodies to *Brucella* spp. have been detected in a wide range of cetacean and pinniped species around the United Kingdom, Canada, and the United States; antibodies to *Chlamydia* spp. are common in Steller sea lions (*Eumetopias jubatus*) and Hawaiian monk seals (*Monachus schauinslandi*); antibodies to caliciviruses and herpesviruses are present in most pinniped species tested; and antibodies to *Toxoplasma gondii* are prevalent in most coastal cetaceans and pinnipeds (Braun pers. comm., Barlough et al. 1987, Stenvers et al. 1992, Nielsen et al. 1996, Jepson et al. 1997, Zarnke et al. 1997, Forbes et al., 2000, Mikaelian et al. 2000, Burek et al. 2001). Thus infectious agents shown to cause reproductive failure in individual marine mammals and terrestrial mammal populations exist and are probably widespread, but their direct and indirect effects on the distribution and abundance of marine mammals still need to be determined.

Disease and Marine Mammal Condition and Growth

Experimental studies in birds have shown that parasites may directly limit the growth of individuals (Booth et al. 1993), potentially reducing future survival and delaying the onset of sexual maturity. Macroparasites such as nematodes, trematodes, and cestodes are common in marine mammals, but whether or not the parasites limit growth in these hosts has not been established because research concerning this relationship has been so limited. In two separate surveys of ringed seals (*Pusa hispida*) infected with the lungworm *Otostrongylus circumlitus,* no correlation between intensity of infection and body condition could be detected, although there was a slight reduction in respiratory parenchyma in infected seals (Onderka 1989, Bergeron et al. 1997). The decrease in the prevalence of infection after the first winter, however, suggests that *O. circumlitus* may play a role in decreasing overwinter juvenile survival. We are unaware of further studies directly investigating impacts of disease on the growth of marine mammals.

In a mark-recapture study of gray seals (*Halichoerus grypus*) Hall et al. (2002) found that postweaned pups with high total immunoglobulin G, particularly males in poor condition, had a lower first-year survival probability. It is not known whether this was because the pups were exposed to disease (although all appeared clinically healthy) or because individuals with naturally high IgG had to make a trade-off between resources needed for growth and development against those needed to maintain relatively higher circulating antibody levels. Such studies, which investigate the interplay between growth and exposure/response to infectious disease, may help us understand more about the relative importance of such interactions during vulnerable periods in the animals' life history.

Disease and Host Defenses

The form and function of marine mammal immune systems have been the focus of attention and research only in the last few years, and then largely because many persistent ocean contaminants have been shown to be immunotoxic to laboratory animals and other species (De Swart et al. 1994, De Guise et al. 1998). How immunity is affected by toxic contaminants has thus been the driving force behind much of the research in this field, often with the underlying assumption that the immune systems of marine and terrestrial species must be similar (see also O'Hara and O'Shea this volume). However, immunity to disease is clearly affected by many factors, including age, sex, species, and season (Nelson et al. 2002). Relatively little is known about the immune systems of marine mammals, particularly cetaceans (De Guise 2002) and how they have evolved compared to the systems of terrestrial species. The integrity and maintenance of immunity are central to the prevention of disease in individuals and populations. The system's main purpose is to control infection by pathogens. If we are to understand more about the impact of disease on marine mammal populations, then host defense mechanisms should be considered in their widest context, from the viewpoints of both comparative and functional immunology.

INTERACTIONS OF DISEASE WITH OTHER FACTORS THAT AFFECT MARINE MAMMAL POPULATIONS

Changes in disease prevalence and, in turn, increased host morbidity or mortality may result from increases in host susceptibility, pathogenicity of the disease agent, or transmission of pathogens to hosts. A number of discrete or interacting factors can influence each of these processes, as discussed later in the chapter.

Nutrition

Food limitation may alter the prevalence and incidence of disease by decreasing the host's immune response to infection (Seth and Boetra 1986) or by altering host behavior (Scott 1988). Interactions between food limitation and parasitism can determine which individuals die during population crashes of ungulates (Gulland 1992). Food limitation, nutritional deficiencies, and infectious diseases all occur in marine mammals, but the interactions among them have not been examined. It was suggested that poor nutrition among seals in some areas of the North Sea resulted in higher mortality from PDV during the 1988 epidemic, although evidence for nutritional immunosuppression in this case was lacking because of the immunosuppressive effect of the virus itself (Heide-Jorgensen et al. 1992b). Stranded California sea lions that are suffering from food deprivation or are feeding on unusual prey species during El Niño years have heavier burdens of the gastric nematode *Contracaecum corderoi* than do stranded animals in other years, and the former are more likely to die from peritonitis caused by perforated parasitic ulcers (Fletcher et al. 1998). Mortality from peritonitis following migration of parasites from the intestine also occurs in California sea otters (*Enhydra lutris nereis*), in this case following infection with the acanthocephalans *Profilicollis altmani, P. kenti,* and *P. major.* Young male otters are more frequently affected by these parasites than are other otters (Mayer et al. 2003). The acanthocephalans are acquired by the consumption of crabs (*Emerita* spp. and *Blepharipoda* spp.) that are intermediate hosts of these parasites but are not the preferred food of most otters. Mayer et al. (2003) hypothesize that young males are more susceptible to infection by acanthocephalans because their lack of feeding experience and low social status lead them to forage on less desirable prey species.

Population Density

Population density can have dramatic effects on disease dynamics, as it can affect the transmission of diseases directly as well as indirectly via effects on host nutrition and predation (Hudson et al. 2002). The spread of PDV during the 1988 epidemic among North Sea harbor seals was limited in some areas by low host density (Thompson and Hall 1993). Recent changes in the mortality of pinnipeds associated with hookworm infestation suggest that this macroparasite affects host mortality in density-dependent ways. The hookworm *Uncinaria lucasi* was an important cause of northern fur seal (*Callorhinus ursinus*) pup mortality on the Pribilof Islands in the 1960s (Olsen and Lyons 1965). Since then, pup numbers and parasite-associated mortality have decreased (Fowler 1990), suggesting that lower pup densities are reducing parasite transmission. Interestingly, initial work suggests that, as California sea lion pup numbers increase on San Miguel Island off California, mortality associated with hookworm infection is increasing (Lyons et al. 1997).

Host Movement and Distribution

Movement of a pathogen-infected host into a previously unexposed host population can result in severe epidemics and can contribute to the competitive success of the invading host (Daszak et al. 2001). Anthropogenic movement of an infected host is called "pathogen pollution" (Daszak et al. 2000). Movements of uninfected hosts into areas with endemic pathogens can also result in disease outbreaks in the new range of the host (Daszak et al. 2001). Movements of marine mammals prompted by environmental changes, such as El Niño events and global warming, can lead to movement of pathogens into susceptible host populations. For example, as noted earlier, the 1988 PDV epidemic in the North Sea harbor seal population was likely triggered by the southward movement of harp seals from the Barents and Greenland seas, carrying the virus from where it was endemic into an area with a naïve population (Heide-Jorgensen et al. 1992a). A dramatic reduction in the fish stocks in the Barents Sea in 1987 probably caused the harp seals to forage farther south in search of prey. The common lungworm of Arctic seals, *Otostrongylus circumlitus,* has only recently been reported in the increasing northern elephant seal (*Mirounga angustirostris*) population in California, where it kills the host before the parasite reproduces (Gulland et al. 1997). This is probably a new host-parasite association, judging by the high pathogenicity and lack of transmission of the parasite before the death of the host. It may be the result of expansion of the elephant seal population leading to increased range overlap with Arctic phocids.

Parasites can affect the coexistence of closely related host species if the parasites have a differential impact on the demographic rates of the hosts (Holt and Pickering 1985). Hookworm (*Uncinaria* spp.) infections of South American sea lions (*Otaria byronia*) and fur seals (*Arctocephalus australis*) along the coasts of Chile and Uruguay may be examples of such parasite-mediated competition, as intensity of infections was greater, while prevalence of lesions was lower, in fur seal pups compared to sea lion pups (George-Nascimento et al. 1992). The demographic effects of the parasite on the two host species are still unclear, and molecular techniques are needed to clarify whether or not the *Uncinaria* sp. in the two hosts is indeed the same species.

Population density may also influence haul-out behavior of harbor seals, which, in turn, influences their degree of aggregation and thus the transmission rates of contagious diseases such as PDV. Outbreaks of PDV in Europe occurred when the population density of seals was high relative to historical levels.

Host Genetics

Species differences in susceptibility to different diseases presumably reflect genetic differences among hosts. For example, gray seals (*Halichoerus grypus*) are more resistant to PDV than sympatric harbor seals (Thompson and Hall 1993). Within the few mammalian species studied to date (mice and sheep), parasite resistance and reproductive success covary with parental similarity (Smith et al. 1999). Polymorphisms at certain alleles controlling immune responses correlate with parasite resistance, and polymorphism at the major histocompatibility complex (MHC) is associated with improved immune response (Slate et al. 2000, Coltman et al. 2001, Messaudi et al. 2002). No studies have yet investigated the correlation between MHC variability and disease resistance in marine mammals. In a recent study of stranded California sea lions, parental relatedness as measured by polymorphism at eleven microsatellites was greater in animals stranding with infectious diseases than in those stranding as a result of trauma (Acevedo-Whitehouse et al. 2003). Among the sea lions with infectious diseases, animals with helminth infections were more inbred than animals with bacterial infections, and there was a significant positive regression of parasite diversity on inbreeding. These results demonstrate the importance of host genetics in modulating the host-parasite interaction, but do not identify mechanisms by which such effects occur.

Predation and Fisheries Bycatch

In terrestrial systems it has been shown that the interaction between parasitism and predation can magnify apparently small effects of parasites on host morbidity. For example, free-ranging snowshoe hares (*Lepus americanus*) that had their worm burdens reduced by anthelmintics were significantly less likely to be caught by predators than a similar group of controls (Murray et al. 1997). Predation rates were reduced still further in hares that received food supplementation, suggesting an interaction between parasite-induced susceptibility to predation and the host's plane of nutrition. As a result of logistic difficulties in conducting such studies, little is known about interactions between predation and parasitism in marine mammals. It is likely that infectious diseases predispose marine mammals to both predation and fisheries bycatch. For example, beach-cast sea otters in California have been observed to suffer from both protozoal encephalitis and great white shark wounds, suggesting that neurological disease associated with protozoal infection increased their vulnerability to predation (Thomas and Cole 1996). Disease studies using bycaught, stranded, and harvested animals could address these issues further.

Increased Urbanization of the Coast

As human populations increase along coastlines, so does contact between marine mammals and humans, their pets and livestock, and their associated pathogens. Molecular analyses of morbillivirus isolates from tissues of Baikal and Caspian seals collected during two successive epidemics in these populations revealed wild-type CDV (Mamaev et al. 1995, Kennedy et al. 2000). Transmission of CDV probably occurs via aerosols from domestic or feral dogs although aerosol transmission of CDV from adjacent susceptible terrestrial wildlife species such as foxes is also possible (Lyons et al. 1993).

Influenza B (B/Seal/Netherlands/1/99) was isolated in 1999 from a juvenile harbor seal with respiratory signs (the first time that influenza B had been isolated from a nonhuman; Osterhaus et al. 2000). Sequence analysis of the isolate showed it to be closely related to strains present in the human population in 1995. Moreover, retrospective serosurveys showed no antibodies to influenza B in the seal population around The Netherlands prior to 1995. Seals presumably were infected by a virus transmitted from humans in 1995 and may now serve as reservoirs for influenza B viruses that had previously circulated only among humans.

Contact between humans and marine mammals is also increased by the practice of rehabilitating sick or injured marine mammals in urban areas. This also increases the exposure of the animals to pathogens of terrestrial mammals as they are often housed in rehabilitation facilities accessible to wildlife such as small rodents. Kidney failure and mortality associated with *Leptospira interrogans var. gryppotyphosa* have been observed in harbor seals infected during rehabilitation, probably as a result of exposure to terrestrial rodents or skunks (Stamper et al. 1998).

Sewage containing pathogens may also influence disease frequency in marine mammals. *Toxoplasma gondii,* an obligate parasite of felids that is shed as infective oocysts in cat feces, appears to be common in southern sea otters, causing varying degrees of meningoencephalitis (Cole et al. 2000). Sea otters may be infected through ingestion of the oocyst stage, either directly from the water or by consuming filter-feeding invertebrates that concentrate the oocysts. Environmental contamination by feral and domestic cat populations, either directly or from the human disposal of cat feces in sewage, might play a significant role in epidemiology of sea otter toxoplasmosis (Cole et al. 2000). *Salmonella bovis morbificans,* a bacterial serotype associated with cattle, has been isolated from harbor seals around the United Kingdom. This may indicate contamination of seal habitat with agricultural runoff (Baker et al. 1995).

Pathogen antimicrobial resistance has been recognized in marine mammals admitted to rehabilitation facilities (Johnson et al. 1998). Frequent use of antibiotics for the treatment of humans, domestic pets, and livestock, combined with the contamination of the environment with resistant bacteria through raw sewage spills, municipal water dumping, and agricultural and storm/flood runoff, has resulted in antibiotic-resistant bacteria in the environment. The distribution of these bacteria in marine mammal populations is unknown, but it is likely influenced by urbanization and is probably an important indicator of pathogen movement between terrestrial and marine mammals.

Runoff from urban and agricultural areas may also influence the prevalence of infectious diseases by altering the frequency of harmful algal blooms, which appears to be increasing (Harvell et al. 1999, Van Dolah this volume). Some toxins produced by harmful algal blooms, such as brevetoxin, may alter the immune system of mammalian hosts, increasing their susceptibility to infectious diseases (Bossart et al. 1998). As brevetoxin-producing blooms are common in areas where die-offs of bottlenose dolphins associated with DMV have occurred, it is possible that brevetoxin modulated the immune response of the dolphins to the virus, thereby increasing the severity of the die-off (Geraci et al. 1999).

Pathogens may also be transported around the marine environment by the global movement of ballast water in ships (Ruiz et al. 2000). A recent study showed that *Vibrio cholerae* can often be delivered to estuaries with commercial ports and that some bacteria in ballast tanks are viable upon arrival.

Contaminants

Experimental studies on captive harbor seals fed fish with varying levels of organochlorine contaminants and on cetacean cells *in vitro* have shown that organochlorines have immunotoxic properties, decreasing natural killer cell activity as well as a series of mitogen- and antigen-induced T-cell responses (De Swart et al. 1994, De Guise et al. 1998). The effect of this alteration of the immune response on resistance to infectious disease in the wild is unclear, as most field studies have a number of confounding variables other than contaminants and disease status. Field studies investigating relationships between organochlorine exposure and effects on the immune system or deaths from disease have not shown consistent patterns, probably because of these variables and small sample sizes.

Blubber organochlorine levels in about 100 harbor porpoises that died around the United Kingdom in 1989–1992 were not significantly different in animals that died from infectious disease and those that died from trauma (primarily in fisheries) (Kuiken et al. 1994). However, animals that died of trauma were not necessarily free of infectious disease. In contrast, a second study of this same population based on samples from fresh carcasses collected from 1990 to 1996 showed significantly higher concentrations of total PCBs, and of sixteen out of twenty-five PCB congeners, in harbor porpoises that died from infectious disease (Jepson et al. 1999). Organochlorine concentrations in blubber of harbor seals that died in the 1988 phocine distemper virus (PDV) outbreak in the United Kingdom had higher levels than those that survived, but the relationship was confounded by the weight loss in sick animals, which may have concentrated lipid soluble compounds in blubber (Hall et al. 1992). Concentrations of PCBs in blubber of striped dolphins from the Mediterranean Sea that died during a morbillivirus epidemic in the early 1990s were extremely high compared to other marine mammals and to striped dolphins sampled by biopsy in the area before and after the epidemic, suggesting that PCBs may have influenced susceptibility to morbillivirus (Aguilar and Borrell 1994). Other studies have found no statistical associations between organochlorine concentrations in juvenile harbor seals collected before and during a PDV outbreak, and the addition of PCBs to the diet of captive harbor seals did not affect susceptibility to morbillivirus challenge (Blomkvist et al. 1992, Harder et al. 1992).

Exposure to organochlorines has also confounded investigations into the cause of premature parturition in California sea lions on San Miguel Island. Sea lions that gave birth to pups prematurely had higher levels of organochlorines in their blubber than did females carrying pups to term, but the former also had antibodies (indicating exposure) to *Leptospira interrogans* and caliciviruses (Gilmartin et al. 1976). These pathogens are capable of causing abortion, and differences in blubber organochlorine levels may be the consequence rather than the cause of differences in pupping date. Thus, from the available data, organochlorines could have: (a) modified immune responses to the pathogens that caused the early births, (b) caused the premature parturition, or (c) had no effect on parturition date. The etiology of premature parturition in this species remains unclear, and the interactions among infectious disease, immunity, and exposure to contaminants remain elusive, mostly because of the large number of confounding variables in field studies and the lack of experimental studies determining effects of exposure to contaminants on marine mammals (O'Shea and Brownell 1998).

Further discussion of contaminants can be found in the chapter by O'Hara and O'Shea in this volume.

Stress

Marine mammals encounter stressors daily; predators, intra- and interspecific aggressors, and adverse oceanic or weather conditions can challenge homeostatic processes. Anthropogenic factors such as vessel traffic, fishing, and underwater noise may increase the homeostatic response to a level that elicits an adverse stress response, which can modulate resistance to infectious diseases. The subtle physiological changes involved in the stress response of marine mammals are still poorly understood but are likely to decrease the immune response through a number of endocrine changes and direct effects on the lymphoid system (reviewed by St. Aubin and Dierauf 2001).

Few studies have investigated the effects of stress on disease epidemiology in marine mammals. Wilson et al. (1999) found that epidermal lesions in bottlenose dolphins were more common in animals potentially stressed by low water temperatures and salinity. Outbreaks of phocine herpes virus 1 in harbor seals occur in rehabilitation centers, where stress is likely a precipitating factor (Visser et al. 1991).

Multiple Pathogens

Interactions among pathogens, particularly between micro- and macroparasites or among different macroparasite species, are often reported in wildlife disease studies (Krecek et al. 1987). These co-occurrences and associations have implications for the impact of the primary disease on the host population as the behavior and infectivity of each determines its distribution and transmission characteristics. Modeling and predicting the effects of these complexities on host population dynamics may cause other problems (Gupta et al. 1994). Some examples of associations have been reported in marine mammals. Geraci et al. (1981) reported that the seal louse (*Echinophthirius horridus*) is an important intermediate host for three developmental stages of the heartworm (*Dipetalonema spirocauda*) in the harbor seal. *Brucella* infection has been found in *Parafilaroides* sp. lungworms in harbor seals (Garner et al. 1997), and phocine herpesvirus 1 is often secondarily associated with PDV in harbor seals (Osterhaus et al. 1985).

REVIEW OF SCIENTIFIC APPROACHES

Trends in Prevalence of Marine Mammal Diseases

Reports of emerging and resurging marine mammal diseases are increasing in frequency and severity (Harvell et al. 1999). This increase may be due, in part, to improved observation and record-keeping following opportunistic examinations, increased numbers of necropsies performed by pathologists rather than biologists, and multidisciplinary investigations of recent epizootics. Stranded animals, fisheries bycatch, subsistence-harvested animals, and animals caught for research purposes are being more closely examined by veterinarians and pathologists. Additionally, various novel technologies have enhanced our ability to identify and measure pathogens and toxins so that agents previously undetected or unidentified can now be assayed even in small or decomposing tissue samples.

It is difficult to determine whether there has been a real increase in the occurrence of marine mammal disease or, rather, whether increased reporting reflects increased interest in marine mammals generally and, perhaps, marine mammal disease in particular. To investigate and assess trends in marine mammal disease over the past 50 years, we reviewed 500 papers published in peer-reviewed journals since 1966. As with other scientific literature, the number of papers on marine mammal disease published each year has increased (Fig. 4.1). In this sample, the number of papers increased fourfold over a period of around 40 years, with peaks occurring during and after the European distemper outbreaks. A list of the papers reviewed (with their abstracts) can be found at www.smru.st-and.ac.uk.

To assess trends in the occurrence of marine mammal disease, we categorized the published studies by the disease agents investigated (viral, bacterial, parasitic, fungal, protozoal, and harmful algal toxins). Trends in the different dis-

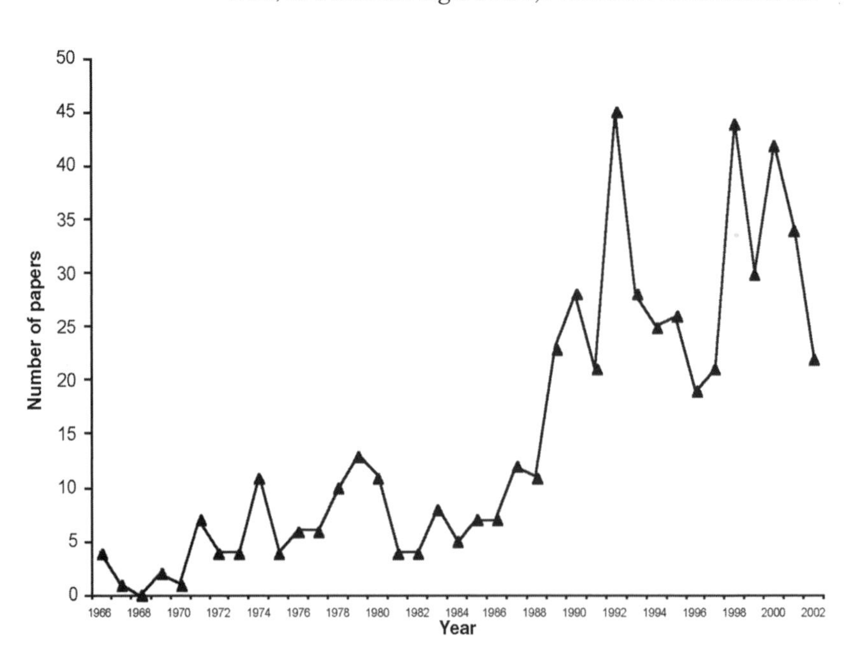

Fig. 4.1. Number of papers published on marine mammal disease, 1966–2002.

ease agents (as proportions of the total number of papers reviewed, by 3-year blocks) are shown in Figure 4.2a,b.

The papers published in the 1960s and 1970s concerned primarily parasitic and bacterial disease. Investigations of viruses emerged in the late 1970s and increased in the 1980s and 1990s. Fungal disease, especially lobomycosis in dolphins, seems to have been studied particularly between the late 1960s and early 1980s, whereas protozoal diseases and the effects of harmful algal toxins on marine mammals were rarely reported until the 1990s. Although biased in some ways (e.g., by delays in completing and publishing analyses), the broad trends seem essentially accurate. They appear to reflect our increasing ability to investigate the effects of the pathogens that are harder to isolate, such as small bacteria, protozoa, and algal toxins.

A number of mass mortalities of marine mammals have been reported in the literature, and this phenomenon may date back at least as far as the 1800s (reviewed in Harwood and Hall 1990). Such high-profile events have been observed and described by naturalists in the past back to Aristotle, but it appears that unusual mortality events have become more frequent worldwide, particularly in the late twentieth and early twenty-first centuries. Certainly, interest in the population and genetic effects of disease on particular marine mammal species has increased since the large die-offs of harbor seals and bottlenose dolphins in the late 1980s. This can be seen in Figure 4.3a, where the papers on mass mortality, compared with studies on individually stranded animals, peak in the early 1990s. Many studies of die-offs that have occurred since the late 1990s have not yet been published, so the recent decrease indicated in the figure may not signal a declining trend in the longer term.

Disease Sampling Strategies

Using the same literature database, we reviewed the different scientific approaches that have been used. Most studies have focused on individuals, and 15% of the publications in the database were case studies reporting disease in a single animal. Few studies have included sufficient numbers of individuals to support inferences at the population level except in the case of mass mortalities.

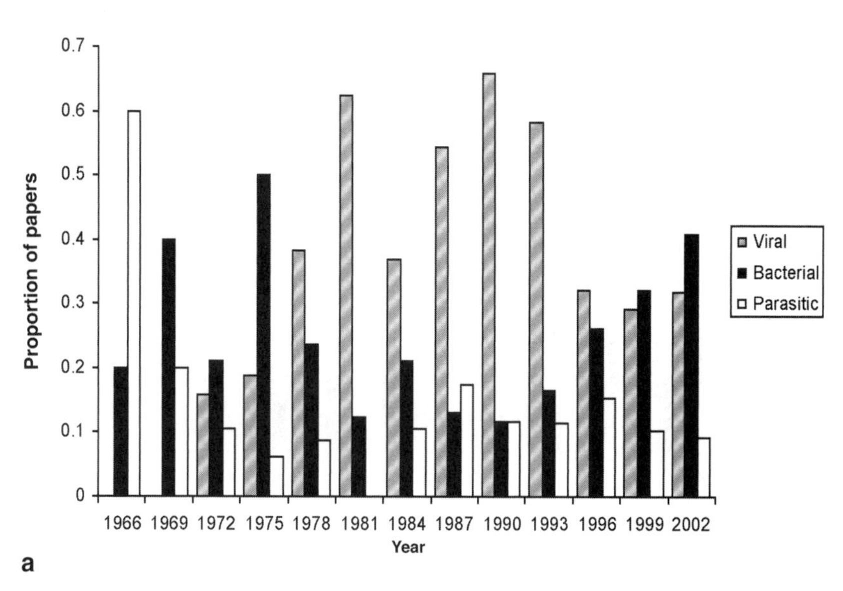

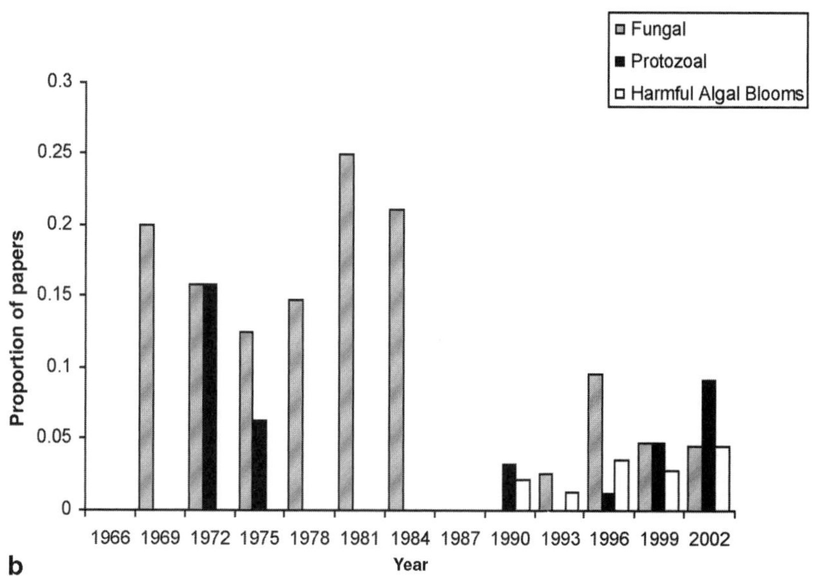

Fig. 4.2. Proportion of disease-related papers published each year, classified by pathogen.

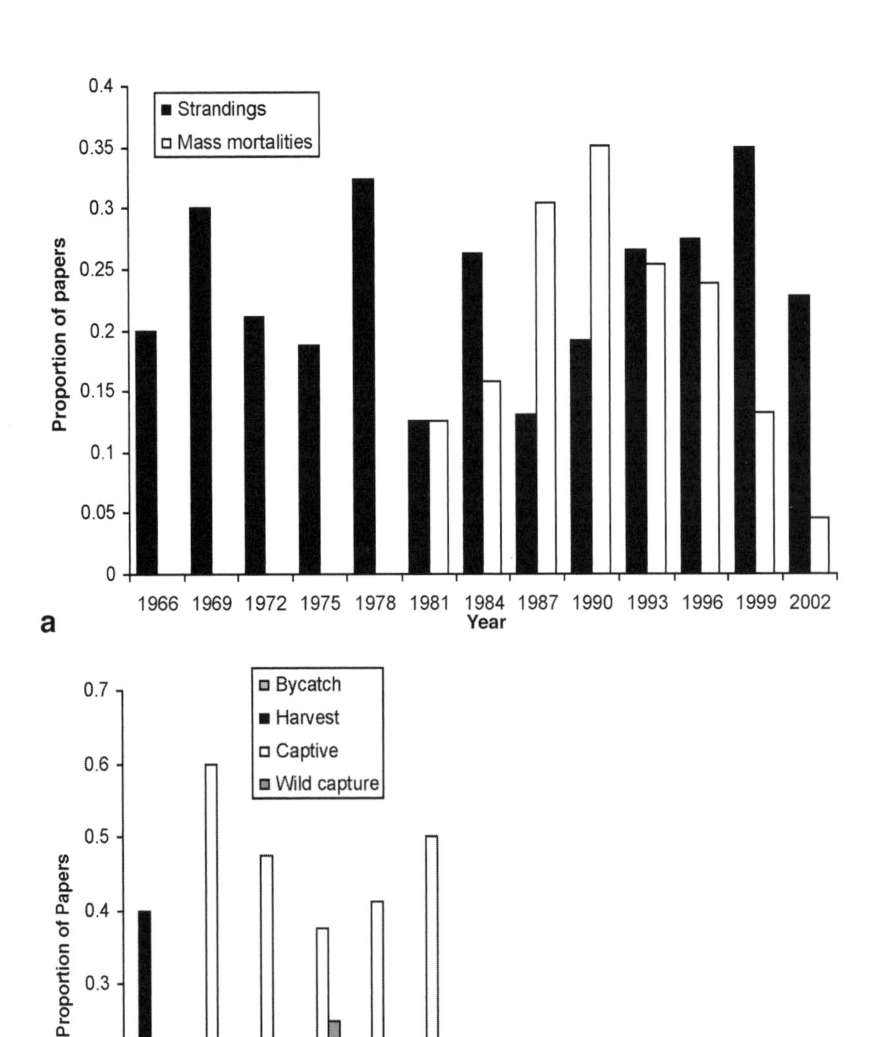

Fig. 4.3. Proportion of disease-related papers published each year, classified by sample source.

Host populations can be sampled by examining stranded, harvested, bycaught, captive, or live-captured-and-released animals. Figure 4.3a,b shows the trends in each of these sampling methods. Studies using harvested and captive animals have declined steadily. Interestingly, the number of studies on wild or live-captured-and-released animals has declined slightly. Studies dependent entirely on animals bycaught in fishing nets did not begin to appear until the late 1990s. This conclusion, however, is biased, as studies of "stranded" animals sometimes include bycaught animals as well as those found washed ashore from natural causes.

Increased concern for animal welfare throughout much of the world has led to many live-stranded animals being taken into captivity for rehabilitation. This in turn has led to the development of centers specifically designed to treat and release animals back into the wild. Such rehabilitation efforts have had important implications for the study of marine mammal disease. Figure 4.4 shows how studies using these animals have been reported in the literature, with no papers before 1965. Although rehabilitated individuals are a biased sample of their populations and only include morbid animals that were salvaged for rehabilitation, they clearly provide a significant sampling route for the study of marine mammal disease.

Each sampling method has advantages and limitations and these are reviewed later in the chapter.

STRANDED ANIMALS AND MASS MORTALITIES. Stranded animals are a useful source of information on diseases in marine mammals. As can be seen in Figure 4.3, they provided the basis for the largest proportion, 53%, of the papers reviewed. A number of infectious agents in marine mammals were first identified in stranded animals, after which their presence in the free-ranging population was confirmed. These include PDV, which caused the deaths of more than 18,000 harbor seals in Europe in 1988 and more than

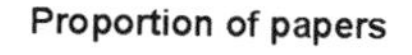

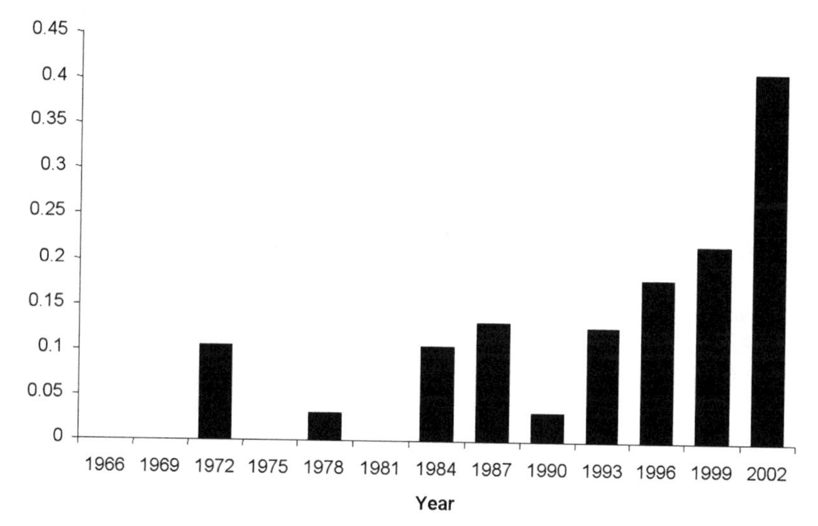

Fig. 4.4. Proportion of disease-related papers published each year on live-stranded animals.

20,000 in 2002 (Osterhaus and Vedder 1988, Jensen et al. 2002), phocine herpes virus (PhHV1) isolated from stranded harbor seals in 1985 (Osterhaus et al. 1985), and *Brucella maritimus* in a variety of species (Garner et al. 1997).

Stranded animals can also be sampled to quantify contaminant levels in tissues, and they can alert us to diseases that are present in the more inaccessible wild animals but would be difficult to detect by random sampling. For example, 17% of the sexually mature California sea lions that stranded and died along the northern coast of California showed neoplasia when examined postmortem (Gulland et al. 1996). In comparison, only one case of neoplasia has been observed in California sea lions at rookeries on San Miguel Island, where there are more than 100,000 of these animals (T. Spraker pers. comm.). Neoplasia pathogenesis is thus more readily studied with stranded sea lions than with those at rookeries, and stranded animals essentially serve as sentinels for their wild counterparts. Stranded animals do not constitute an ideal sentinel system, however, as they do not represent the entire population (Aguilar and Borrell 1994). Also, samples consisting of stranded animals usually have skewed age and sex structures, and biological data such as individual life histories, feeding habits, reproductive success, or disease progression are not typically available. Finally, contaminant levels in tissues collected from animals found dead may be significantly affected by decomposition (Borrell and Aguilar 1990).

It is expected that epidemics, when they occur, cause mortality in animals of all ages and leave survivors that generate an immune response to the agent concerned (Heesterbeek and Roberts 1995). Often a proportion of the carcasses wash ashore and are recovered for study, but this may not always be the case; in areas of low human population densities and convoluted coastlines, those that do come ashore may not always be found and reported. Alternative methods for retrospective detection of the infectious agent that caused an epidemic are to test survivors for antibodies or to examine animals susceptible to the same diseases that share habitat with the species in question. For example, distemper antibodies were found in domestic dogs in areas surrounding African wild dog (*Lycaon pictus*) habitat in Kenya. These latter packs disappeared, leading to the conclusion that the wild dogs died from distemper although no carcasses were found (Alexander and Appel 1994).

Stranded animals, whether alive or dead, can be used to detect disease, but their usefulness is limited when the objective is to assess the prevalence of a disease and its impact on a host population. As noted earlier, stranded-animal samples are skewed toward those carcasses that are likely to wash ashore. Decomposition is a problem for obtaining accurate diagnoses and good histology from dead stranded animals, while human concern and intervention can interfere with research on live stranded animals.

HARVESTED ANIMALS. Disease occurrence in harvested animals has been considered in the past but this source of material has contributed to only a small proportion of the studies reviewed (9%). The sample taken is often skewed and not representative of the population (often pups or juveniles are overrepresented and, in the case of commercial whaling, adults may be overrepresented; e.g., Lambertson 1986, 1992). The number and geographical distribution of species that can be studied in this way are also limited. However, studies of harvested animals can be very good for indicating the prevalence of infection using serological, molecular, and histological methods.

CAPTIVE ANIMALS. Up to the mid-1980s, disease among captive animals was well reported. However, with the decline in the number of display facilities maintaining live marine mammals, there has been a decline in information from this source. There are also major limitations when studying disease in captive animals and making inferences about those in the wild. Studies generally report single cases where intervention and treatment make it difficult to interpret the natural course of any disease. Many captive animals are kept in unnatural environments together with other species that they would not normally encounter and are also fed unnatural diets, so the stress in such situations might exacerbate any disease states.

LIVE-CAPTURED-AND-RELEASED ANIMALS. Carrying out studies on free-living individuals that are captured, sampled, and released within a short time frame has its own limitations. Apart from being expensive, it is often stressful for the animals, particularly if chase/capture scenarios are involved, which can confound the results obtained. Such studies are often logistically difficult to execute, yield relatively small sample sizes, and are limited in the species that are amenable to being sampled (pinnipeds and some small cetaceans). Again, the sample obtained may be skewed and limited in season. However, some unique long-term, longitudinal studies (such as on the Sarasota Bay bottlenose dolphins; Wells et al. 1987) are akin to cohort studies in human epidemiology, which is currently the most powerful study design in which to investigate the prevalence and impact of disease on a population (Rothman 1986).

BYCAUGHT ANIMALS. Disease studies using only bycaught animals are rare, other than studies on dolphins taken in the eastern tropical Pacific tuna industry. Many individuals accidentally caught in fishing nets are removed and discarded; they may later wash ashore and be reported as strandings. The main limitations of bycaught samples are that the number of species that can be sampled is limited, only small numbers are usually obtained, and samples are often not representative of the population. Because entanglement is generally fatal, the types of studies that can be carried out are limited to those that involve dead animals. Nevertheless, bycatch provides opportunities for some useful research regarding disease prevalence, and it can provide controls for other disease studies.

LICENSED EXPERIMENTAL RESEARCH FACILITIES. Relatively few research facilities throughout the world are licensed within their country to keep marine mammals in captivity for short periods of time for research purposes. Whereas most studies carried out in such facilities address questions of basic physiology, some disease studies have also been conducted (Harder et al. 1992, De Swart et al. 1994). Research into health status and immune function using these semiexperimental systems could help fill some of the gaps in knowledge, although it would be limited in scope (e.g., age, gender, species) and applicability.

Each of the sampling sources described previously has advantages and disadvantages, so that the choice of method may depend on the questions being addressed as much as on the logistical difficulties involved in sampling animals using the different methods. For example, if the objective of a study is to detect a particular disease within a population, then stranded animals might be the best sources. However, if the aim is to determine the point or period prevalence of a disease, live-captured or bycaught animals would probably be more suitable. The latter are also likely to be the most appropriate sources when a random sample of controls is required from a population. Most work on diseases of marine mammals to date has been opportunistic, taking advantage of animals available from a variety of sources. To develop clean hypothesis testing and clarify some of the important interactions between infectious agents and other factors, increased use of experimental facilities with clearly designed studies is needed.

Review of the Disease Literature

Table 4.2 identifies the different types of studies that have been reported in the literature surveyed. These have been categorized into broad classes that reflect the main method used in the paper to detect, diagnose, investigate, or model the disease agent or process being studied. Because many of the studies used multiple methods, they are given in the table as the total number of papers using each method type, by 3-year classes.

As can be seen, most publications include information about the pathology and clinical signs of the disease, with some histology and virology reported, particularly in the 1990s. Very few papers have included epidemiology or mathematical modeling, and recent advances in molecular techniques only begin to appear in the late 1990s and early 2000s. Interestingly, there are still few published studies of host genetics and immune function and even fewer that employ or describe new diagnostic methods.

Table 4.3 shows the number of papers by species. The vast majority of studies have been carried out on the harbor seal, with the bottlenose dolphin the most commonly studied cetacean. This probably reflects the relative abundance and nearshore distribution of these two species and the fact that bottlenose dolphins have often been kept in captive display facilities where case studies have been carried out. Harbor seals can also be studied in captivity and have been subject to mass mortalities in recent years; harbor porpoises, which also are high on the list, are often subject to stranding or bycatch. Interestingly, the distribution is about equal between seals and cetaceans, but it is clear that almost nothing has been reported about diseases in some species.

KEY ISSUES AND RECOMMENDED FUTURE RESEARCH

There is a lack of essential information needed to predict and prevent diseases that negatively impact marine mammal populations. Our understanding of diseases in marine mammals is poor compared to our knowledge of such processes in terrestrial animals. This is due to a lack of resources to routinely analyze and interpret disease in marine mammals, as well as to logistic difficulties that have hampered opportunistic investigations to date. As a result, we still do not know which diseases are normal components of a healthy marine ecosystem and which are novel and a consequence of anthropogenic factors. We also do not know if there is a real increase in diseases in marine mammals, nor do we know how to control or prevent these diseases. To overcome these hurdles, we have to invest in an infrastructure that will change how we study diseases in

Table 4.2 Number of papers categorized by type of study (discipline) and year

Year	Serology	Bacteriology	Immunology	Virology	Histology	Pathology/ clinical signs
1966	1	1	0	0	0	4
1969	1	1	0	0	2	6
1972	2	3	1	1	1	9
1975	1	5	1	3	4	7
1978	2	3	0	10	6	18
1981	0	3	0	9	0	6
1984	1	3	1	7	3	9
1987	8	6	2	6	4	20
1990	12	11	3	27	19	32
1993	8	11	4	24	17	23
1996	10	18	2	15	12	28
1999	12	22	9	15	21	45
2002	7	2	1	3	7	10

Year	Molecular	Hematology/ biochemistry	Vaccine development	Epidemiology/ mathematical modeling	New diagnostic methods	Host genetics
1966	0	0	0	0	1	0
1969	0	0	1	0	0	0
1972	0	0	0	0	0	0
1975	0	0	0	0	1	0
1978	0	1	0	0	0	1
1981	0	1	0	0	0	0
1984	0	0	0	0	0	0
1987	0	2	2	2	0	0
1990	8	1	2	8	3	1
1993	13	0	0	1	8	0
1996	19	2	0	4	1	0
1999	26	1	2	2	9	0
2002	6	1	0	0	0	0

marine mammals. The designation of a specialized marine mammal disease laboratory as a centralized resource would allow for diagnosis and pathogenesis research (a schematic diagram of how such a system would address various questions in this field is shown in Figure 4.5). Samples could be screened against a standard panel of diseases, and collaborations would be encouraged to investigate emerging and previously unstudied pathogens. Data and samples could be made available for collaborative research. Such an approach requires a long-term commitment, transgenerational studies, longitudinal studies, and baseline monitoring, all necessary elements to determine the true impact of disease on a marine mammal population.

Key Issues

1. Multidisciplinary approaches are currently rare.
2. There are few laboratories with expertise and resources to analyze samples specifically from marine mammals.
3. There are few statisticians with appropriate expertise and there is a severe lack of good experimental design.
4. Measuring disease or changes in disease in free-ranging animals in a marine system is logistically difficult.

Future Research Areas

1. Standardize data collection and integrate disease data collection with other marine mammal population data to develop baseline information. This is needed to give a solid baseline with high-quality accessible data on the identity of endemic diseases and their effects in marine mammals.
   - Develop a centralized marine mammal diagnostic laboratory: (a) Develop reagents that function across species (e.g., biomarkers, monoclonal and polyclonal antibodies, primers); (b) develop culture techniques for infectious agents, including viruses, bacteria and rickettsia; (c) standardize serology tests for diseases in marine mammals; (d) improve field methods (e.g., RNA preservation media such as "RNA Later") that allow the use of molecular techniques (e.g., for RT-PCR and viral DNA/RNA detection).

Table 4.3 Study species reported in papers reviewed

Species		1–5	6–10	11–30	31–50	>50
Antarctic phocids			x			
Arctocephalus australis gracilis	South American fur seal	x				
Arctocephalus forsteri	New Zealand fur seal	x				
Arctocephalus pusillus doriferus	Australian fur seal	x				
Arctocephalus pusillus pusillus	South African fur seal	x				
Arctocephalus townsendi	Guadalupe fur seal	x				
Arctocephalus tropicalis	Antarctic fur seal	x				
Balaena mysticetus	Bowhead whale	x				
Balaenoptera acutorostrata	Minke whale	x				
Balaenoptera borealis	Sei whale	x				
Balaenoptera musculus	Blue whale	x				
Balaenoptera physalus	Fin whale	x				
Callorhinus ursinus	Northern fur seal			x		
Cystophora cristata	Hooded seal	x				
Delphinapterus leucas	Beluga			x		
Delphinus delphis	Common dolphin		x			
Enhydra lutris	Sea otter	x				
Erignathus barbatus barbatus	Bearded seal	x				
Eschrichtius robustus	Gray whale	x				
Eumetopias jubatus	Steller sea lion	x				
Globicephala melaena	Long-finned pilot whale	x				
Grampus griseus	Risso's dolphin	x				
Halichoerus grypus	Gray seal			x		
Hydrurga leptonyx	Leopard seal	x				
Inia geoffrensis	Boto	x				
Kogia breviceps	Pygmy sperm whale	x				
Lagenorhynchus acutus	White-sided dolphin	x				
Lagenorhynchus albirostris	White-beaked dolphin	x				
Lagenorhynchus obliquidens	Pacific white-sided dolphin		x			
Lissodelphis borealis	Northern right whale	x				
Lobodon carcinophagus	Crabeater seal	x				
Megaptera novaengliae	Humpback whale	x				
Mirounga angustirostris	Northern elephant seal		x			
Mirounga leonina	Southern elephant seal	x				
Monachus schauinslandi	Monk seal			x		
Neophoca cinerea	Australian sea lion	x				
Neophocaena phocaenoides	Finless porpoise	x				
Obobenus rosmarus divergens	Pacific walrus	x				
Obobenus rosmarus rosmarus	Atlantic walrus	x				
Orcinus orca	Killer whale	x				
Otaria byronia	Southern sea lion		x			
Phoca caspica	Caspian seal	x				
Phoca groenlandica	Harp seal		x			
Phoca hispida hispida	Ringed seal		x			
Phoca sibirica	Baikal seal			x		
Phoca vitulina	Harbor seal					x
Phocarctos hookeri	Hooker's sea lion	x				
Phocoena phocoena	Harbor porpoise				x	
Phocoena spinipinnis	Burmeister's porpoise	x				
Physter macrocephalus	Sperm whale	x				
Platanista gangetica	Ganges river dolphin	x				
Sousa chinensis	Indo-Pacific humpback dolphin	x				
Stenella attenuata	Pantropical spotted dolphin	x				
Stenella coeruleoalba	Striped dolphin			x		
Trichechus manatus latirostris	Florida manatee	x				
Tursiops truncatus	Common bottlenose dolphin					x
Zalophus californianus	California sea lion				x	

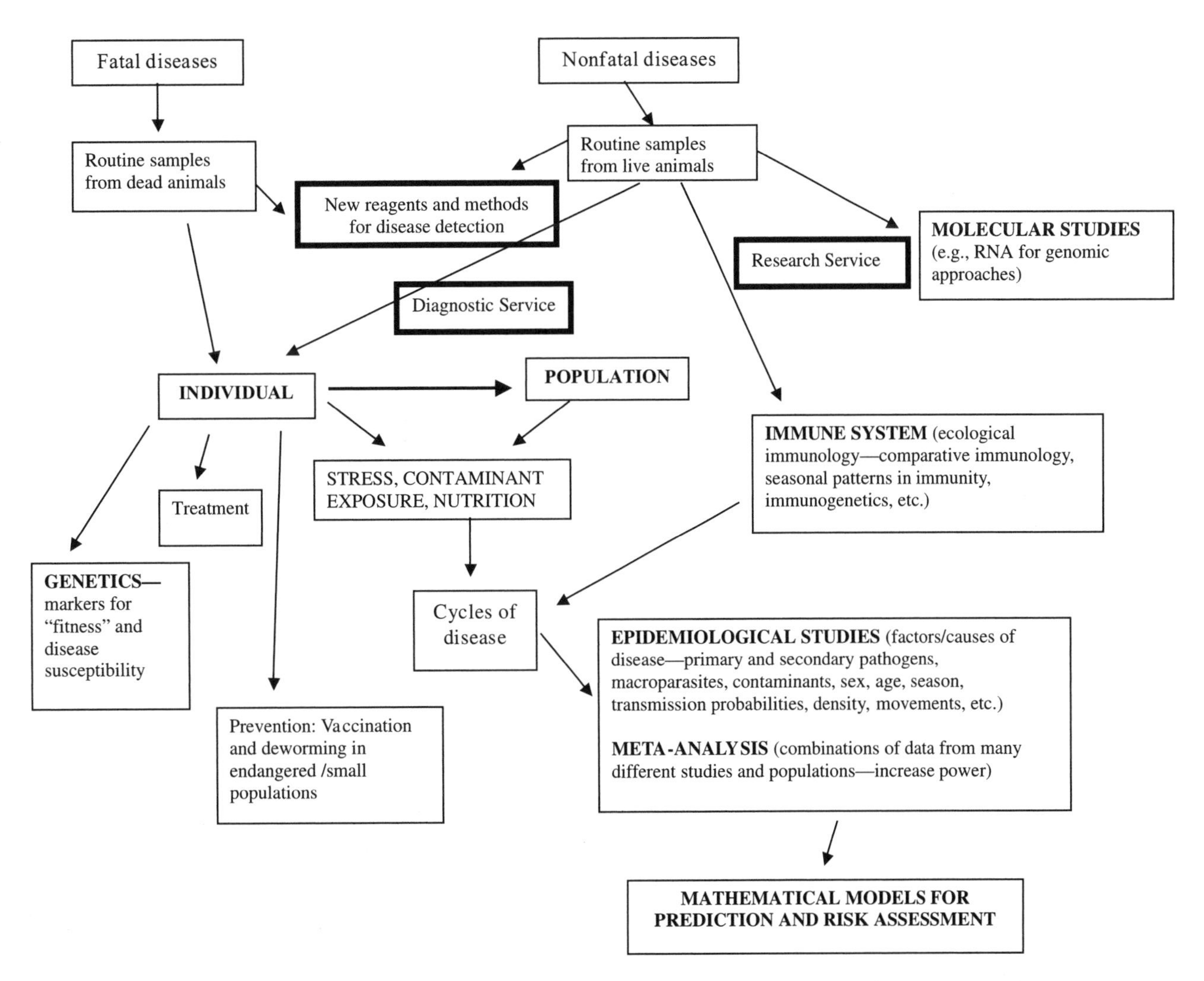

Fig. 4.5. Schematic diagram suggesting how a specialized marine mammal disease laboratory could be used to address questions involving marine mammal disease.

- Standardize and centralize collection of additional marine mammal disease and population data (e.g., serology data, bacterial culture data, parasite identification).
- Provide routine training to relevant sites regarding standardized data collection, reporting, and epidemiological analysis.
- Thoroughly analyze and report the National Oceanic and Atmospheric Administration's (NOAA) Marine Mammal Health and Stranding Program Level A retrospective data.

2. Identify unusual disease outbreaks. This is needed to identify epidemic and novel diseases that can reduce survival of marine mammals.
   - Enable rapid reporting of NOAA's Level A data to the centralized database; ensure that current data are routinely reported to target audiences.
   - Develop a rapid-response team with skills to diagnose disease and conduct epidemiological investigation.
   - Develop multidisciplinary teams to identify factors leading to disease outbreaks.
3. Identify risk factors and evaluate interventions. This is needed to be able to control diseases and factors precipitating them in marine mammals.
   - Support for multidisciplinary teams that can prioritize case-control and cohort studies related to abnormal disease processes and interventions.
   - Establish the nature of the interactions among disease, immunity, and nutrition by experimental and field studies using robust epidemiological study designs.
   - Develop a standardized, centralized marine mammal population database that is linkable to outside environmental databases (e.g., water temperature, salinity, harmful algal blooms).
4. Determine associations between health effects and population parameters, especially in regard to nonfatal diseases currently largely ignored. This is needed to determine the impact of diseases on populations of marine mammals, and thus the significance of disease in conservation and management of marine systems.

TODD M. O'HARA
AND THOMAS J. O'SHEA

Assessing Impacts of Environmental Contaminants

THE PROBLEMS OR PERCEIVED PROBLEMS (unanswered questions) regarding chemical contaminants are very simply that these substances (i.e., contaminants, biotoxins, pharmaceuticals, industrial spills, and wastes) can kill animals or change reproductive, metabolic, and immune functions, which alters population dynamics. These impacts threaten individuals and populations and are well documented for laboratory and domestic animals and, in rare cases, for marine mammals as well. Conclusive cause-and-effect relationships linking chemicals to toxicosis or population-level effects are seldom documented for free-ranging marine mammals as these cases are difficult to diagnose or research. Scientists have assessed body burdens of selected toxicants in marine mammals, but have yet to answer the fundamental questions being asked by policy makers and managers: "Are toxicants killing or impairing the health of marine mammals?" If so, to what extent (i.e., is there an impact on a few individuals, on a population, or on a species)? The answers are certainly more readily available for cases involving acute, high-dose (high-exposure) episodes and large numbers of individuals (i.e., many dead animals on the beach covered with oil or associated with an algal bloom) as compared to the more insidious, relatively lower rate of mortality (i.e., immune-compromised animals dying over months, not hours) or reproductive dysfunction scenarios (i.e., decreased neonatal survival over years or decades) where dramatic events (i.e., bodies on the beach) are not evident. The more insidious effects could eventually have more of an impact at the population level and are thus worthy of consideration. Obviously, simply measuring residue concentrations in tissues of marine mammals, their prey, and habitat does not provide the needed information for assessing subtle but important potential impacts.

Compounding the basic difficulty of studying toxicants in marine mammals is that many new chemicals (sometimes referred to as "emerging" chemicals) are being produced, used, and released into the environment on a daily basis. These releases occur at unknown rates for many compounds from broad-ranging (nonpoint) sources such as, for example, volatilization of flame retardants from household items or pesticides from fields, lawns, orchards, and gardens; forest fires; internal combustion engine exhausts; urban and agricultural run-off. Considering the difficulty in conducting (and the resultant dearth of) studies that provide cause-and-effect information and the economic importance of various industries, policy makers have not become convinced of the need to institute precautionary or preventative measures based on "potential" impacts on marine mammals. Thus we strongly suggest that more direct linkages of chemically induced lesions in individuals and impacts on populations are needed to provide convincing information to managers and policy makers, as negative economic impacts have to be justified beyond the precautionary principles commonly suggested. This does not reflect a lack of concern on our part (as we are very much concerned about current trends in chemical contamination of marine mammal habitats), but rather reflects what we consider an economic-political reality. Presenting agencies with an overabundance of contaminant residue data has not had much effect, as the accompanying data usually lack any impact assessment; a very fair rebuttal is "So, what—everything is contaminated!" This chapter briefly describes the status quo regarding contaminants in marine mammals in a general sense. However, our major intent is to prescribe changes that will provide some of the data needed to better understand the impact of chemicals on marine mammals, so as to bring about the policy changes that are necessary to improve the management of chemical releases.

MARINE MAMMALS AND CONTAMINANTS: WHY BE CONCERNED?

As noted previously a toxicant can cause an animal literally to drop dead, clearly linking exposure and death in a very easily described time sequence (i.e., an oil spill occurs and the next day oiled animals appear and die on a beach). Other mechanisms operate with some chemicals that are teratogenic (birth defects) or "transgenerational." An example in humans would be thalidomide, which was first introduced in the 1950s as a sedative and was prescribed for nausea and insomnia in pregnant women. However, the drug was eventually found to be the cause of severe birth defects in children whose mothers had taken it in the first trimester of pregnancy. Diethylstilbestrol (DES) taken during pregnancy has been shown to increase the incidence of clear-cell adenocarcinomas of the vagina or cervix and vaginal adenosis in daughters exposed in utero (during pregnancy). These two examples highlight the delayed response to a toxicant exposure at a critical phase in development that can have significant effects in mammals. In general, many of the environmental contaminants (organohalogens, heavy metals) are known to affect endocrine systems and cross the placenta or are transmitted in milk and possibly act in a delayed fashion.

Pinnipeds, cetaceans, and other marine mammals share many of the detoxification (defense) mechanisms and the susceptibilities (vulnerability) to toxicants with the experimental mammal models (oddly enough, humans and laboratory and domestic mammals). We know that many of the biochemical (e.g., enzymes, receptors) and physiologic systems (e.g., immune, central nervous system, thyroid) impacted by toxicants exist in marine mammals. In some cases, however, marine mammals have unique morphologic or physiologic conditions that alter toxicant behavior and the animals' response. It is well known, especially for polar-adapted species, for example, that marine mammals have proportionately much larger lipid stores than most terrestrial mammals. This lipid store clearly accumulates many lipophilic chemicals, such as the polychlorinated biphenyls (PCBs) and 2,2-*bis*-(*p*-chlorophenyl)-1,1,1-trichloroethane (DDT), a widely used pesticide. A marine-based diet and adaptations for deep and prolonged dives also alter the physiology of marine mammals and may create more susceptible or more tolerant systems depending on the toxicant of interest and the particular species. These potentially differing aspects of marine mammal physiology and biochemistry are poorly understood and weaken our ability to extrapolate the toxic effects of contaminants at the concentrations at which we detect them (i.e., what is the equivalent of harbor seal blubber in laboratory rodent models?). For biomagnifying chemicals like the PCBs for which concentrations in tissues of animals increase with increasing trophic level, animals feeding in a particular habitat at a higher trophic level have a greater body burden. An unfortunate strategy of the females is to excrete some of the toxic compounds via their lipid-rich milk, which results in high exposures of neonates to those chemicals; this is due to the interaction of mobilization of lipid-associated toxins from the blubber of the dam, the concentration of the toxins in the milk, and the relatively large intake by an animal of much smaller body size, resulting in a rather large dose. We should focus our impact assessment on species and specific cohorts (neonatal exposure and effects assessment) of most concern. However, not all toxicants of concern act in this lipophilic manner and may require very different approaches (i.e., petroleum hydrocarbons can be of acute or chronic concern and for the most part do not biomagnify in marine mammals).

Based on these considerations, it should be clear that relatively low-level exposures to some chemicals at critical life stages can result in dramatic effects on individuals and/or subtle but important population-wide impacts affecting population growth or maintenance and health. One should also develop an appreciation of the fact that marine mammals tend to accumulate certain contaminants and that reproduction is a vulnerable process because mothers harboring high body burdens of those contaminants produce relatively highly contaminated milk for their young. Further, the young are physically and developmentally quite vulnerable during the life phase that is concurrent with this usually intensive suckling. The effects may not be expressed until well beyond adolescence (consider human examples presented previously), maybe as poor reproductive performance.

Considering the high-profile nature of marine mammals and the emotional responses of the public to them, one can appreciate that reports of "highly contaminated" charismatic marine mammals and the impact these have socially, in particular as a "poster child" for environmental activism or fundraising, need to be addressed. In part because of this perception, many groups have selected marine mammals as potential sentinels of the "health" of coastal habitats and the ocean (Aguirre et al. 2001b, Reddy et al. 2001). Because of the various niches occupied and the numerous marine mammal feeding strategies, this concept of "sentinel" has some merit. However, complicating this monitoring value is the fact that many biological factors (e.g., age, gender, reproductive status, body condition, feeding behavior, other stressors) can alter the concentrations of contaminants present and the response measures assessed (i.e., biomarkers). For migrating animals (i.e., gray whales) or marine mammals with unknown distributions, it is difficult to indicate which geographic regions are responsible for the contamination, which thus limits their utility as "sentinels."

The use of marine mammals for food is worldwide (e.g., North America, Scandinavia, North Atlantic, Russia, Japan, Antarctica, Caribbean, Africa), and contamination of these food products has led to advisories on consumption of various tissues. Generalizations about food safety and quality cannot be made because the contamination varies more than a thousandfold depending on the species and region considered for some of the contaminants. Whatever one's view of consuming food products from marine mammals may be, neither animal rights activism (or environmentalism) nor financial or political interests in maintaining commercial or subsistence use should be the driving force for food consumption advice. What is needed is a balanced scientific and public health approach that recognizes the benefits and risks of these food items. Unfortunately, we cannot cover this topic in more detail in this chapter, but we feel mutual respect and careful science and policy can best develop the advice needed in regard to consuming or not consuming various products from marine mammals.

The scientific study of contaminants in marine mammals began slowly some 40 years ago with the first documentation of the presence of anthropogenic chemicals in the tissues of whales, pinnipeds, and small cetaceans (Osterberg et al. 1964, George and Frear 1966[**AQ1**], Koeman and Van Genderen 1966, Sladen et al. 1966, Holden and Marsden 1967). Attention of scientists and policy makers has focused on this topic ever since, and interest in the subject has grown tremendously in recent years (for a history of the field, see O'Shea and Tanabe 2003). Much of this increase in attention has been due to a suspicion that contaminants contribute to several observed phenomena: (1) alarming die-offs or population declines of some conspicuous species of marine mammals; (2) reproductive anomalies and other pathological conditions in marine mammals; (3) cross-sectional population declines to levels prompting designation of species or stocks as threatened or endangered; and (4) provocative reports of unsafe food supplies for those consuming marine mammals. Simultaneously, there has been growth in the number of contaminants detected in the tissues of marine mammals and large increases in the numbers and quantities of new contaminants (including hydrocarbons, pharmaceuticals, and metals) that are being synthesized by industry and utilized and discharged by society. Research in environmental chemistry teaches us that many of these contaminants inevitably enter marine ecosystems, where marine mammals may be exposed to and affected by them.

Since the 1960s scientists have discovered much about concentrations of a subset of anthropogenic chemicals in the tissues of some marine mammals (O'Shea 1999), but clear answers to basic questions about the effects of these contaminants remain elusive. For example, there are still no definitive studies to clearly infer that dichlorodiphenyldichloroethylene (DDE), perhaps the most ubiquitous and well-studied organochlorine contaminant in marine mammals, has any effect on their reproduction, as was established long ago for raptors, as reviewed in Blus et al. (1996). The scientific study of contaminants in marine mammals to date has allowed few definitive positions and therefore has had minimal effect on policy. Are contaminants killing or impairing the health, reproduction, and immune systems of marine mammals? How extensive are such impacts? What are the more subtle but potentially insidious effects of chemical contaminants on marine mammals? Strong science (i.e., appropriate experimental design; careful hypothesis testing to answer cause-and-effect questions), rigorous diagnostic procedures, and credible risk-assessment protocols are needed to create more reliable foundations for inference. Otherwise the true nature of the impacts of contaminants on marine mammal mortality and health will remain unknown and arguable. We support a precautionary approach to management practices that would mitigate or prevent likely increased exposure of marine mammals to contaminants in an absence of specific information on likely effects of such exposure. However, we emphasize that scientific research in this field must employ methods that allow solid, direct inferences whenever possible. Such methods can establish a more reliable body of knowledge about the effects of contaminants on marine mammals than is currently available.

SCOPE AND AIMS OF THIS REPORT

We describe needed future directions in research, policy, management, diagnostic, event response, and professional disciplinary support for improving the understanding, diagnosis, and management of the impacts of contaminants on marine mammals. In particular, we provide a general synopsis (not a literature review of marine mammal contaminants) of the status and trends of the science of marine mammal toxicology, research recommendations (including another look at those made by others in recent scientific meetings), a discussion of methods of mitigation, and suggestions for new approaches to resolve issues related to contaminants and the conservation of marine mammals. Our concern is broad in potential scope and policy ramifications. We make no attempt here to review all the facets of the topic comprehensively. Our primary goal is to provide a blueprint of recommended actions to ensure that future contaminant research is carried out effectively and in ways that optimally inform policy makers.

Even though many thousands of chemicals are introduced into the world's oceans, most information about contaminants and marine mammals is weighted toward organohalogens (especially organochlorines [OCs]) and metals and, to a lesser extent, radionuclides and petroleum hydrocarbons (O'Shea 1999). As a result, much of the material presented here is pertinent primarily to those substances. This chapter considers all marine mammals as defined under the Marine Mammal Protection Act (MMPA) but does not attempt to address each species individually. Unfortunately we must, at times, make generalizations that we recognize have exceptions and limitations. Our geographic foci are the coastal areas of the United States and U.S. territories. In some instances, the geographic scope may be broader because marine mammals are international resources and subject to treaties and agreements to which the United States is a party and research and management for certain aspects of this field are more advanced in other nations. Nonetheless, our "blueprint" for better assessing contaminants and contaminant effects in marine mammals is not specific to the United States; as the problems transcend political boundaries, so do our recommendations.

RECENT SOURCES OF INFORMATION ON CONTAMINANTS AND MARINE MAMMALS

Growth of interest in contaminants and marine mammals is exemplified by several recent overviews, syntheses, workshop reports, and books devoted to this field. Over the past 5 years, review chapters on contaminants and toxicology have appeared in various volumes (O'Shea 1999, O'Hara and O'Shea 2001, O'Shea and Aguilar 2001, Ross and Troisi 2001, de Wit et al. 2004). Papers from a workshop dealing primarily with cetaceans and contaminants appear in a monograph edited by Reijnders et al. (1999), and a recent workshop report provides recommendations for research on contaminants and sea otters (Reeves 2002a). A workshop report edited by Ross and De Guise (1998) addresses research needs regarding contaminants in the context of marine mammal health. Papers on the toxicology of marine mammals have been published in a book edited by Vos et al. (2003a), which also suggests avenues for future research. Perhaps the most germane source of material for our chapter is the volume that resulted from the Marine Mammal Commission–sponsored Workshop on Marine Mammals and Persistent Ocean Contaminants, held in Keystone, Colorado, in 1998 (O'Shea et al. 1999). Findings of that workshop provide a partial foundation for our recommendations in this chapter. They are excerpted in the appendix, along with new comments and annotations from our current perspective. Among the additional pertinent reviews are those by Walker and Livingstone (1992), Loughlin (1994), Hellou (1996), Kamrin and Ringer (1996), Law (1996), Arctic Monitoring and Assessment Programme (2002), and de Wit et al. (2004). A collection of reviews of cellular and molecular tools in marine mammal management, health, and disease (Pfeiffer 2002) could also prove useful in advancing marine mammal toxicology if applied to address individual and population health of marine mammals in a rigorous and meaningful way (also see Gulland and Hall and Van Dolah in this volume). Marine ecosystem health assessment or "Conservation Medicine" (Aguirre et al. 2001a) using marine vertebrates as monitoring tools or sentinel species represents another emerging approach (Aguirre et al. 2001b).

THREATS TO MARINE MAMMALS

Exposure of Marine Mammals to Contaminants

Marine mammals are exposed to contaminants via the foods they eat, the water in which they swim, and the air they breathe. This exposure is not simple, and varies in each of these external compartments as a function of many complex factors. Nonetheless, understanding and mitigating or removing exposure forms the basis for diagnostics and treatment of toxicosis, regardless of whether it is based on contaminant poisoning of an individual or an ecosystem. A simple descriptive understanding of "exposure" is inadequate. A quantitative assessment is required because "the dose makes the poison," as demonstrated by the fact that many human medicines are highly therapeutic at recommended doses but can be deadly when such doses are exceeded. The principle that an understanding of dosage is important is a basic tenet of toxicology. Inadequate knowledge concerning dose-response relationships remains a major limitation to assessing threats of contaminants to marine mammals and consumers of marine mammal products. Determinations are required for the actual degree of exposure, the expected response, and methods for "treating" or mitigating the exposure.

These are not easy determinations to make for the multitude of chemicals to which marine mammals are exposed, and they are certainly more difficult to address in the absence of reliable data on feeding ecology (including food intake) of marine mammals. Prey is probably the most significant source of marine mammal exposure. The point where contaminants first enter the marine environment before they reach prey in the food chain is where policy makers and managers can have the most significant and positive impact, as once most contaminants are released into the marine ecosystem, there is little chance of chemical recovery. Effective management must address point (e.g., catastrophic oil spill) and nonpoint (e.g., atmospheric transport to remote regions) inputs of contaminants to the aquatic and marine environments, including chemical use inland. Pesticides used in equatorial and temperate areas have effectively reached the poles (e.g., Sladen et al. 1966, Holden and Marsden 1967, Arctic Monitoring and Assessment Programme 2002, de Wit et al. 2004). Exposure of marine mammals to contaminants is a global issue.

Biases in Classes of Contaminants Investigated in Marine Mammals

The most commonly studied contaminants in marine mammals include the organochlorines such as the polychlorinated biphenyls (PCBs) and the insecticide 2,2-*bis*-(*p*-chlorophenyl)-1,1,1-trichloroethane (DDT) and its metabolites, and heavy metals such as mercury, cadmium, and lead. Exposure of marine mammals to anthropogenically derived essential and nonessential elements (metals) has been and will continue to be of concern. However, organochlorines and a few selected metals seem to represent a default list of chemicals when "pollutant studies" are carried out or proposed. In some cases these are contaminants of great relevance. However, analyses of contaminant residues in the tissues of marine mammals or their prey limited to these groups greatly diminish the potential for analyses of a broader range of chemicals. Chemicals studied should be based on well-reasoned probabilities of exposure and possible adverse health effects. Contaminants of possible importance depend on geographic location, season, prey selection, trends in use, disposal practices, and physical-chemical properties of the compounds. A list of "emerging" chemicals around the world that are seldom evaluated in the context of marine mammal exposure would be overwhelming (but see Arctic Monitoring and Assessment Programme Trends and Effects Program: 1998–2003, Section B, Pages 30–31 Revision: November 2000, and de Wit et al. 2004). Some additional emphasis has recently been placed on butyltins, brominated organics, naturally occurring halogenated organics, phthalates, perfluoro-octane sulfonate (PFOS), polyaromatic hydrocarbons (PAHs), and metabolites of organochlorine contaminants (hydroxylated, methylsulfonated).

The focus on exposure to the more traditionally assayed contaminants such as organochlorines as opposed to testing hypotheses about exposure to emerging compounds may delay recognition of new threats. Unless integrated with other areas of investigation, simply screening tissues and prey of marine mammals for a familiar roster of contaminants because of past precedents emphasizing these substances is no longer a very fruitful exercise for advancing our scientific understanding about the significance of contaminants to the health of marine mammals.

PERSISTENT AND BIOMAGNIFYING OCEAN CONTAMINANTS AS SOURCES FOR POTENTIAL MARINE MAMMAL TOXICOSIS

Sources of Contaminants in Marine Mammals. Human activities contribute tons of many types of chemicals to the air, to the surface waters (inland and coastal), and directly to the marine environment (e.g., Giam and Ray 1987, Aguirre et al. 2001a,b). Sources are varied and include industry, agriculture, energy production and use, waste disposal, and other large-scale activities. These sources can affect marine water and habitat quality around the globe. Many represent point sources (direct discharges) that can be easily identified and managed as compared to the large-scale nonpoint sources that cross continental boundaries and travel many thousands of kilometers (e.g., volatile components of pesticides and mercury, and biotransport mechanisms are important for "source" determination). Contamination of marine mammals is widespread and the anatomical, physiological, and biochemical adaptations of these animals (particularly the presence of fat or blubber and high trophic levels of some species) predispose them to accumulating persistent lipophilic contaminants and other chemicals not using fat-based pathways (e.g.., mercury). Many contaminants are pervasive in marine waters, whereas contamination with some chemicals can be heaviest in more restricted near-shore bays, estuaries, and coastal waters.

The Stockholm Persistent Organic Pollutants (POPs) Protocol was signed in 2001 and ratified by fifty countries (although not by the United States) in May 2004 and provides a global agreement for the phase-out of POPs. This Stockholm Convention on POPs included an assessment report on DDT, aldrin, dieldrin, endrin, chlordane, heptachlor, hexachlorobenzene, mirex, toxaphene, polychlorinated biphenyls, dioxins, and furans (the so-called "dirty dozen"). The United Nations Environment Programme (UNEP) Chemicals Program (http://irptc.unep.ch/gmn/default.htm) is launching the "Global Network for Monitoring of Chemicals in the Environment," a project that aims to create an electronic forum and working group to coordinate methodologies and analyses of chemicals in the environment. For further discussion of "sources," readers should consult websites for the United Nations Environment Programme (http://irptc.unep.ch/pops/) and the Arctic Monitoring and Assessment Programme (2002) and de Wit et al. (2004).

Some POPs have tremendous past or current social and economic value for controlling disease agents or promoting agriculture. Replacing POPs in an economically feasible manner around the globe (in particular for developing countries) is key to preventing further contamination. This requires a clear understanding of the "replacement technologies" for these pollutants (Mörner et al. 2002) and mitigation needs (e.g., for existing transformers and capacitors containing PCBs, see Inter-Organization Programme for the Sound Management of Chemicals 2002). It is encouraging that UNEP is monitoring mitigation measures (United Nations Environment Programme 2002a).

The next step is to apply this rigorous approach to other chemical classes of concern (i.e., heavy metals, oil, radionuclides). Policy and mitigation measures for the POPs, if effective, require decades to be reflected in chemical residue monitoring programs for marine mammals. For chemicals that may be used in the future, government and industry need to be active (or, more accurately, proactive) in estimating whether marine mammals and other wildlife are likely to be unharmed before full-scale manufacture and release. This requires extensive educational programs, lobbying, and changes in approval processes for pesticides, chemical usage, and waste disposal, which creates a large economic burden. Justification of "costs" will be needed—what is marine mammal health worth? Marine mammals are some of the most high-profile (charismatic megafauna) creatures on Earth. A justification of costs should not be confined to economic or ecological values but should explicitly also include the nutritional, cultural, and spiritual values of marine mammal resources. Because marine mammals are international resources and marine pollution is a global issue, we stress the importance of full and meaningful U.S. participation in international initiatives.

Persistence and Biomagnification in Primary Producers and/or Lower-Trophic-Level Organisms. For many compounds, this first step in bioaccumulation (water to an organism) can be of the greatest magnitude. Typically, the process is expressed in terms of bioaccumulation factors (BAFs), and a variety of methods for calculating BAFs have been published. Hoekstra et al. (2002b) used the following equation:

$$\text{BAF (for } OC_x) = [OC_x \text{ in } \textit{Calanus} \text{ spp.}]/[OC_x \text{ in water}]$$

where [OC_x in *Calanus* spp., a copepod] are the mean lipid-adjusted OC concentrations in these zooplankton samples (ng/g lipid) and [OC_x in water] is the mean OC concentration in water samples (ng/ml). This or similar equations could be used for phytoplankton or other primary producers and organisms at trophic levels lower than marine mammals.

Comparison of PCB congener profiles in zooplankton and water samples suggests that biotransformation by cytochrome P-4502B isozymes is low in *Calanus* and that limited phase I metabolism may occur; thus the OCs are persistent in this group. These results suggest that hydrophobic OCs (log K_{ow} 3–6) in *Calanus* species are at equilibrium with the water concentrations and that physical partitioning, rather than biotransformation, is the major factor governing OC profiles in marine zooplankton. Persistence and lipophilicity set the stage for biomagnification in marine mammals (e.g., baleen whales) and marine mammal prey (fish and cephalopods for most odontocetes and pinnipeds, and for some mysticetes). It is important to recognize that compounds that are "very lipophilic" (log $K_{ow} > 6$) tend to have a reduced potential of accumulation because they are not as readily available for absorption from the gut (Hoekstra et al. 2002a,b). This exemplifies the need to investigate multiple congeners to assess ecological and biological "persistence." Furthermore, considering the structure activity relationship (SAR) and vulnerability of some OCs to metabolism (biotransformation), one can estimate metabolic or biotransformational ability (isozyme presence and/or capacity) within a biological system.

Residue concentrations of contaminants in tissue or body fluids (such as blubber, whole blood, liver, plasma) collected at a single time are of limited value for assessing exposure to contaminants in marine mammals, especially when developing a risk assessment. Estimation of oral exposure over time, via quantitative feeding ecology studies, would elucidate the rate of exposure to contaminants ("cumulative" dose) and provide a better approximation of true exposure (i.e., toxicokinetic considerations). Typically, toxicologists and pharmacologists assess the response of a biological system (an animal) to a set exposure (certain concentration or dose over time) and determine risk based on the amount of exposure (oral, inhaled, injected) as determined from experimental laboratory tests or well-documented intentional or accidental exposures (epidemiological evidence). Marine mammal toxicology to date has unfortunately relied too much on concentrations of contaminant residues in tissues, which do not in themselves allow accurate assessments of risk or impacts.

Persistent Contaminants in Marine Mammals of Lower Trophic Levels. There are no experimental data on the toxic effects of persistent contaminants in marine mammals occupying lower trophic levels, although many studies describe concentrations of OCs in blubber and metals in livers and kidneys of baleen whales (see reviews by O'Shea and Brownell 1994, O'Shea 1999, and O'Shea and Aguilar 2001). However, justifiable inferences regarding effects based on this literature are very limited. O'Shea and Brownell (1994) concluded that no study provided evidence of harm to baleen whales and that reported concentrations were lower in mysticetes than in most piscivorous marine mammals. Krahn et al. (1997, 1999, 2001) and Tilbury et al. (2002) came to similar conclusions in analyses of contaminants in the tissues and stomach contents of gray whales (*Eschrichtius robustus*) and other marine mammals. Conversely, Colborn and Smolen (1996) used a conservative "weight of evidence" approach to suggest that negative impacts of organochlorine contaminants were playing a role in the lack of population growth in endangered North Atlantic right whales (*Eubalaena glacialis*), although Weisbrod et al. (2000) found no evidence of harmful exposure for this species. In contrast, we consider the less persistent, nonbioaccumulative biotoxins to have a well-demonstrated potential for harm to baleen whales, sirenians, and other marine mammals and to be worthy of investigation (Van Dolah this volume, Durbin et al. 2002, and many others).

Once xenobiotics enter biological matrices, the chemicals can be acted upon by numerous highly selective enzymatic (biotransformation) systems that can alter some of the "environmentally persistent" compounds, even those that biomagnify. These transformation processes may vary among marine mammals to an important degree at the "receptor" level. Some studies have measured chiral PCBs in marine mammals and associated prey. The use of enantiomer fractions (EFs) of chiral PCBs or potential metabolism of PCB congeners by marine mammals from prey items has not been well studied. As an example, Hoekstra et al. (2002b,c) evaluated contaminant enantiomers in blubber and liver from the bowhead whale (*Balaena mysticetus*). The accumulation of several chiral PCBs (PCB-91, 135, 149, 174, 176, and 183) in bowhead blubber was enantiomer specific relative to bowhead liver and zooplankton, suggesting that biotransformation processes in the bowhead whale are enantioselective. Despite evidence for enantioselective biotransformation, all three congeners accumulated in the bowhead relative to PCB-153.

The change in the EF values of chiral PCBs from *Calanus* (zooplankton prey) to the bowhead was less than the change in EF values of chiral OCs observed between other prey and predators (Wiberg et al. 1998, 2000, Fisk et al. 2002). Results for EFs from studies of cetaceans generally demonstrate that interspecies differences exist in accumulation of chiral PCBs. The EF values for PCB-136 and PCB-174 were similar for the fin whale (*Balaenoptera physalus*) and the bowhead whale, which suggests similar enantioselective biotransformation between these two mysticetes. Bowhead whale EF values were significantly different from those reported for odontocetes, such as the long-finned pilot whale (*Globicephala melas*), Risso's dolphin (*Grampus griseus*), bottlenose dolphin (*Tursiops truncatus*), and striped dolphin (*Stenella coeruleoalba*) (Reich et al. 1999, Jimenez et al. 2000). The differences between baleen and small-toothed whales may be attributed to the lengthy period of evolutionary divergence of mysticetes and odontocetes, species-specific stereochemically sensitive processes, the degree of contamination, and the different food chains occupied by piscivorous cetaceans (odontocetes and fin whale) and the bowhead whale, which feeds almost exclusively on invertebrates.

Nonracemic proportions of chiral PCBs have been documented in some fish species (Wong et al. 2001). Therefore the accumulation of nonracemic congeners from prey items has to be taken into account with any interpretation of chiral biotransformation process (or processes) in higher-trophic organisms. This analytical chemistry technique allows investigators to propose and examine mechanisms of PCB "bioprocessing," and should be combined with studies addressing congener profiles based on known metabolic fates in mammals and assays for specific categories of metabolites (e.g., hydroxylated, methylsulfonated).

Sirenians (three species of manatees and the dugong) are completely herbivorous and thus would not be expected to be exposed to significant quantities of persistent contaminants that biomagnify in the food chain. Concentrations of most organochlorines measured in sirenians have been low in comparison with those in piscivorous marine mammals (see review by O'Shea 2003). However, sirenians may have unique exposure patterns because they feed at the sediment level and may have greater exposure to contaminants that are taken up from sediments by vascular plants (certain metals, see O'Shea et al. 1984) or that accumulate in or adsorb to sediments in contaminated areas (e.g., certain dioxins or dibenzofurans, as described by Haynes et al. 1999 and McLachlan et al. 2001). Thus, this group (all species of which are endangered or threatened under U.S. law) includes species that provide examples of the need to assay for nonconventional compounds beyond the routine quantitation of typical OC pesticides and PCBs. Florida manatees are also susceptible to certain biotoxins (Van Dolah et al. 2003, Van Dolah this volume).

Persistence and Biomagnification of Contaminants in Marine Mammals at Higher Trophic Levels. Studies of thousands of individual piscivorous pinnipeds and odontocetes have shown that OCs accumulate in tissues, particularly blubber, of these higher-trophic-level marine mammals (e.g., Tilbury et al. 1997, 1999). Some of the highest concentrations of OCs and certain other substances reported in any species of mammal, terrestrial or marine, have been found in pinnipeds and odontocetes, verifying that their exposure can be significant. However, there have been few experimental, interpretive, and ancillary studies that would allow conclusions about the effects of particular exposures (see review by O'Shea and Tanabe 2003).

Dynamics and variability in transfer of contaminants upward in food chains are illustrated by examples from the Arctic. Kucklick et al. (2001) collected blubber from ringed seals (*Pusa hispida*) and subcutaneous fat from polar bears (*Ursus maritimus*) near Barrow, Alaska, for POP analyses. Apparent biomagnification factors indicated lower POP concentrations in samples taken from bears in the Barrow area than in those taken from certain locations in the Canadian Arctic. This is not surprising because some Alaskan polar bears prey upon or scavenge marine mammals from lower trophic levels (such as bowhead whale and walrus [*Odobenus rosmarus*] carcasses) more than is the case for Canadian bears, which feed predominantly on ice seals.

The PCB congener patterns seen by Kucklick et al. (2001) demonstrate the metabolism and elimination of certain PCB congeners in the polar bear as compared to the ringed seal, similar to previous studies (Letcher 1996, Letcher et al. 1996, 1998). Thus, "persistence" in the food chain is variable, with polar bears appearing to have a higher biotransformation and elimination capacity for the OCs. However, studies in Alaska are confounded in that feeding ecology is not well described, and it is essential that polar bears from coastal areas be evaluated to determine the degree to which they scavenge on marine mammals. Recent studies of polar bears of the Svalbard region of Scandinavia have caused heightened concern about the potential impacts of POPs on polar bears (Arctic Monitoring and Assessment Programme

2002, de Wit et al. 2004), including potential endocrine and immune system disruption.

Concentrations of PCBs and DDTs in blubber of Alaskan killer whales (*Orcinus orca*) are much higher than in blubber of various other cetaceans and pinnipeds that reside and feed in Alaskan waters (Ylitalo et al. 2001), and they are similar to those in the transient killer whales from coastal waters of British Columbia (Ross et al. 2000). Ylitalo et al. (2001) found that concentrations of OCs in transient (marine mammal-eating) killer whales were much higher than those found in resident (fish-eating) animals from the same geographic regions, apparently owing to the differences in diet (again underscoring the importance of understanding feeding ecology to evaluate the impact of contaminants).

Assessment of sequence in birth order is important in identifying individuals with the highest in utero and lactational exposure. Young mammals provide an avenue for mothers to excrete lipophilic compounds, and first-time mothers have larger loads built up over their long, multiyear periods of growth and sexual maturation. Wells et al. (2004) reported some interesting findings for Sarasota, Florida, bottlenose dolphins. Organochlorine concentrations in blubber, milk, and blood were evaluated relative to age, sex, body condition, birth order, and health parameters. As would be predicted, prior to sexual maturation, males and females displayed similar OC concentrations; males accumulated PCBs as they aged, whereas females did not after they had given birth to and nursed their first calf. Wells et al. (2004) reported that high rates of first-born calf mortality were correlated with higher concentrations of OC contaminants in blubber and plasma of primiparous females. Subsequent calves exhibited higher survivorship. First-born calves had higher concentrations than subsequent calves of similar age. Concentrations in young adult females were lower than in older females, reflecting a progressive decrease in PCB transfer rates to offspring as a consequence of the age-related increase in calving intervals. This study does not conclude that OCs result in mortality of first-born calves, but like studies addressed later in the chapter points to a critical highly exposed cohort for recruitment: the first-born calves/pups. Similar bottlenose dolphin studies are underway at Indian River Lagoon, Charleston, South Carolina, and other regions along the East Coast of the United States (P. Fair and P. Becker pers. comm.).

In killer whales, mean concentrations of selected PCBs and DDTs in blubber of first-recruited animals were roughly fivefold higher than those in later-recruited whales. Persistent OC concentrations in sexually mature resident killer whales were more highly correlated to birth order than to age. This indicates the remarkable "persistence" of OCs not only within an individual, but also in a maternal line, and that the first-born individuals represent a critical cohort to evaluate exposure and effects of contaminants. Thus fine-tuning field studies to include addressing the "highest risk" cohort (first-born) for intensive evaluation could be very informative for interpreting effects or impacts of OCs on marine mammals of higher trophic levels. This high-risk cohort was documented for pups of northern fur seal primiparous dams by Beckmen et al. (1999, 2003). However, health risk models should also include demographic calculations of the population consequences if first-born young are more likely to suffer impacts of contaminant exposure. Depending on the dynamics of other population processes, these consequences may be severe or trivial, depending on diverse factors including, but not limited to, the underlying population dynamics and other stressors that the population faces.

ASSESSMENTS FOR POTENTIAL TOXICITY AT MULTIPLE LEVELS AND SCALES. Potential toxicosis must be assessed at multiple levels and scales, beginning with measurements of contaminant concentrations in environmental and organismal matrices and scaling up to population, species, ecosystem, and global levels. Within individual organisms, contaminants can affect developmental processes (in utero, growth and development, maturation, transgenerational) and reproduction; induce pathologies (gross, histologic, molecular); and make organisms more susceptible to infection, predation, and other stressors. Increases in morbidity and mortality are possible. Direct and indirect effects of contaminants on populations and ecosystems could be manifested as changes in recruitment, prey quality and availability, or reproductive performance.

Contaminant measures should include chemical concentrations, congener or isomer descriptions, organic versus inorganic forms, racemic states, metabolites (reduction-oxidation, changes in major functional groups), and other quantifiable information. Careful interpretation of high-quality analytical chemistry data can lead to testable hypotheses concerning bioaccumulation, biomagnification, toxicokinetics, and biotransformation and can involve evaluations at the molecular, organelle, cellular, or tissular levels (Letcher et al. 1998, Wiberg et al., 1998, 2000, Reich et al. 1999, Jimenez et al. 2000, Fisk et al. 2002, Hoekstra et al. 2002c, Pfeiffer et al. 2002).

Molecular (genetic, biochemical, or receptor) responses or "lesions" indicate potential exposure through changes in enzyme activity (e.g., cytochrome P450), changes in levels of mRNA or protein (metallothionein), molecular derangement (DNA adducts, oxidative damage to macromolecules), nutrient or endocrine imbalances (thyroid, as in Rolland 2000), and alterations of receptors or critical proteins (e.g., inhibited acetylcholinesterase). Organelle responses can be very specific for some toxicants that attack specific "receptors" and alter subcellular function; thus, assessing such responses can be highly diagnostic. Cellular changes display an integrative response to a toxicant in which some response to the toxicant is being measured rather than the specific alteration. This balance of response and the level of true "injury" that occurs is weighed among the multiple "defense" mechanisms (e.g., antioxidants) and cellular targets attacked by the toxicant.

Cellular changes can be monitored as cell injury or death (e.g., leakage of cellular constituents as measured in serum), cytology or histopathology (visible anatomic/ultrastructural changes), and in situ or in vitro as functional assays (e.g., of immune function, blastogenesis, phagocytosis; behavioral or physiological changes). Development of modern biomolecular techniques such as microarrays focused on specific marine mammal genes would be very useful to marine mammal toxicology. Tissular responses integrate cellular, organelle, and molecular responses and appear as gross or histopathological change (e.g., fibrosis, changes in mass) and organ-system-based functional loss or change (e.g., lack of clearance of normal metabolic wastes). These changes can be general and may not necessarily involve specific or pathognomonic changes or clinical signs.

Organ function can be significantly altered without observable gross or histological change. This is especially true for the many neurotoxins, where clinical signs (e.g., altered motor control, alertness) may be the only indicators of poor health. The "organismal adverse response" is a conglomeration of the many interacting systems that affect and respond to exposure to contaminants sufficient to cause injury. This response can be subtle in regard to the health and survival of an individual but quite important to the population (e.g., decreased reproductive performance) or can be as devastating to the individual as death. It is essential that knowledge of these biochemical/molecular, cellular, and tissular responses be available for proper diagnostics and research into the effects of contaminants. These levels of understanding allow investigators to link particular chemical exposures to impacts on marine mammals. Generally these data or approaches are not available or are seldom taken in studies of marine mammals. However, obtaining this type of solid evidence will improve policy making and management because it will reduce uncertainty and silence many critics of the "weight of evidence" approach that is more commonly employed.

Stock or population effects interest most managers but are even more difficult to ascertain if the health of individuals in the population has not been properly assessed. Sampling an adequate number of animals in the cohorts and gender of greatest concern is essential. At higher levels, species effects can be variable in mammals, with some species being quite sensitive, whereas closely related species may be tolerant of the adverse effects of specific chemicals. For this reason, extrapolation from one species to another is risky. Even so, no cetacean or pinniped model species has been selected and properly evaluated for effects of exposure to contaminants. The closest approximation of this desideratum involved feeding fish with different concentrations of mixtures of compounds to harbor seals and assessing aspects of reproduction, physiology, and immunology (e.g., Reijnders 1986, review by Vos et al. 2003b).

Beyond the species level, ecosystem effects can occur even if the marine mammal under consideration is not directly affected by the chemical of concern (no toxicosis) if the prey base or water quality is so altered that an area becomes uninhabitable or the carrying capacity is significantly reduced (e.g., see Ragen this volume). Understanding the feeding ecology of marine mammals is therefore important not only for assessment of chemical exposure, but also because marine mammal populations need healthy prey and habitat to survive (Aguirre et al. 2001b).

Global impacts can occur because many contaminants are pervasive throughout the biosphere. The widespread atmospheric and oceanic transport of contaminants, in combination with the fact that many marine mammals are migratory and many of their populations represent shared, or straddling, stocks, makes the issue of contamination international in scope even though some sources and effects may appear to be local or regional. For some contaminants, effective mitigation will only be possible via global management agreements that require multilateral economic and political engagement. Such problems and solutions are, perhaps, most apparent in the high Arctic (de Wit et al. 2004). The high public profile of some marine mammals, such as polar bears and cetaceans, can contribute to global awareness of marine pollution issues, but campaigns in this regard have to be grounded in science and free of unnecessary or unwarranted alarmism, emotion, and extremism.

The previous hierarchical summary of exposure and effect levels emphasizes the importance of establishing the linkages (plausible and investigated mechanisms) between documented exposure to a chemical and an observed effect (potential toxicosis or population change). By not including these linkages, observations of necessity remain simple associations or correlations. Mechanism-based approaches provide the most convincing evidence of effects and the most compelling arguments for policy makers.

Routes of Exposure and Current Understanding of Effects

ROUTES OF EXPOSURE. Our overview assumes that the primary route of exposure of marine mammals to contaminants is food. We do not cover other routes of exposure in detail, although there is evidence that they can be important [e.g., pulmonary exposure to natural dinoflagellate toxins in manatees (Bossart et al. 1998)]. Exposure to oil is the major exception to our generalization. Such exposure can include oiling of baleen (but see Geraci 1990), eyes, fur, and skin, as well as pulmonary exposure to inhaled oil or vapors. Investigations of oiling incidents require careful examination and sampling to assess impacts properly. This capability is lacking for most of the U.S. coastline, especially in remote areas. Yochem et al. (www.hswri.org/whatsnew.htm) are establishing baseline biomedical data for pinnipeds in California that include the measurement of background petroleum exposure and information on normal physiology and pathology of systems susceptible to petroleum exposure (e.g., reproductive and immunological systems). Marine mammals are vulnerable to petroleum exposure including natural seeps and oil processing, refining, and transportation activities (e.g., Kucklick et al. 1997). Similar efforts to those of Yochem et al. (previous website) in other regions of high oil release or spill potential (e.g., Alaska, Gulf of Mexico) should be a priority, although preliminary results suggest that bowheads and bearded seals (*Erignathus barbatus*) harvested by subsistence users on Alaska's North Slope have undetectable levels of PAHs in their tissues (D. Wetzel and J. Reynolds pers. comm.). The lack of detectable PAHs makes it even more urgent that biochemical and histologic indicators (or biomarkers) be established for exposure to PAHs and its effects for marine mammals as we cannot rely on simple comparisons of tissue concentrations. Again, not all contaminants and chemicals of concern accumulate in tissues, so other means of assessing exposure and impacts are needed.

DIFFICULTIES IN CONDUCTING PROPER EFFECTS ASSESSMENT. Management-based questions related to the possible effects of contaminants have typically focused efforts and resources on endangered or threatened species or stocks or populations or individuals considered "highly" contaminated. This makes research more difficult, politically and logistically. Two major themes have predominated in investigations concerning contaminants in marine mammals: (1) speculation that unexplained lesions or effects (population decline) with no obvious cause are due to contaminants, and (2) findings of "high" levels of contaminants in individuals or populations followed by speculations or "seeking" a lesion or an effect. Because of a multitude of confounding variables, these approaches are unlikely to reveal causes and effects in a convincing way.

Inability to Properly Assess Effects. As noted previously, the field of marine mammal toxicology is desperately in need of well-designed, control-based studies. Assessment of effects remains the most elusive component of contaminants research on marine mammals, the major reasons being:

1. A general lack of understanding of marine mammal health and physiology.
2. Historic overreliance on extrapolation from unrelated species and unrelated tissues for toxicological assessments using single time-point concentrations of contaminant residues in tissues, coupled with a continued emphasis on what have become trivial questions and uninteresting hypotheses regarding contaminant concentrations found in tissues (e.g., reconfirming patterns well known to vary by sex and age).
3. Hesitation to perform studies using capture-release and captive animals to relate contaminant exposure to effects or a lack of effects (generally citing expense and ethics as reasons).
4. Experimental designs of studies of free-ranging animals without the appropriate statistical power, often failing to use proper epidemiological approaches to determine the presence or absence of effects with reasonable confidence.
5. Dependence on correlative analyses and associations versus more appropriate mechanism-based interpretive studies (true physiologic responses and adverse health impacts) to ascertain cause-effect relationships.

Improvements Needed. Unless the restrictions just listed change, marine mammal toxicologists will continue to struggle with effects assessment. Improvements are needed in several areas of research and diagnostic evaluations, as follows:

1. Current understanding of the toxicological significance, if any, of some biological responses (such as cytochrome P450, metallothionein, and histology) to OCs and other contaminants is vague. There is a need to move beyond correlative studies (e.g., some PCBs were higher in the blood of animals with a lower immune function test) to more interpretive studies that assess mechanisms of action (linkages of exposure and observed effects), as has been done for some biotoxins associated with specific lesions and clinical signs (Van Dolah et al. 2003).
2. Many biochemical and cellular assays currently being used in other biological systems have to be applied to and validated for marine mammals to allow mechanism-based evaluations of impact (e.g., molecular epidemiology).
3. The limited inferences from correlations between possible effects (e.g., population declines, immune function, reproductive failure) and contaminant concentrations in tissues have to be advanced by including more rigorous interpretive studies.
4. More rigorous interpretive studies should include region-based, intensified, consistent, and coordinated diagnostic approaches that are integrated with research on (and collaboratively make use of) "unhealthy" (i.e., some stranded) and "presumed healthy" (captive, bycaught or hunted, captured-and-released) individuals.
5. Research has to be multi- and transdisciplinary, long-term, and problem based rather than basic-science or purely single-investigator driven.
6. Future investigations must be designed to give greater consideration to ruling out alternative agents and explanations as causes of declines in health or population abundance (avoid trying to prove or re-prove a single cause).
7. Diagnostic and research biomedical paradigms should be merged into single programs or initiatives of regional relevance.
8. The scientific basis for extrapolation from terrestrial to marine mammals and for extrapolation across marine mammal species has to be improved. The no-observed-effect level (NOEL), lowest-observed-adverse-effect level (LOAEL), and effect concentration (EC_{50}) from unrelated species and tissues are sometimes used, with weak or no justification, to extrapolate "toxicity" in marine mammals (examples can be found in Arctic Monitoring and Assessment Programme 2002, de Wit et al. 2004, and numerous other publications).

9. Marine mammal toxicology has to stay in the mainstream of comparative toxicology and comparative medicine to ensure automatic access to the latest technology and recruitment of leading talent into the field.

Factors Hampering Improved Understanding of the Current Status of Marine Mammals in Relation to Exposure to Contaminants

UNRELIABLE MONITORING. Monitoring is required to determine whether management changes (such as restrictions on chemical use, changes in oil consumption or shipping, changes in disposal processes) result in decreased, unchanged, or increased burdens in tissues and, by inference, corresponding health effects in marine mammals over time and space. Effective monitoring is difficult and requires samples collected according to strict quality-control procedures from living or recently expired marine mammals. Such samples are often difficult to collect owing to geographical remoteness, and funding for on-site groups that can obtain the samples may be limited. Sampling and shipment of samples can require lengthy approval periods from multiple authorities and the cooperation of several individuals, agencies, or organizations. To be meaningful, monitoring also requires specimens with comprehensive ancillary data on biological variables such as age, sex, stock, season, and reproductive status. Samples collected from animals with little additional biological and health data cannot be interpreted adequately, and efforts to obtain such samples should be of lower priority (if conducted at all). Sampling over large temporal (decadal) and spatial (intercontinental) scales may be required to detect and understand changes or differences in contamination. Archiving samples and data requires funding and coordination, and laboratory analyses must be carried out using interlaboratory quality-assurance (QA) and quality-control (QC) protocols.

In the face of such challenges, adequate funding must be targeted to establish data and sample collection programs and archives for marine mammals that are not sporadic and intermittent. Teamwork and vision should be fiscally rewarded. Currently, the system is dependent on groups outside the federal government (local and state governments, nonprofit organizations, universities) for much of the examination of marine mammals and sample collection related to contaminants research and monitoring. Therefore, efforts that do not appear cooperative and that provide limited data on the individuals sampled should be of lowest priority for funding. The 1998 Marine Mammal Commission workshop report (O'Shea et al. 1999) suggested particular species for priority monitoring of trends in contamination and related studies. These include harbor seals (*Phoca vitulina*), gray seals (*Halichoerus grypus*), California sea lions (*Zalophus californianus*), harbor porpoises (*Phocoena phocoena*), bottlenose dolphins, beluga whales (*Delphinapterus leucas*), bowhead whales, minke whales (*Balaenoptera acutorostrata*), sea otters (*Enhydra lutris*), walruses, and polar bears. Tabulations of references with available trend information can be found in other sources (e.g., O'Shea 1999, Vos et al. 2003a, de Wit et al. 2004)

KNOWLEDGE OF FEEDING ECOLOGY. Addressing the ecological connections of marine mammals with prey that is or could be directly affected by contaminants requires a different perspective than dealing with the direct impacts of contaminant toxicosis on marine mammals. Understanding feeding ecology is relevant to both direct and indirect effects of contaminant exposure. More is known about the direct effects of contaminants on marine microbes, invertebrates, and fish because they are accessible, both politically and logistically, for study in the laboratory. Critical water and tissue concentration levels for adverse effects on fish and invertebrates have been published (e.g., Giam and Ray 1987 and Niimi 1996 for PCBs; Hellou 1996 for PAHs; Rainbow 1996 for heavy metals; Lemly 1996 for selenium, and Walker and Livingstone 1992 for persistent pollutants). Virtually nothing is known, however, about the response of marine mammal populations to changes in prey abundance caused by contaminants. Regulatory agencies have established permissible concentration thresholds for many contaminants in water, sediments, fish, and invertebrates, but such thresholds have not been established for marine mammals.

THREATS TO HUMAN CONSUMERS OF MARINE MAMMALS

Consumption of Marine Mammals by Subsistence Users

The consumption of marine mammal tissues continues today in numerous countries. Many groups claim a special relationship with marine mammals based on nutrition, culture, and spiritual importance (Reeves 2002b, Hansen et al. 2003, de Wit et al. 2004). Marine mammals are important resources in some northern coastal communities and island nations (see websites for IWC, International Convention for the Regulation of Whaling, and World Council of Whalers; Reeves 2002b). Indigenous people in Alaska use marine mammal products for food, clothing, and culturally important activities. There is a recognized benefit to the individual hunter and the entire community in preparing for the hunt and in the pursuit, acquisition ("catching"), and processing of these animals (Huntington 1992, Braund and Moorehead 1995, Reeves 2002b, Alaska Eskimo Whaling Commission).

Subsistence users can be advocates of habitat protection for marine mammals in the face of powerful economic and political forces promoting industrial exploitation of the coastal and offshore environment (Freese 2000, Balch 2002). Despite the importance of marine mammals to the local diet, the "dread" of exposure to contaminants has made many Alaska Native communities anxious about continuing to consume whales, seals, walruses, and polar bears. This clearly impacts the management paradigm and contaminants research field for marine mammals in this region and in some cases tissues have been assessed for contaminants at the behest of local management groups (*Alaska Bulletin* 1986, O'Hara et al. 1999, Woshner 2000, Brower et al. 2002, Willetto et al. 2002a,b).

Deficiencies in Knowledge and Resulting Controversies

Most available studies address contaminant concentrations only in certain marine mammal tissues, some of which are not typically part of the human diet or prepared for human consumption. Researchers may acquire and analyze samples simply because they are available from hunter-killed animals, with little regard as to whether, or how, the tissue is relevant to human consumption. Thus, some projects that claim to address species of concern for contamination do not involve analyses of tissues that are important to consumers, which is a weakness of many long-term archival projects and agency-based studies.

At present, marine mammal food chains in most of North America are contaminated but do not appear to be affected by radionuclides (Efurd et al. 1996, Cooper et al. 2000). Mercury concentrations in marine mammals of northern Alaska vary a great deal (Woshner 2000, Woshner et al. 2001a,b, 2002), within and among species, and it is likely that epidermis and muscle represent the largest source of mercury for consumers of cetacean tissues; liver and muscle may be the major source of mercury for consumers of pinniped tissues, but this requires confirmatory investigation.

Recently, Verbrugge and Middaugh (2004) reviewed the extent to which PCBs and related contaminants in wildlife posed a threat to Alaska Natives who consumed such wildlife as part of a subsistence diet. Their conclusion was revelatory: "The potential risks associated with POP exposure through subsistence food consumption are smaller than the risks associated with a decreased use of traditional foods, or the risks associated with many other aspects of Alaskan life." The issue of contaminants, thus, is complex, sometimes overstated, and requires a balance in terms of understanding nutrition and other factors (see later in the chapter).

Even in cases where a commonly consumed tissue has been analyzed for contaminants, there has rarely been an accompanying assessment of its nutritional value (e.g., calories, protein, omega-3 fatty acids, essential elements, vitamins). Such assessments are critical for balancing the risks and benefits of consuming marine mammal products (Hansen et al. 2003, Reynolds et al. in press). It is misleading to present a statement of risk (concentrations of contaminants in tissues) without also addressing nutritional value (*Alaska Bulletin* 2003).

Controversies Regarding Contaminants in Marine Mammals as Foods

Recent reports (Hansen et al. 2003, de Wit et al. 2004) indicate that marine mammals are the most likely sources of POPs and mercury for northern residents. The State of Alaska publication, *Alaska Bulletin* (2003) highlights the anxiety generated by recommendations regarding "safe" and "unsafe" levels of contamination, as local sources in Alaska have become the cause of great concern on the part of residents. Nonetheless, agencies and nongovernmental organizations (NGOs) have sometimes made inappropriate and improperly documented assessments that have increased anxiety levels in subsistence users. There is a need to determine which authorities are responsible for advising on human health issues related to consuming marine mammals and to ensure that contaminant data are interpreted and explained in a scientifically rigorous but understandable manner to human consumers.

This has resulted in the so-called "Arctic dilemma" in that the main source of contaminants for Native people is traditional foods, such as marine mammals, but that these foods also provide beneficial nutrients as well as cultural and spiritual identity and are known to promote health (Egeland et al. 1999, Hansen et al. 2003). The loss of nutritional value as people reduce their consumption of marine mammals is a serious health problem (Bjerregaard et al. 2003).

We re-emphasize a note of caution about advice concerning contaminants in traditional foods (Arctic Monitoring and Assessment Programme 2002): "Most health problems are caused by a combination of factors. Compared with the role of lifestyle and inheritance, contaminants alone may play a *modest* role." This statement needs to be communicated clearly and in a manner that is not offensive, that is, that well-known causes of mortality and morbidity, unrelated to contaminants (e.g., alcohol, tobacco, lifestyle, western foods), are likely affecting human health on a much greater scale (e.g., see Verbrugge and Middaugh 2004). This message should also include the caution that cultural impacts of not hunting and using marine mammals will likely be severe. For example, there has been a 71–142% increase in the prevalence of diabetes throughout Alaska. Most of this is a result of the loss of a traditional lifestyle (Alaska Native Tribal Health Consortium 2002). In comparison, it is possible that the effects of contaminants are nonexistent or very subtle. An Arctic Monitoring and Assessment Programme (2002) report states: "Moreover, changes in smoking, drinking, and dietary habits can make it difficult to single out contaminants as the cause of an observed change in health." These confounding variables provide a daunting challenge for human health experts investigating the effects of contaminants in Arctic human populations.

The issue of marine mammals as a source of contaminants in human consumers is not unique to Alaska (Kinloch et al. 1992): similar concerns have been raised in Canada (Dewailly et al. 1993, Wormworth 1995, Jenssen et al. 1996), Greenland (Hansen 1990), the Faroe Islands (Simmonds et al. 1994), and other locations (Simmonds and Johnston 1994). In whaling countries, there is disagreement about the safety of whale meat in the human diet. Information on the risks of consuming marine mammals as food has to be presented in a way that distinguishes between what is claimed and what is substantiated by scientific evidence. Insofar as there is use of and demand for marine mammals as food in other nations and the United States is a signatory to international treaties and agreements (e.g., the International Convention for the Regulation of Whaling), U.S. scientists need to be involved in collaborative studies with researchers from other countries where humans consume marine mammal products or at least be conversant with the issue of dietary exposure to contaminants in areas other than Alaska.

INFRASTRUCTURE ISSUES FOR RESEARCH AND MANAGEMENT OF MARINE MAMMALS IN RELATION TO CONTAMINANTS

Research and management responsibilities related to marine mammals and environmental contaminants are fragmented. Effectiveness could be improved by: (1) placing research and management authority for marine mammals under one agency; (2) mandating that this single agency conduct and coordinate research, management, and diagnostics for marine mammals; and (3) developing regional teams that are not species based but are required to work with designated "core laboratories" (and thus must indirectly work with other regions) and experts (archivists, statisticians, chemists, pathologists, biological oceanographers, toxicologists). Many of these core laboratories exist within fields of human, domestic animal, and some wildlife research and diagnostics (e.g., the Centers for Disease Control and Prevention, the Department of Agriculture's National Veterinary Services Laboratory, the U.S. Geological Survey's National Wildlife Health Center, the Southeastern Cooperative Wildlife Disease Study, the Canadian Cooperative Wildlife Health Center, many state diagnostic laboratories). The science of marine mammal toxicology is limited by the lack of a diagnostic infrastructure and basic health and biological baseline data. This situation could be rectified by the formation and support of a centralized group to consult for diagnostic or research support, as exists for humans, domestic livestock or pets, and some wildlife. At a minimum, the existing laboratories for domestic animals and wildlife should be encouraged to be a part of the marine mammal health research and diagnostic laboratory infrastructure.

THE RECOMMENDATIONS

We divide our recommendations into ten points that primarily describe needed research and ten that describe needed actions in policy and infrastructure in support of research and scientific investigations. These recommendations are not assigned priority. We feel that all the recommendations are important and that decisions regarding priority of action should rest with the responsible agencies. Our recommendations overlap many of those from the 1998 Marine Mammal Commission workshop (O'Shea et al. 1999, reprinted in appendix). Specific overlap is noted where applicable. This overlap underscores the need for a formal evaluation of progress in meeting the recommendations of the 1998 workshop (our Recommendation [B.1], later in the chapter).

A. Recommendations for Research and Scientific Investigations

1. *Design and implement a series of experiments on the effects of sublethal exposure to selected organochlorine contaminants on captive California sea lions.* Based on knowledge from prior research, impacts of exposure of marine mammals to contaminants are likely to manifest as: (a) reproductive, (b) immunological, and (c) other pathological (including behavioral and neurologic) effects. A clear and major step forward in understanding such effects can be accomplished through well-designed experiments on captive marine mammals. The 1998 Marine Mammal Commission workshop identified these three areas as among the most critical scientific uncertainties, and working groups outlined approaches to their investigation, including the suggestion to implement experiments on captive otariids. The California sea lion is the best species on which to conduct these experiments because it is abundant, many aspects of its biology are well understood, methods for captive maintenance and husbandry are well established, and the scientific literature suggests that the species has suffered ill effects from exposure to contaminants (as well as biotoxins).

Organochlorines (beginning with exposure experiments using single compounds) should be the first contaminants used for these studies because of past findings regarding these substances in California sea lions and other wildlife, their ubiquitous occurrence in marine mammals around the globe, and the intense current focus on organochlorines in chemical-residue surveys of marine mammals and in the development of policy regarding marine pollution. A multidisciplinary working group should be formed to design the research protocol in coordination with the group described in research recommendation (2) (following). This series of experiments would directly respond to many of the needs identified by working groups of the 1998 workshop on reproduction, immunology and disease, and risk assessment. It would also directly respond to the following previously identified principal conclusions of that workshop: (1), (7), (10), (14), (17), and (18) (see appendix), as well as shed light on other identified needs.

2. *Develop and conduct a formal risk assessment for organochlorines in California sea lions in conjunction with the experimental study recommended previously in (1).* This assessment should start before the above experiments are conducted and begin by convening a group of experts to review the risk assessment for California sea lions initiated at the 1998 workshop (see Working Group Report on Risk Assessment in O'Shea et al. 1999). The expert group should

use previous work as a starting point to develop a thorough framework and model for risk assessment. Such a framework would define the gaps in knowledge that these experimental exposure studies must be designed to fill. A final risk assessment would be completed once experimental findings are available and other pertinent data have been gathered. This is a specific application of the previous recommendation to make formal risk assessments in the study of marine mammals and contaminants (recommendation [7] in appendix).

3. *Review, endorse as appropriate, and provide support for the implementation of other previously identified specific studies of contaminant effects (including experimental approaches) on other marine mammals, particularly small cetaceans and sea otters.* Detailed justification and suggested approaches for these studies can be found in the reports of other workshops (Reijnders et al. 1999, Reeves 2002a) as well as the 1998 Marine Mammal Commission workshop (O'Shea et al. 1999) and the IWC's Pollution 2000+ program (Reijnders et al. 1999). Although logistics may prove difficult for small cetaceans, we urge that logistics and animal rights concerns not be allowed to overshadow the policy and scientific needs for reliable information from experimental field and captive studies. This recommendation embodies aspects of recommendations (3), (10), (14), and others from the 1998 Marine Mammal Commission workshop.

4. *We suggest that the Marine Mammal Commission and other involved agencies endorse intensified contaminant studies of polar bear feeding ecology and health assessment using both captured (research) and hunter-killed (subsistence) bears, but not studies intended simply to report contaminant burdens in tissues.* This need is underscored by recent studies of polar bears in the Svalbard region that have heightened concern about the potential impacts of contaminants on polar bears (reviewed in Arctic Monitoring and Assessment Programme 2002 and de Wit et al. 2004), including potential endocrine disruption that could jeopardize reproduction. This recommendation is consistent with recommendations (1), (12), (13), (14), and (18) of the 1998 Marine Mammal Commission workshop.

5. *Change the emphasis of surveys of chemical contaminant concentrations in marine mammals.* Several suggested avenues for such new investigations are noted below. These also follow from recommendations (1), (5), (6), (13), and (15) of the 1998 Marine Mammal Commission workshop.

5a. *Include new, emerging, or little studied chemicals in analyses of contaminants in tissues of marine mammals.* Analytical chemistry has great potential for revealing poorly known compounds that may be of hidden significance as contaminants of marine mammals, but this potential is typically underused. Instead, emphasis is placed by default on the most commonly studied contaminants, which usually include the organochlorines (PCBs, DDT, and other members of the "dirty dozen") and heavy metals (e.g., Hg and Cd). Analytical chemistry resources that are not devoted to well-designed monitoring or research but instead to "surveys" should refocus from confirming the obvious and now trivial (e.g., DDE is usually higher in blubber than other tissues and is highest in adult males) to more cutting-edge exploratory analyses.

5b. *Base the list of "new" chemicals (e.g., brominated flame retardants) to be surveyed on well-reasoned probabilities of exposure and possible adverse health effects.* These probabilities depend on geographic location, season, prey selection, chemical attributes, trends in use, and disposal practices. The focus on traditionally assayed contaminants delays or prevents exploration of "emerging" contaminants. Agencies should be encouraged to promote screening. Documenting the absence of specific chemicals is important, as long as it is done with a high level of certainty. This would require a broader inclusion of chemists and toxicologists as this expertise tends to be chemical-agent or class based.

5c. *Agencies must define chemicals of concern not only on the basis of their potential for persistence or accumulation, but also on their potential impacts on populations or potential for toxicosis.* A potentially harmful contaminant may have a short half-life (little accumulation), but if input is constant, it is essentially persistent in the environment; some toxicants may be fast acting (cause acute toxicosis) and of immediate concern, but not cumulative.

5d. *Agency research programs should assess the ecological and biological "persistence" and "bioactivity" of multiple congeners of complex groups of contaminants.* Consideration must be given to structure-activity relationships and biotransformation products of some OCs in determining exposure of marine mammals to contaminants. Simple measurement and reporting of a few congeners to estimate the concentration of a class of compounds (e.g., PCBs) is no longer sophisticated or progressive.

5e. *Use biomagnification factors, bioaccumulation factors, congener profiles, enantiomer measures, and metabolite determination across age and sex cohorts of marine mammals to improve understanding of "bioprocessing" of contaminants.* Understanding chemical fate and accumulation or magnification (i.e., BAFs and BMFs) in an organism greatly assists in determining toxicosis and possible effects.

5f. *Assign priority to analytical chemistry studies that use samples accompanied by detailed biological information.* Such information should be related to reproduction, age, feeding ecology, and health over time. Good chemistry (congener profiles, enantiomers, metabolites) should be combined with good biology. An example includes the long-standing research with bottlenose dolphins and contaminants (Wells et al. 2004) and more recent efforts by investigators at Indian River Lagoon, Florida, and the area around Charleston, South Carolina.

5g. *Maintain quality programs for monitoring temporal and spatial trends in chemical contaminants found to be of concern in relation to marine mammal exposure.* Consider multiple environmental and biological factors in designing spatial and temporal monitoring of contaminant concentrations. Develop and adopt standard QA/QC procedures for collection, chemical analysis, interpretation, and archiving of samples and data on contaminants in marine mammals. Improve the infrastructure for existing marine mammal tissue archival programs with statistically based sample designs to answer some of the questions outlined previously.

6. *Research on biomarkers in marine mammals should move beyond correlative studies to more interpretive studies that assess mechanisms of action (linkages of exposure and observed effects).* Such a change in emphasis (e.g., a difficult-to-interpret finding that some PCBs were higher in the blood of animals with a lower immune function test versus a more definitive finding as for the effects of DDT and associated metabolites on reproduction for many avian species) is more likely to allow quantifiable assessments of health impacts on individuals and populations. Documentation of mechanistically based linkages increases the level of certainty that contaminants are or are not involved in observations of unhealthy or declining marine mammal populations. It increases the confidence in policy and management decisions beyond "weight of evidence" approaches. This recommendation is consistent with recommendations (11), (12), and (17) of the 1998 Marine Mammal Commission workshop and could be incorporated in part with our recommendations (1) and (3) above.

7. *Conduct research to improve understanding of marine mammal feeding ecology in the context of exposure to contaminants and to assess potential impacts of contaminants on prey of marine mammals.* Although direct toxicosis is at the forefront of attention, it is equally important to understand the magnitude of the response of marine mammal populations to contaminant-caused changes in both abundance and nutritive quality of their prey. Regulatory agencies have established permissible concentrations of many contaminants in water, sediments, fish, and invertebrates; this has not been done for marine mammals. Feeding-ecology studies are important and of direct use to those studying marine mammal population dynamics and exposure to contaminants.

8. *Conduct a formal evaluation, including new research as needed, of the toxicological components, benefits, and risks of humans consuming marine mammals.* This evaluation and research must provide a clearer understanding of the actual consumption patterns of people and actual exposure to these contaminants. It should provide basic information on the nutritional composition of marine mammal tissues. This will enable both an assessment of the health of the individual marine mammal and the quality of marine mammals as sources of nutrients to consumers.

9. *Establish their feasibility and, if possible, implement retrospective and future epidemiological analyses of the health histories and contaminant status of marine mammals in captivity.* Management of marine mammals in captivity includes maintenance of detailed records of reproduction history, health, survival, and other factors thought to vary with contaminant exposure in wild populations. The efficacy of using these records in conjunction with biopsies and analyses of contaminants in rations to conduct epidemiological analyses and to establish a database for comparisons with field studies should be evaluated. If feasible, such studies should be carried out in cooperation with the public display industry. This repeats recommendation (8) of the 1998 Marine Mammal Commission workshop.

10. *Review the literature related to interspecies and intertissular extrapolation for assessing the effects of environmental contaminants, with a focus on quantitatively expressing the uncertainty in this methodology.* In the absence of effects data for marine mammals, efforts to interpret contaminants data must depend on reference to unrelated species and tissues (e.g., Arctic Monitoring and Assessment Programme 2002). The limitations and uncertainty of this approach have not been clearly communicated and appreciated. The disciplines of comparative medicine or comparative toxicology could prove quite valuable in this regard, and experts in those fields should be commissioned to develop a thorough review of the topic as applied to studies of contaminants in marine mammals. This recommendation calls into question the desirability of following recommendations (9) and (12, in part) of the 1998 Marine Mammal Commission workshop without a more formal evaluation. It is consistent, however, with portions of other recommendations and conclusions of that workshop.

B. Policy and Infrastructure Recommendations to Facilitate Research and Scientific Investigations

1. *A working group should be designated to advise the Marine Mammal Commission, the Department of Commerce, the Department of the Interior, and other relevant organizations on means to implement research recommendations from the Commission's 1998 Workshop on Marine Mammals and Persistent Ocean Contaminants.* Those recommendations remain critical to the advancement of scientific understanding of this issue, but progress in implementing them has been piecemeal at best. Progress has not been formally evaluated. This advisory group should make evaluations, assign priorities, critically evaluate steps taken to achieve the recommendations, and assess their feasibility.

2. *A centralized diagnostic and research support facility for studies of contaminants and marine mammals should be developed. The diagnostic and research biomedical paradigms in marine mammal toxicology must be merged to improve the interpretations of critical health and toxicosis assays to approximate those typically used in humans or other animals.* The science of marine mammal toxicology is limited, in part, by the lack of a diagnostic infrastructure and the lack of basic health and biological data for marine mammals. This could be rectified by formation of a centralized group and facility to consult for diagnostic and research support. Such facilities have been successfully implemented for humans, domestic livestock or pets (numerous state and federal laboratories), and

some terrestrial wildlife (e.g., National Wildlife Health Center [www.nwhc.usgs.gov/]; Southeastern Cooperative Wildlife Disease Study [www.scwds.org/]; Canadian Cooperative Wildlife Health Center [http://wildlife.usask.ca/english/frameWildlifeTop.htm]). Potential for developing this support infrastructure within existing laboratories for domestic animal or wildlife health should be evaluated. Currently there is far too much reliance on using disjointed in-kind or poorly funded efforts of nonagency groups to carry the burden of investigating, researching, and diagnosing causes of adverse health effects on marine mammals. These groups should be tied together and coordinated in a well-funded consortium as part of the developing infrastructure within a strong collegial environment. This recommendation is consistent with (but more specific than) recommendations (1) and (2) of the 1998 Marine Mammal Commission workshop and parallels a recommendation of Gulland and Hall (this volume).

3. *Federal marine mammal research and management programs should be reorganized and consolidated.* Currently, most of these programs are mandated to two different cabinet-level departments (Commerce and the Interior) and multiple agencies and subdivisions within those departments. Regulatory and policy decisions regarding contaminants in the environment rest with several other cabinet-level departments. This fragmentation hampers communication, duplicates bureaucracies, and retards proactive research and policy on marine mammals, including questions regarding the effects of environmental contaminants. We recommend placing marine mammal research and management under one agency that will coordinate relevant research, diagnostics, management, and policy (including investigations of contaminants) and provide for a single liaison with other agencies and groups. This recommendation is consistent with those of the U.S. Commission on Ocean Policy (2004).

4. *For the conservation of marine mammals and the welfare of consumers, areas of emphasis should include legislation, agreements, and funding initiatives aimed at assembling contaminants data and eliminating continued use of chemicals known as the "dirty dozen" and those having similar characteristics.* Details of such efforts can be found at websites (www.iisd.ca/chemical/pops5/) and in specific documents (United Nations Environment Programme 1999a,b). The interests and role of subsistence users (see www.ienearth.org/pops_inc5_un.html) should be acknowledged, because Native organizations have been closely involved in the treaty negotiations. The results of the fifth session of the Intergovernmental Negotiating Committee (INC-5) for an International Legally Binding Instrument for Implementing International Action on Certain Persistent Organic Pollutants (POPs) that met in 2000 in Johannesburg, South Africa, is awaiting ratification by the United States. We encourage appropriate agencies and groups to recognize the vulnerability of marine mammals to contamination and to address the U.S. Congress on this issue. This recommendation follows numbers (4) and (5) of the 1998 Marine Mammal Commission workshop.

5. *Agencies responsible for marine mammal research and management should join the chemical manufacturing industry in developing procedures to demonstrate that marine mammals and other wildlife will be unharmed before there is full-scale manufacture and release of newly developed chemicals.* The burden of proof should be on those who potentially will profit financially to demonstrate that their activities will not cause undue harm, rather than on the government (and taxpayers) to demonstrate that those activities are harmful. This will require extensive coordination and changes in approval processes for registration of pesticides, chemical use plans, municipal waste discharging, and other activities that lead to sanctioned release of potentially dangerous chemicals into the environment. Collaborations with and perspectives from industry are needed in finding solutions to current and future issues regarding marine mammals and contaminants. This recommendation is consistent with several of the principal conclusions and recommendations of the 1998 Marine Mammal Commission workshop.

6. *Agencies in the United States responsible for marine mammal research and management should explore joint funding initiatives with a wider group of potential cooperating agencies and foundations to develop proposals for interdisciplinary studies of marine mammal health, contaminants, and ecology.* Agency managers and administrators should explore such initiatives with government science-funding agencies such as the National Institutes of Health and the National Science Foundation and with major private foundations. This recommendation is consistent with recommendation (2) of the 1998 Marine Mammal Commission workshop.

7. *Agencies in the United States should support research that follows the basic tenets proposed by the Pollution 2000+ program of the International Whaling Commission (IWC).* The IWC has attempted to address the transboundary issue of contaminants via their Pollution 2000+ program (Reijnders et al. 1999), but has made little progress because it lacks funding and a focused organizational basis. (The project relies primarily on in-kind participation.) Some elements of this proposed program are in common with recommendations of the 1998 Marine Mammal Commission workshop. Agencies should endorse and promote this effort so that it receives needed funding and institutional commitment to accomplish its well-designed intentions and plans.

8. *U.S. interests should ensure that the United Nations includes marine mammals within the scope of efforts related to contaminants policy.* The UNEP Chemicals Programme (http://irptc.unep.ch/gmn/default.htm) is launching the "Global Network for Monitoring of Chemicals in the Environment," a project that aims to create an electronic forum and working group to coordinate methodologies and analyses of chemicals in the environment. Exposure of marine

mammals to contaminants should be included in these impact and trend assessments.

9. *Agencies should be encouraged to develop a more multidisciplinary workforce (with more disciplines and perspectives represented) in investigations of marine mammals and contaminants.* This can be accomplished in part by assigning liaisons with the Department of State, the U.S. Environmental Protection Agency, and other organizations that can "tap into" needed data, information, and expertise. Diversifying expertise will discourage agencies from becoming autocratic and dealing with contaminants issues from the perspective of mostly internal expertise. This is consistent with recommendation (1) of the 1998 Marine Mammal Commission workshop.

10. *Marine mammal research and management agencies in Alaska collecting tissues from species taken by subsistence hunters should be mandated to provide nutrient data and to work closely with local public health agencies in developing risk assessments.* Marine mammals are sources of nutrients that may play a role in resulting toxicosis or protection from contaminants. Investigators and responsible marine mammal research and management agencies (including the Marine Mammal Commission) should consult with Alaska user groups (see Table A.1) and local public health organizations in Alaska to develop a realistic relative risk assessment of human consumption of contaminants in tissues of marine mammals. A consortium of environmental and public health agencies (e.g., Food and Drug Administration, National Institutes of Health, and the Environmental Protection Agency at the federal level) should be involved in this assessment. Such an assessment should integrate consideration of the social importance and benefits of marine mammal hunting to coastal communities in Alaska.

APPENDIX: PRINCIPAL CONCLUSIONS AND RECOMMENDATIONS ACROSS DISCIPLINES FROM THE MARINE MAMMAL COMMISSION 1998 WORKSHOP ON MARINE MAMMALS AND ENVIRONMENTAL CONTAMINANTS, WITH ANNOTATED COMMENTS

At a meeting sponsored by the U.S. Marine Mammal Commission and cooperating agencies (National Marine Fisheries Service, Environmental Protection Agency, U.S. Geological Survey, and National Fish and Wildlife Foundation), a group of 54 scientists representing multiple disciplines and drawn from seven nations gathered in Keystone, Colorado, in 1998 to review the field of study of marine mammals and contaminants and to suggest directions and approaches for future research. To achieve progress toward this goal, the attendees identified twenty principal conclusions and recommendations for future directions in research on marine mammals and contaminants (O'Shea et al. 1999). We summarize these in Table A.1. We also include additional commentary of our own following each comment, which does not necessarily reflect the views of the Keystone participants. A broad range of opinions can be expected on the relative importance of some of these topics, and they were therefore not listed in any particular order of priority. The full working group reports (O'Shea et al. 1999) provide detailed explanations and greater specificity.

Comments on the Workshop Recommendations

1. *Integration of multiple approaches.* Integration requires overcoming the resistance imposed by individual scientist, laboratory, or agency-based predilections to investigate narrowly focused topics in isolation. We recommend that efforts be made to overcome this by (a) developing specific grant initiatives that encourage or demand integration and multidisciplinary approaches; (b) establishing formal linkages among multiple agencies and organizations that have partial purviews over the numerous areas of scientific research and policy that are involved in the study of contaminants and marine mammals, including human health concerns; (c) creation of central, interdisciplinary facilities devoted to diagnostics and research on the health of marine mammals and the ecosystems of which they are a part. These goals may be accomplished through federal legislation (perhaps through amendments to Title IV of the Marine Mammal Protection Act), government reorganization, and formal interagency agreements. Effective integration may also require shifts in philosophies and reward systems of individual scientists and the institutions that employ them.

2. *Stable support for critical long-term programs with increased emphasis on wider collaborations.* We concur with the assessment of this topic as detailed in the workshop report, and recommend that similar measures as we noted for our first comment be considered to assist in achieving the needed long-term support and wider collaboration.

3. *Long-term interdisciplinary studies of local populations.* We concur with the assessment of this topic as detailed in the workshop report. Wells et al. (2004) is an example of a United States–based long-term effort for bottlenose dolphins and other local efforts should be encouraged.

4. *Compilation and dissemination of information.* This action is needed for anthropogenic chemicals in general and in particular for marine ecosystems and marine mammals. However, there is no marine mammal toxicology center, institute, brain trust, or similar organized core of expertise anywhere in the United States that keeps a reliable reference database on chemical inputs, responds to events or newly raised issues of concern, or adequately evaluates new chemicals that come on-line. Enhanced collaboration and establishment of agreements between industry groups and public organizations, and perhaps enabling legislation, reorganization and centralization, will be required to achieve this recommendation.

5. *Monitoring environmental loads and ongoing inputs of persistent contaminants.* We agree with this assessment, but note that monitoring should include well-designed monitoring of marine mammals and their prey that follows other recommendations of the 1998 workshop regarding protocols, sampling designs, and integration with other investigations on marine mammals and contaminants.

6. *Universal protocols.* This effort is underway by the Marine Mammal Health and Stranding Response Program (MMHSRP) (i.e., the Alaska Marine Mammal Tissue Archival Program [AMMTAP]) yet suffers from lack of field personnel, consistent communication, complicated interagency negotiations over funding, and lack of collection of health-related samples (e.g., formalin-fixed tissues, serum). Management of marine mammals by a single agency with no conflicts of interests (i.e., objective science not influenced by commercial fisheries management or environmental group interests) and with a dedicated research and monitoring division would likely overcome this and several other problems.

7. *Use formal risk assessment procedures to evaluate threats.* We agree with this recommendation and specifically suggest that such an assessment be initiated for California sea lions and certain organochlorines in conjunction with well-designed experiments on captive individuals.

8. *Use of rehabilitated and captive marine mammals and associated databases.* The U.S. government agencies such as the National Marine Fisheries Service and the Animal and Plant Health Inspection Service (U.S. Department of Agriculture) currently provide oversight and record-keeping for marine mammals in captivity and thus have pre-existing linkages that can help facilitate and support studies in voluntary cooperation with oceanaria. These should strive to follow rigorous epidemiological procedures. Such studies might not provide insights comparable to those based on true experimentation or dosing, but would improve understanding of basic toxicokinetics and toxicodistribution and provide correlations and associations to improve hypothesis testing through dosing experiments. Examples of the potential resources for the study of contaminants represented by bottlenose dolphins currently held in captivity have been provided by Reddy and Ridgway (2003).

9. *Use of surrogate animal models.* Although there are many examples in which studies of surrogate species have been used in place of humans, humans are indeed used in experimentation (i.e., clinical trials, drug approval processes) to better understand processes beyond the limitations of extrapolation. We still lack the luxury of results of direct experimental study of contaminants in marine mammals. Furthermore, although studies of surrogate models can provide insights as described in the workshop report, intensive rigorous examination as to the actual utility of these models as applied to marine mammals will still be required. We discourage overreliance on surrogate models and favor more direct approaches.

10. *Controlled experimental studies to address critical questions.* The absence of controlled experiments on marine mammals critically hampers management and sound policy-making decisions. The so-called "weight of evidence" approach is easily criticized and discredited on the strictest scientific grounds. This leads to a lack of consensus in marine mammal toxicology, and thus policy makers, the public, and others are left to choose whom they will "believe." Well-designed experiments with replication can provide real knowledge to move beyond a reliance on belief.

11. *Understanding processes linking exposure to effects.* This is not simply a basic research issue. The study of processes linking exposure to contaminants to effects in marine mammals must also allow for a diagnostic perspective. In our opinion, a wide gap continues to exist between research and diagnostic paradigms in marine mammal toxicology. There has been an absence of both diagnostic and research toxicology, in part because of a preponderance of investigators who lack formal credentials in toxicology and an overreliance on chemical residue studies by chemists and field biologists without input from toxicologists.

12. *Destructive sampling of tissues for biomarkers and contaminants.* In our opinion, many biomarkers are currently of low utility for studies of marine mammals. They tell us little about the health of individual marine mammals and are a long way from assessing the health status of populations. There can be a perplexing overreliance on biomarkers in the absence of basic biological and health (disease) data. Biomarkers cannot replace basic health assessments. Measuring contaminant concentrations in prey or in marine mammal tissues provides straightforward evidence of exposure, and use of biomarkers of exposure seems more indirect and less reliable. Biomarkers can be affected by confounding variables unrelated to contaminants, and their responses and adverse effects are not necessarily directly connected. We feel that validation experiments are necessary before biomarkers can be acceptable for application to assessments of contaminant exposure or effects in marine mammals.

13. *Expansion of sampling and monitoring programs to include histopathology, immunotoxicity, and life history information.* This recommendation is of fundamental importance and helps move beyond biomarkers.

14. *Selecting model species of marine mammals.* These recommendations are fundamentally sound. However, because of the potential for coordinated experiments and epidemiological approaches with captive individuals as well as historical and ongoing studies, we favor moving forward with an emphasis on California sea lions and bottlenose dolphins.

15. *Selecting model contaminants.* We agree with these principles, but for monitoring and surveys, greater focus is needed on little-studied and potentially emerging compounds. For experimental studies, we favor an initial focus on a small number of the most ubiquitous organochlorines.

16. *Complex mixtures.* We underscore the importance of the need to study specific contaminants in isolation in order to assist managers and policy makers regarding decisions about production, use, and fate of specific chemicals. We also agree with an emphasis on first understanding dose-response relationships for individual chemicals of greatest concern. Priorities for experimental exposures of marine mammals should begin with single substances or two-way interactions before moving to complex mixtures.

17. *Dose-response relationships.* We agree that these are fundamental considerations.

18. *Assessing endocrine disruption.* Documentation that an endocrine measure (such as T3/T4 level) is correlated with a contaminant concentration does not indicate an adverse effect. Contaminant and hormone measures can vary for many reasons that are not indications of adverse effects (e.g., lipid levels, binding proteins, reproductive or nutritional status). Contaminants have multiple targets in addition to the endocrine system, but a focus on endocrine disruption ignores other possible toxicoses and dysfunction. See Rolland (2000) for a review of information and deficiencies in knowledge related to contaminant exposure and thyroid function. Objective deduction is important for future studies of endocrine disruption, which must move beyond multiple correlative studies lumped as "weight of evidence," which ignore numerous studies that show no effect. These assessments require well-balanced approaches.

19. *Understanding blubber physiology and estimating total body burdens of lipophilic contaminants.* This recommendation is appropriate for understanding certain specific, lipophilic contaminants.

20. *Statistical power in experimental designs and sampling designs.* This issue is a major overriding problem. There is much concern about potentially subtle toxic effects of contaminants in marine mammals. Such effects occur within a "noisy" marine mammal health environment (i.e., an environment influenced by, for instance, nutrition, bacterial and viral exposure, trauma, cyclic changes and rhythms). Large sample sizes are usually needed, and determination of sample size requirements must enter into planning for all studies of contaminants in marine mammals. Additionally, there is increasing recognition that a priori biological hypothesis testing in field studies can be better served by model selection processes based on information-theoretic approaches rather than statistical null hypothesis testing (e.g., Anderson et al. 2000, Burnham and Anderson 2002). Such valuable alternative approaches should be considered in design and analysis of field sampling of marine mammals for contaminant studies.

Table A.1 A summary of the recommendations from the 1998 Marine Mammal Commission Workshop on Marine Mammals and Persistent Ocean Contaminants

Recommendation from 1998 Workshop	Key elements of recommendations
1. Integration of multiple approaches	Integration of laboratory, captive animal, and field studies Integration of physiological, behavioral, reproductive, clinical, pathological, and toxicological data Link immune status, health, reproduction, and survival of individuals to trends at population and ecosystem levels
2. Stable support for critical long-term programs with increased emphasis on wider collaborations	Ongoing long-term research should be a high priority and given firm support Marine mammal specialists should collaborate with scientists in human medicine and environmental toxicology on an international basis
3. Long-term interdisciplinary studies of local populations	Continue and expand long-term studies of key wild populations that allow measurement of contaminant exposure, health, and immune responses of individuals in the context of life history, distribution, demographic, and other ecological information
4. Compilation and dissemination of information	Basic information about poorly known lipophilic, persistent, bioaccumulative, or toxic chemicals must be made widely available U.S. government agencies and relevant international bodies must actively cooperate toward reaching this goal to anticipate future marine contaminant problems
5. Monitoring environmental loads and ongoing inputs of persistent contaminants	Production and environmental inputs of contaminants of proven concern continue in many areas of the world, creating an ongoing need for global-scale monitoring Contaminant monitoring should encompass other ecosystem components as well as tissues of marine mammals Other needs include identification of emission sources, transport mechanisms, transport pathways, and calculation of environmental loads

continued

Table A.1 continued

Recommendation from 1998 Workshop	Key elements of recommendations
6. Universal protocols	Universal protocols should be adopted for sample collection and storage, laboratory analytical procedures, and data reporting Protocols should stipulate quantitation of a minimum number of persistent contaminants to enhance the comparability of data sets for purposes of risk assessment and trend analysis Analytical techniques should be used to the fullest to identify all anthropogenic chemicals in tissues, and thus expand the number of existing and new chemicals known to accumulate in, and pose potential threats to, marine mammals
7. Use formal risk assessment procedures to evaluate threats	Formal risk assessment procedures should be adapted and used to evaluate threats of contaminants to specific marine mammal populations
8. Use of rehabilitated and captive marine mammals and associated databases	Marine mammals held in captivity are likely exposed to contaminants in routine daily rations of marine products, at least at low concentrations, and are an underused resource for the study of contaminant impacts on physiology and health Use of these resources may provide insight on the variability in and relationships among contaminants and health and physiological processes, biomarkers, reproduction, and survival
9. Use of surrogate animal models	Experiments with surrogate model species can provide insights about mechanism of action, comparative risks of different chemicals, and dose-response relationships Selection of an appropriate surrogate depends on the questions posed; the usefulness of the information generated will depend upon certain assumptions and extrapolations, which will require critical evaluation
10. Controlled experimental studies to address critical questions	Critical questions about impacts of the most commonly observed persistent contaminants in marine mammals have been asked for more than 25 years, without clear answers; cause-and-effect relationships have never been established Without convincing experimental evidence, uncertainty and debate regarding cause-and-effect relationships and necessary mitigation will continue, while health of individuals and the persistence of species and populations may be in jeopardy Evaluation of effects of "new" contaminants may also require experiments with captive marine mammals Experiments should be limited to nonlethal exposures to doses of contaminants similar to those experienced by some wild populations, and should be well justified and well designed
11. Understanding processes linking exposure to effects	Understanding linkages between contaminants and the health, immune system, or reproductive fitness of marine mammals at the cellular or molecular level is most likely to come from laboratory studies Cell culture and in vitro techniques can provide species-specific data on topics such as toxic equivalency using cell lines of marine mammals Other laboratory mammals can be used for detailed, invasive experimentation needed to improve understanding of the biochemical processes linking exposure and effects Semifield and epidemiological studies of wild populations can help validate inferences Some marine mammal populations under protection or well-regulated exploitation are recovering and experiencing rapid population growth, which can confound the evidence for contaminant-related effects on populations
12. Destructive sampling of tissues for biomarkers and contaminants.	Biomarkers can be used to assess chemical exposure and with further development may be capable of predicting effects in marine mammals; biomarkers may prove useful in studies of effects of short-lived, nonbioaccumulative contaminants Biomarker studies should be included with other contaminant-related research on marine mammals and closely linked to information on contaminant burdens, exposures, and physiological and life history traits Biomarkers for marine mammals should be sensitive, rapid, inexpensive, and field-adaptable through nondestructive sampling. Validation studies are necessary to increase the value of nondestructive samples as surrogates for target organs of marine mammals
13. Expansion of sampling and monitoring programs to include histopathology, immunotoxicity, and life history information	It is uncertain to what extent some contaminants may affect the immune systems of marine mammals, but there is sufficient reason for concern Long-term programs for sampling marine mammals should support investigations of histopathology and immunotoxicology given the associated life history information
14. Selecting model species of marine mammals	Model species must be identified and employed in studies of contaminants and marine mammals The most appropriate species are marine mammals for which considerable information is already available about population dynamics, life history, and physiology; that are in captivity or otherwise easily accessible; and that are involved in related ongoing studies Other factors to consider are the feasibility of obtaining sufficient numbers of high-quality samples; whether the species occurs across a gradient of habitats; and conservation status

Table A.1 continued

Recommendation from 1998 Workshop	Key elements of recommendations
	Possible model species include bottlenose dolphins, harbor porpoises, belugas, bowhead whales, minke whales, harbor seals, gray seals, California sea lions, polar bears, and Florida manatees
15. Selecting model contaminants	Criteria for selecting model compounds for priority monitoring and study should include levels of production, potency of bioaccumulation, and toxicity High priority should be placed on chemicals with known adverse effects and that marine mammals are known to be exposed to: PCBs, DDT and metabolites, other organochlorines, butyltins, and a few trace elements are obvious candidates; mutagenic and genotoxic effects of PAHs warrant study in exposed populations
16. Complex mixtures	All marine mammals have body burdens of many different contaminants, in variable amounts, stored or circulating in different organ systems; marine mammals in the wild are exposed to complex mixtures rather than to single chemicals Effects from exposure to multiple contaminants may be synergistic or antagonistic, thus experimental exposures should include complex mixtures that mimic as closely as possible the exposure in nature Exposures to complex mixtures should not be conducted until dose-response relationships for individual chemicals involved are understood; care should be taken to avoid conclusions based solely on in vitro exposures Some contaminants will have to be studied in isolation in order to inform managers and policy makers in decisions about production, use, and fate of specific chemicals
17. Dose-response relationships	Effects of contaminants should be understood well enough to predict responses from specific doses, including likely no-effect levels of exposure Analyses of the dose-response relationship must be pursued on a species-by-species, contaminant-by-contaminant basis Experiments and sampling designs should require that exposure be controlled or measured, and that potentially confounding factors (e.g., age, sex, reproductive status, nutrition) be taken into account
18. Assessing endocrine disruption	Some common contaminants of marine mammal tissues are known to affect the endocrine systems of other species; the potential for endocrine disruption in marine mammals should be regarded as a serious possibility and be subject to aggressive research and evaluation Systematic appraisal of a number of morphological and other endpoints of endocrine disruption should be incorporated in routine marine mammal stranding and health evaluations
19. Understanding blubber physiology and estimating total body burdens of lipophilic contaminants	Physiological condition affects the distribution of lipophilic contaminants in the bodies of marine mammals; further research on lipid dynamics is required to improve understanding of the processes determining this distribution Estimation of total body burden is important for risk assessment, and also requires information on the total amount of blubber, muscle, and other tissues
20. Statistical power in experimental designs and sampling designs	In both laboratory and field studies of contaminants in marine mammals it is essential that hypotheses be clearly formulated and that a statistical model be developed to determine the appropriate sample size before the experimental or sampling protocol begins

Note: Recommendations were not ranked in any order of priority by the workshop.

ACKNOWLEDGMENTS

The authors thank the editors of this volume, and we highlight the dedicated efforts of John E. Reynolds III and two anonymous reviewers. The initial effort to construct this report was a consultation (Portland, Oregon, August 2003) coordinated through the Marine Mammal Commission and the National Fish and Wildlife Foundation (NFWF), especially Whitney Tilt. The funding for that consultation and for follow-up efforts was provided by the U.S. Congress to the Marine Mammal Commission. Editing associated with the Portland consultation was handled most notably by R. R. Reeves. O'Hara thanks the North Slope Borough and the University of Alaska Fairbanks for the time and support to work in marine mammal toxicology.

FRANCES M. VAN DOLAH

Effects of Harmful Algal Blooms

COASTAL WATERS AROUND THE WORLD periodically experience harmful algal blooms (HABs) that adversely affect marine animal health and reproductive success, coastal ecosystem integrity, human health, and economies that depend on coastal resources. Acute morbidity or mortality of marine mammals often correlates with the presence of toxic algal blooms, and to date four classes of algal toxins have been associated with marine mammal mortality events. A marine mammal can suffer exposure to algal toxins either directly through its respiratory system or indirectly via food web transfer. Their susceptibility is therefore dependent not only on the occurrence of toxin-producing algae within the habitat of the mammals in question but also, in the case of food web transfer, on the co-occurrence of appropriate prey species during an algal bloom to serve as vectors to higher trophic levels.

Marine mammal mortality events associated with HABs appear to have increased in recent years. This may reflect, at least in part, the increased attention paid to such events and our improved ability to identify algal toxins in marine mammal tissues. However, it is probably also an indication of the increases in both frequency and geographic distribution of HABs that have been documented over the last quarter-century (Smayda 1990, Hallegraeff 1993, Van Dolah 2000). Management of the impacts of algal toxins on marine mammals requires an understanding of the causes and consequences of HABs, the reasons for their apparent increase, and their relationships to large-scale oceanographic and climate changes.

Most of the examples and much of the discussion in this chapter are rooted in North America, but the problems associated with HABs are global. To a considerable extent, my North American bias reflects the fact that my direct personal experience has been centered there. Also, rigorous investigations of the links between HABs and marine mammal health require particular expertise and professional interest, substantial commitments of money and other resources, and a relatively sophisticated logistical infrastructure, all of which have been uniquely available in North America over the last several decades.

OVERVIEW OF ALGAL TOXINS

The origins of marine algal toxins are unicellular algae that respond to favorable conditions in their environment by proliferating or aggregating to form dense concentrations of cells called "blooms." In many cases toxic species are normally present in low concentrations with no evident environmental health impacts; their toxicity generally depends on their presence in high cell concentrations. Toxin-producing marine phytoplankton are primarily members of two algal groups, the dinoflagellates and diatoms. Many phytoplankton within these groups produce novel compounds that have potent biological activity. However, less than 5% (<100 species) of the known dinoflagellate species and fewer than twenty-five species in one genus of diatoms are known to produce compounds that are toxic to mammals. Most of these compounds are potent neurotoxins that target either ion channels in the brain or neuromuscular junctions or components of the cell-signaling pathways that interact with these channels. Although the reasons toxins are produced by these algal species are not fully established, they most likely provide a competitive advantage over other algal species or serve as antipredation mechanisms targeting zooplankton or small herbivores. Their impacts on higher-trophic-level species, such as marine mammals or humans, may thus be incidental.

Five major classes of algal toxins have been well studied because of their toxicity to humans through the consumption of seafood: saxitoxins, brevetoxins, domoic acid, ciguatoxins, and okadaic acid (Fig. 6.1a–e). Of these, the first four are neurotoxins that have been associated with marine

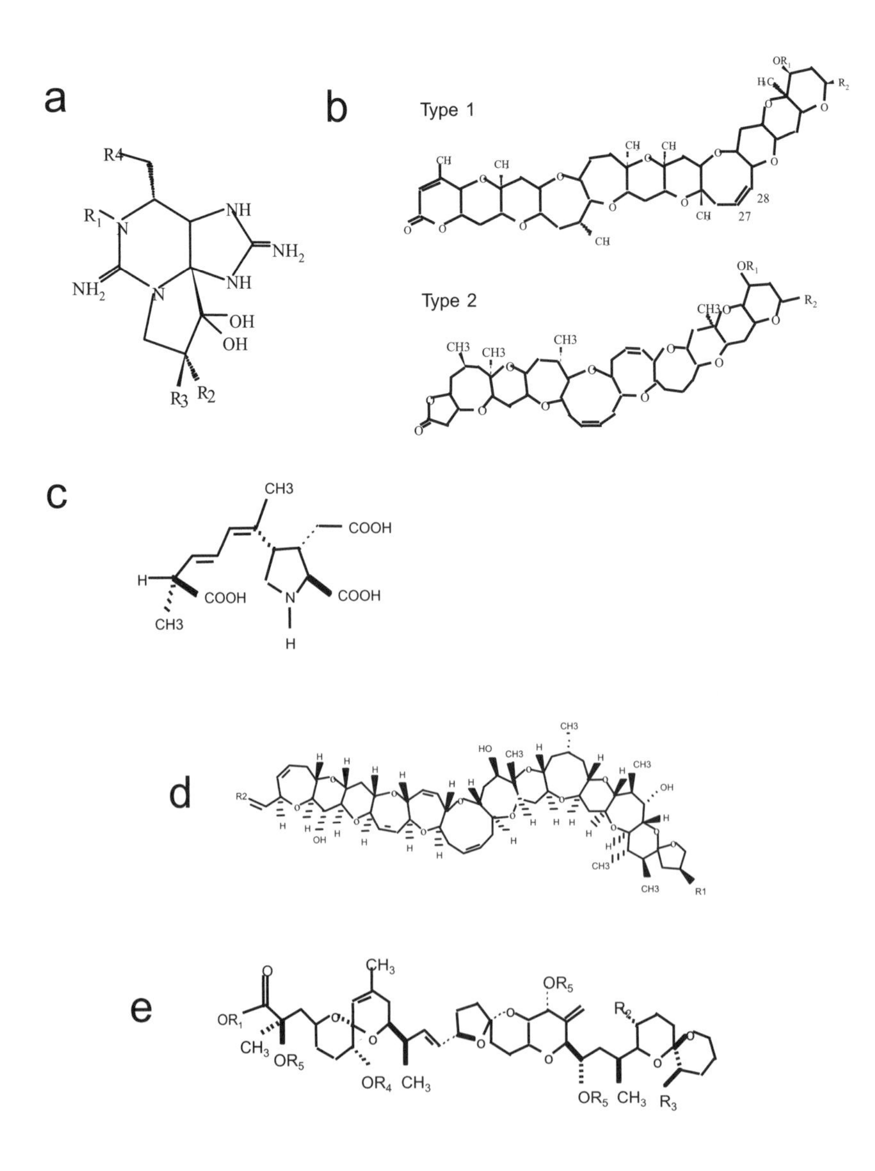

Fig. 6.1. Structures of major classes of algal toxins with adverse human health effects: (a) saxitoxin, (b) brevetoxin, (c) domoic acid, (d) ciguatoxin, and (e) okadaic acid.

mammal mortality events. The fifth, okadaic acid, is a protein phosphatase inhibitor responsible for diarrhetic shellfish poisoning in humans and is also a tumor promoter that has been proposed (but not proven) to be involved in fibropapilloma disease in sea turtles (Landsberg et al. 1999). In addition to these comparatively well-studied classes, several novel algal toxins that cause adverse human health effects have been identified over the past decade, including *Pfiesteria* toxin, azaspiracid, yessotoxins, and spirolides, and these should be kept in mind when investigating marine mammal mortality events with no obvious cause. The following sections provide brief descriptions of each of these toxin classes.

Saxitoxins

Saxitoxins (STXs) are water-soluble, tetrahydropurine neurotoxins of which more than twenty-one family members are currently recognized. These toxin congeners are produced in varying combinations by several dinoflagellate species belonging to three genera (*Alexandrium, Gymnodinium,* and *Pyrodinium*) in different regions of the world and also by several species of cyanobacteria found in fresh water. In humans, STXs are responsible for paralytic shellfish poisoning (PSP), with symptoms that include tingling and numbness of the perioral area and extremities, loss of motor control, drowsiness, incoherence, and, in extreme cases, death from respiratory paralysis. The STXs bind to site 1 of the voltage-dependent sodium channel and block neurotransmission (R. E. Levin 1992). However, because STX is water-soluble, it is rapidly cleared from the blood (<24 h in humans) so victims generally survive if they are put on life support. In the United States, STXs historically posed a threat primarily on the northeastern and western coasts, but they have recently also been found in the Indian River Lagoon, Florida, where their occurrence was unprecedented (Landsberg et al. 2002).

Domoic Acid

Domoic acid (DA) is a water-soluble amino acid that mimics the neurotransmitter glutamate. A potent activator of certain subtypes of glutamate receptors present in the brain, DA is produced by several species of diatoms in the genus *Pseudonitzschia;* other species in this genus are non-toxic. The first insight into DA as an algal toxin came about through a human intoxication event on Prince Edward Island in eastern Canada in 1987, when some 100 people became ill after consuming contaminated mussels (Perl 1990). A hallmark of this event was the permanent loss of short-term memory in certain victims, which led to DA intoxication in humans being termed amnesic shellfish poisoning (ASP). The symptoms of ASP in humans include gastrointestinal problems (e.g., nausea, vomiting, diarrhea) and neurological difficulties, including dizziness, disorientation, lethargy, seizures, and permanent loss of short-term memory. Neurotoxicity caused by DA results in brain lesions in areas of the brain where glutamate receptors are heavily concentrated, particularly in regions of the hippocampus responsible for learning and memory processing (Perl et al. 1990). In the United States, DA poses a threat primarily on the West Coast, although toxin-producing diatom species are also present on the northeast coast and in the Gulf of Mexico.

Brevetoxins

Brevetoxins (PbTx) are a suite of polyether toxins best known as the Florida red tide toxin, produced by the dinoflagellate *Karenia brevis.* However, PbTxs have also been shown to be produced by newly identified *K. brevis*–like species in New Zealand (Haywood et al. 1996) and by raphidophytes of the genus *Chattonella* in Japan (Khan et al. 1995a,b). They have also been identified recently in a fish-killing *Chattonella* bloom in Delaware Bay, raising the possibility of broader environmental and public health issues in mid-Atlantic estuarine waters (Bordelais et al. 2002).

Brevetoxins, like STX, target the voltage-dependent sodium channel, but in this case their binding to site 5 causes the channel to open under conditions in which it is normally closed, resulting in inappropriate neuronal transmission. Brevetoxin exposure in humans is primarily through the consumption of toxic shellfish and is thus termed neurotoxic shellfish poisoning (NSP), with symptoms that include nausea, tingling and numbness around the mouth, severe muscular aches, loss of motor control, and, in particularly severe cases, seizure. A second route of exposure to PbTxs is through aerosolization of the toxin caused by the lysis of fragile *K. brevis* cells by wind or wave action. In humans, exposure to aerosolized toxin results in coughing, gagging, and burning of the upper respiratory tract (Pierce et al. 2003). Fish are particularly sensitive to PbTxs, probably because the fragile *K. brevis* cells are lysed as water passes through their gills, permitting passage of these lipophilic toxins across the gill epithelium directly into the blood stream. Consequently, Florida red tides are generally associated with extensive fish kills.

Ciguatoxins

Ciguatera fish poisoning (CFP) is an intoxication caused by a suite of polyether toxins known as ciguatoxins (CTXs). Ciguatoxins are produced by the dinoflagellate *Gambierdiscus toxicus,* which grows on filamentous algae associated with coral reefs and reef lagoons. They enters the food web when these algae are grazed upon by herbivorous fishes and invertebrates, and the toxins are biotransformed and bioaccumulated in the highest trophic levels of reef fishes. Although ciguatera occurs persistently at certain locations, outbreaks are sporadic and unpredictable at others. Reef disturbance from storm damage or human activities frequently precedes ciguateric conditions (Ruff 1989, Kaly and Jones 1994), and the overgrowth of corals by macroalgae, caused by coral bleaching, overfishing, or nutrient enrichment may also promote ciguateric conditions by providing increased substrate for the dinoflagellate to grow (Bagnis 1987, Kohler and Kohler 1992).The CTXs are structurally related to the PbTxs and bind to the same site on the voltage-dependent sodium channel. However, CTX is much more potent than PbTx, and the symptoms, while initially similar to those of PbTx intoxication, may be followed by additional neurological problems, including reversal of temperature sensation, tachycardia, hypertension, blurred vision, paralysis, and, in extreme cases, death. Following acute intoxication, chronic symptoms in humans often persist for weeks, months, or even years. In the United States, ciguatera is a problem primarily in Hawaii and southern Florida.

Okadaic Acid

Okadaic acid (OA) is one of a family of polyethers that cause diarrhetic shellfish poisoning (DSP) in humans, a comparatively mild intoxication that generally resolves within 2 or 3 days. Diarrhetic shellfish poisoning is widespread, with essentially seasonal occurrences in Europe and Japan; it has also been documented in South America, South Africa, New Zealand, Australia, and Thailand. The first confirmed case of DSP in North America occurred in 1990 in Nova Scotia, Canada (Marr et al. 1992). The causative organism in this outbreak was a dinoflagellate, *Prorocentrum lima,* which was found in the nutrient-rich environment associated with filamentous algae growing on raft cultures of mussels. The diarrhetic shellfish toxins are inhibitors of protein phosphatases, which are critical components of cellular processes involved in metabolism, ion balance, and neurotransmission. Diarrhea associated with DSP is probably caused by the hyperphosphorylation of proteins, including ion channels, in the intestinal epithelia (Cohen et al. 1990), resulting in impaired water balance and loss of fluids. However, OA-like polyether toxins have been identified as tumor promoters (Suganuma et al. 1988), raising the question of what effect low levels of chronic exposure to DSP toxins might have in humans, as well as in wildlife such as marine turtles (Landsberg et al. 1999).

Pfiesteria Toxin

Pfiesteria piscicida, a dinoflagellate first identified in North Carolina coastal waters in 1990, has been linked to lesioned fish kills in the mid-Atlantic region of the United States and to a human condition known as estuary-associated syndrome, with symptoms that include fatigue, headache, respiratory irritation, skin lesions or burning sensations on contact, disorientation, and memory loss. A study of people exposed to *Pfiesteria* or *Pfiesteria*-associated water demonstrated severe impairment in cognitive function, as compared to unexposed individuals from the same occupational, educational, and geographic area (Grattan et al. 1998). *Pfiesteria* differs from the previously discussed dinoflagellates in that it is nonphotosynthetic, heterotrophic, and more animal-like than plantlike. There is compelling evidence that its occurrence at toxic levels coincides with the eutrophication of coastal waters through intensive swine and poultry raising in areas of North Carolina and Maryland (Burkholder 1998).

The toxins responsible for fish lethality or human neurological symptoms have not yet been fully characterized, although some insight into its toxic mode of action suggests interaction with P2X7 purinerigic receptors (Kimm-Brinson et al. 2001) found in the brain and immune system. There is currently no evidence that toxicity is transferred through the food web to top predators such as marine mammals; however, this syndrome might be considered in the event of marine mammal morbidity or mortality in estuarine waters of the U.S. mid-Atlantic states.

Azaspiracid

Azaspiracid is a newly identified nitrogen-containing polyether toxin, reported for the first time in 1995 and considered responsible for an outbreak of gastrointestinal illness in eight people following the consumption of shellfish from the west coast of Ireland (Furey 2003). It has since been shown to be produced by the dinoflagellate *Protoperidinium crassipes,* not previously known to be toxic, and to have a wide distribution in northern Europe (James et al. 2003). Unlike the other diarrhetic shellfish toxins, azaspiracid causes necrosis in the lining of the small intestine and lymphoid tissue and fatty changes in the livers of mice; it has therefore been added to the list of algal toxins tested for in shellfish for human consumption in the European Union. The mode of action of this toxin has not yet been elucidated.

Yessotoxins

Yessotoxin (YTX) is one of several related sulfated polyether toxins produced in a number of dinoflagellate species, including *Protoceratium reticulatum* (= *Gonyaulax grindleyi* in part) and *Lingulodinium polyedrum.* Yessotoxin, originally thought to be one of the diarrhetic shellfish toxins, has recently been shown to have no diarrhetic activity and little toxic potency when administered orally to mice (Tubaro et al. 2003). However, it does have potent lethality to mice when injected, inducing neurological symptoms. Blooms of both *Lingulodinium polyedrum* and *Gonyaulax grindleyi* have been implicated in fish and shellfish mortality events although the role of YTX in those events was not investigated.

Spirolides

Spirolides are a group of macrocyclic amines produced by the dinoflagellate *Alexandrium ostenfeldii* and implicated in shellfish toxicity in northern Europe (Cembella et al. 2000). The mode of action of this toxin is not clear, but injection into mice causes neurological symptoms followed by rapid death. The distribution of *Alexandrium ostenfeldii* in North

America is limited to eastern Canadian waters and the Gulf of Maine.

MARINE MAMMAL MORBIDITY AND MORTALITY EVENTS ASSOCIATED WITH ALGAL TOXINS

Saxitoxins

Saxitoxins were implicated in the mortality of humpback whales (*Megaptera novaeangliae*) in Cape Cod Bay, Massachusetts, between November 1997 and January 1998 (Geraci et al. 1989). During the late fall, humpback whales usually prey on sandlance (*Ammodytes* spp.) on Stellwagen Bank off the Massachusetts coast (Anderson and White 1989). However, in the fall of 1987 sandlances were largely absent from this area and humpbacks were apparently feeding on Atlantic mackerel (*Scomber scombrus*). Fourteen whales died within 5 weeks in Cape Cod Bay and northern Nantucket Sound.

Baleen whales had not previously been reported to mass-strand, and therefore the mortality of fourteen animals was highly unusual. The whales affected in this event were robust, unlike many mass-stranded animals. Furthermore, death appeared to have occurred quickly, based on the observation of one whale that had exhibited normal behavior 90 min before it was found dead. The stomachs of six of nine carcasses examined contained incompletely digested fish, indicating that the whales had been feeding not long before death, so an acutely toxic substance was suspected. As the STX-producing dinoflagellate *Alexandrium tamarense* blooms annually in this region, STXs were investigated as a potential causative agent. Saxitoxin-like activity was detected in the stomach contents, liver, and kidney, as determined by mouse bioassay; however, the presence of toxin could not be corroborated by instrumental analysis. Thus, identification of STX as the causative agent remains circumstantial. Nonetheless, STX was found by high-performance liquid chromatography (HPLC) in the viscera and livers of mackerel caught during local waters in the same time frame. Considering the toxin concentration found in the mackerel, Geraci et al. (1989) estimated that a whale consuming 4% of its body weight daily would have ingested 3.2 μg STX/kg body weight. By comparison, the lethal dose of STX in humans is estimated at 1 to 4 mg (ca. 6–24 μg/kg; R. E. Levin 1992). Physiological adaptations that may make humpback whales more susceptible to the toxic effects of STXs include: (1) approximately 30% of their body weight is blubber in which the water-soluble STXs do not partition, so these toxins are more concentrated in metabolically sensitive tissues; (2) the diving physiology of humpback whales concentrates blood in the heart and brain and away from those organs that function in detoxification, further concentrating a neurotoxin in sensitive tissues (Geraci et al. 1989); and (3) respiration is not under autonomic control in cetaceans (W. F. Perrin pers. comm.).

Saxitoxins were also suspected in the mortality of endangered Mediterranean monk seals (*Monachus monachus*) on the coast of Mauritania, West Africa, in May and June 1997. More than 100 animals, representing more than 70% of the local population and approximately a third of the world population of the species died in this event. A previously undescribed morbillivirus (WAMV-WA) was isolated from three of these monk seal carcasses (Osterhaus et al. 1997a,b), indicating that this virus was active in the population at the time. Morbilliviruses have been identified as causative agents of mass mortality events involving other marine mammal species (see Gulland and Hall this volume) and were thus identified as a likely causative agent in this case. However, unlike in previous morbillivirus-associated events, these monk seals appeared to die quickly, with no signs of long-term illness. Dying animals exhibited lethargy, lack of motor coordination, and paralysis in the water, symptoms similar to those associated with STX exposure. Hernandez et al. (1998) identified three toxic dinoflagellate species present at moderate concentrations in waters near the seal colony, and fish collected from seal feeding grounds were positive for STX. The seals' livers also tested positive for STX but at low concentrations. As in the case of the humpback whales off Cape Cod, without further insight into toxin effect levels in these animals, it was difficult to determine whether the levels of STXs observed were sufficient to have caused the monk seal mortality.

Episodic mass mortality from PSP toxins could be extremely important to the population biology of long-lived mammals such as the Mediterranean monk seal (Forcada et al. 1999). The 1997 event was age specific, with mostly adults being affected, and it resulted in a significant change in the age structure of the population (12% juveniles prior to the event versus 29% after). As the event reduced the breeding population to fewer than seventy-seven individuals, it may have reduced the population's genetic variability and ultimately compromised the survival of the species.

Investigations into a possibly more subtle role of STXs in an unexplained decrease in reproduction rate of the endangered North Atlantic right whale (*Eubalaena glacialis*) have been spurred by evidence of trophic transfer of STX to right whales from blooms of the dinoflagellate *Alexandrium fundyense*. The North Atlantic right whale population currently consists of only about 350 individuals and may be declining, with the average calving interval increasing from 3.7 years in the 1980s to 5.1 years between 1993 and 1998 (Kraus et al. 2001). Blooms of STX-producing *A. fundyense* occur almost annually in Grand Manan Basin in the lower Bay of Fundy, a major feeding ground for North Atlantic right whales. The copepod *Calanus finmarchicus*, which dominates the right whale's diet in the bay, is known to feed on *A. fundyense*. During a bloom of this dinoflagellate, *C. finmarchicus* there were found to contain an average of 0.41 ng of STX per copepod (Durbin et al. 2002). Concentrations of copepods in the area averaged 5,742 copepods/m^3, and right whales present during the study period were observed feeding on them. Based on an ingestion rate of

4×10^8 copepods/day, it is estimated that the whales would have ingested between 4.73 and 9.65 μg STX/kg per day during the period of the study (Durbin et al. 2002). This is similar to the estimated lethal dose in humans (R. E. Levin 1992). Saxitoxin measured in right whale feces from the area exceeded 20 μg/g (Doucette et al. 2002). The significance of this level of STX in whale feces is not yet clear with respect to the level of the toxin circulating in the blood or with regard to its effects on behavior and physiology. Chronic, sublethal exposure to STX may affect swimming and diving behavior, which in turn could result in decreased feeding rates, poorer fitness, and ultimately reduced calving rates (Durbin et al. 2002). Consistent with this notion, the diving frequency of right whales in the Bay of Fundy is reported to be lower than that of right whales in Cape Cod Bay, and their breathing rate is higher (Durbin et al. 2002). Thus, the effects of chronic exposure to STX on right whale reproductive fitness warrant further investigation.

Saxitoxin is also suspected as a contributing factor in the mortality of bottlenose dolphins in the Indian River Lagoon, Florida, during 2001 and 2002. Prior to April 2002, when a fisherman caught pufferfish, which causes symptoms of paralytic shellfish poisoning after being consumed, STX was unknown in the southeastern United States. An investigation of the toxicity of the suspect pufferfish revealed the presence of high concentrations of STX (Quilliam et al. 2002), and the STX-producing dinoflagellate *Pyrodinium bahamense* was identified as the causative organism in the lagoon (Landsberg et al. 2002). The subspecies *Pyrodinium bahamense* var. *bahamense,* known to occur in the lagoon, has never before been known to be toxic; however, it is related to the STX-producing subspecies *Pyrodinium bahamense* var. *compressum* responsible for extensive STX toxicity for people in Southeast Asia. It is not known at this time if the subspecies responsible for STX toxicity in the lagoon is the endemic subspecies *bahamense,* which has suddenly expressed toxicity, perhaps in response to changing conditions in the lagoon, or rather that the subspecies *P. b.* var. *compressum* has been introduced into Florida waters. However, during the past several years, the northern lagoon has experienced a number of unusual events, ranging from increased numbers of deaths among bottlenose dolphins, manatees, fish, and horseshoe crabs; increased tumor incidence in hard clams; and reduced natural recruitment and hatchery losses of hard clams (Landsberg et al. 2002, pers. comm.).

Bottlenose dolphins that died in the lagoon during this period tended to be emaciated and displayed significant skin lesions, suggesting that multiple factors were associated with their poor health status (G. Bossart pers. comm.). However, at least two dolphins examined had pufferfish in their stomachs (Barros 1993, Barros and Wells 1998), suggesting that STX may have played a role in their deaths. Pufferfish do not appear to be a component of the bottlenose dolphin's normal diet; it is possible that, in the absence of their normal prey, these already malnourished animals fed on toxic pufferfish, which were the immediate source of their demise, but not the underlying cause of poor fitness of bottlenose dolphins in the Indian River Lagoon.

As troubling as the advent of an apparently new algal intoxication of marine mammals may be, it is important to point out that STX-producing dinoflagellate blooms occur naturally in some ecosystems. The long-term coexistence of marine mammals with toxic species is suggested to have influenced the distribution of sea otters (*Enhydra lutris*) in Alaskan coastal waters. Sea otters are voracious predators of bivalves, consuming 20 to 30% of their body weight per day, with the butter clam (*Saxidomus giganteus*) accounting for most of the prey eaten in Alaskan waters (Kvitek et al. 1991). The butter clam is capable of sequestering STXs at high concentrations in its siphon, where they are retained for as long as a year, presumably as a chemical defense against predation (Kvitek and Beitler 1991). In the Kodiak region of Alaska, butter clams are abundant, and sea otters and butter clams are co-distributed both in inside passage waters and along the outer coast. Butter clams are also abundant throughout southeastern Alaska, but they become toxic only in the former, where they remain toxic year-round as a result of seasonal *Alexandrium* blooms. Butter clams on the outer coast of southeastern Alaska are never toxic. Correspondingly, sea otter distribution in southeastern Alaska is limited to the outer coast. To test the hypothesis that the distribution of sea otters in southeastern Alaska is influenced by butter clam toxicity, Kvitek et al. (1991) carried out feeding studies on caged otters. Otters fed live butter clams both decreased their rate of consumption when offered toxic clams and selectively discarded the most toxic tissues (the siphons and kidneys) of clams containing even low levels (<40 μg/100 g) of toxin. These findings indicate that otters have the ability to detect and avoid toxic clams, and such behavioral avoidance of toxic prey is consistent with the hypothesis that sequestration of STXs by the butter clam is a chemical defense against predation that may influence otter distribution.

Domoic Acid

In 1991 the first evidence of DA on the west coast of North America was obtained as a consequence of a mass mortality of pelicans (*Pelecanus occidentalis*) and cormorants (*Phalacrocorax penicillatus*) in Monterey Bay, California. Affected birds exhibited neurological symptoms similar to those reported in experimentally exposed mice, including scratching and head weaving (Work et al. 1993). Domoic acid was present in the stomach contents of the affected birds, with northern anchovies (*Engraulis mordax*) being the main component. Moreover, frustules (siliceous cell walls) of *Pseudo-nitzschia* spp. were found in the stomachs of the fish, confirming the trophic transfer of the toxin through this vector species.

The first confirmed DA poisoning of marine mammals occurred on the California coast in 1998 when seventy California sea lions (*Zalophus californianus*) stranded along central California from San Luis Obispo to Santa Cruz (Scholin et al. 2000). All the animals were in good nutritional condition and displayed clinical symptoms that were predominantly neurological, including head weaving, scratching, tremors, and convulsions (Gulland 2000). Of the seventy animals that stranded, fifty-seven died. Most of the clinically affected animals were adult females, of which 50% were pregnant. Juveniles were of both sexes. No adult males were affected, which may reflect normal differences in general distribution and foraging by adult male and female sea lions along the California coast. Many of the fetuses of pregnant females were found dead, and two pups born during the episode did not survive. Domoic acid was identified in the serum, urine, and feces of many (but not all) of the sea lions that exhibited clinical symptoms, with the highest concentrations found in the urine and feces. Although it is difficult to extrapolate effect levels from mice, the highest concentration of DA observed in serum (0.17 μg/ml) was of similar magnitude to that in mice treated with an LD_{50} dose of toxin (Peng and Ramsdell 1996). No measurable toxin was found in other tissues, consistent with rapid clearance of the toxin via the kidney, as observed in rodents and primates (Iverson et al. 1989, Truelove and Iverson 1994). All sea lions that died within 24 h of stranding had significant brain lesions, primarily in the hippocampus, that were consistent with DA poisoning (Gulland 2000).

The origin of the DA responsible for this mortality event was a bloom of *P. australis* that developed in Monterey Bay in May 1998, reaching a peak abundance of about 200,000 cells/liter (Scholin et al. 2000). Anchovies collected from the bay during the bloom had high levels of DA in their gut, and *P. australis* frustules were visible in their stomach contents. Anchovy vertebrae and otoliths and frustules from *P. australis* were also found in sea lion fecal samples that contained high DA concentrations, confirming anchovies as the primary vector of the toxin to the sea lions.

This mortality event is likely to have affected larger numbers of sea lions than recorded, because of the extensive areas of coastline in central California that are inaccessible for monitoring and the lack of postmortem examinations on animals that stranded dead and were decomposed. Nonetheless, given a population of California sea lions estimated at 167,000 individuals, it is unlikely that an event causing mortality of less than 0.04% constitutes a threat at the population level (Gulland 2000).

Approximately 1 month following the peak of strandings of California sea lions in the 1998 event, an increased number of southern sea otter (*Enhydra lutris nereis*) deaths were observed in the same region. The southern sea otter's distribution is limited to approximately 250 miles of coastline, from Santa Cruz to Purisima Point. The southern sea otter has been federally listed as a threatened species since 1977, and although its population has still not recovered from heavy hunting in the nineteenth century, by 1994 there were approximately 2,400 animals (Ralls et al. 1996). As noted in *Encyclopedia of Marine Mammals* (Estes and Bodkin 2002:856), "For most of the 20th century, the California sea otter maintained a slow but steady increase in numbers and range." Analyses of serum, urine, kidneys, and stomach contents collected from southern sea otters that died from unidentified causes between 1995 and 2000 ($n = 50$) revealed that twenty-eight were positive for DA in one or more fluids/tissues. Unlike California sea lions, southern sea otters feed primarily on benthic invertebrates, including crabs, clams, and mussels. Many of the animals that tested positive for DA had been feeding on the spiny mole crab (*Blepharipoda occidentalis*) and, less frequently, on another mole crab species (*Emerita analoga*). Although there are no data on DA levels in *B. occidentalis, E. analoga* is an intertidal suspension-feeder that can consume *Pseudonitzschia* spp. and was shown to have accumulated DA during the 1998 *Pseudonitzschia* bloom in Monterey Bay. The delayed response of southern sea otters to the DA poisoning event therefore appears to reflect their route of exposure, primarily through the benthic food web following the sinking-out of the decaying bloom.

A similar mortality event occurred in 2000 involving 187 California sea lions stranded on the central coast of California from June through December. Analyses of serum and urine samples confirmed the involvement of DA, and blooms of *P. australis* were found in Monterey Bay during the extended event.

Concurrent with the 2000 sea lion mortality event, abnormally high numbers of gray whale (*Eschrichtius robustus*) strandings occurred, primarily during the northward migration from their Baja California calving grounds to their northern feeding grounds. A total of 350 gray whale deaths was documented along the migration route in 2000, compared with 273 in 1999 and fewer than 50 in earlier years. Many of the gray whales stranded in an emaciated condition, likely caused by poor foraging conditions in the Bering Sea. Domoic acid was confirmed in the serum, urine, and feces of one young animal that stranded alive, but emaciated, near Santa Cruz, and concentrations in its tissues were sufficiently high to implicate DA toxicity as a contributing factor in its death. It is likely that in this case malnutrition was a predisposing factor. This was the first evidence of an effect of DA on a baleen whale. Frustules of *Pseudonitzschia australis* were identified in fecal samples of both gray and blue whales (*Balaenoptera musculus*) during the 2000 *P. australis* bloom season; thus it is apparent that both species were exposed to DA, perhaps at subacute levels. Krill serve as major prey of blue whales, which can consume up to 2 tons a day. Recent studies on krill, *Euphausia pacifica* and *Thysanoessa spinifera,* suggest that they readily consumed toxic *P. australis* during the 2000 blooms in Monterey Bay, accumulating DA to high levels (Bargu et al. 2002). Although DA has not been implicated in blue whale deaths, Lefebvre et al. (2002) estimate that, based on the DA levels accumulated in the 2000 Monterey Bay *P. australis* bloom, krill could convey up to 85 g/day (0.85 μg/kg of DA to a blue whale.

In February 2002 the third DA event in 5 years began on the California coast. This episode proved to be the second-largest documented marine mammal mass mortality event in the United States and affected the largest diversity of marine mammal species. Unlike the previous sea lion events, the 2002 occurrence began during the winter in southern California, when larger than usual numbers of long-beaked common dolphins (*Delphinus capensis*) were found dead or dying. Although there was no evidence for a *Pseudonitzschia* bloom at the time, toxin testing revealed DA in the urine or feces of almost all of the stranded animals. In total, more than ninety cetacean deaths were attributed to DA poisoning during this event. By April 2002 California sea lions had begun to come ashore farther north along the central California coast with symptoms typical of DA poisoning. More than 670 pinnipeds were treated at rehabilitation centers, with the estimated number of dead in the thousands. Altogether nine marine mammal species were involved in this event: two balaenopterids (*Balaenoptera acutorostrata, Megaptera novaeangliae*), two delphinids (*Delphinus capensis, D. delphis*), two phocoenids (*Phocoenoides dalli, Phocoena phocoena*), two pinnipeds (*Zalophus californianus, Phoca vitulina*), and one mustelid (*Enhydra lutris*) (Heyning 2003). The adverse effects of this occurrence went far beyond marine mammals, with hundreds of seabird deaths and a rare mass mortality of Humboldt squid (*Dosidicus gigas*) during the same period, with sampled individuals testing positive for DA.

Retrospective analysis reveals that clusters of stranded animals showing signs similar to those individuals in the recent DA events had been reported in the past along the California coast although no causes of death were determined. In 1978, forty animals stranded in Ventura County displaying similar neurologic effects (Gilmartin 1979); in 1986, 1988, and 1992, sea lions with similar signs were admitted for rehabilitation to the Marine Mammal Center in Sausalito, California, and in 1992, eighteen sea lions stranded in San Luis Obispo County displaying the same symptoms (Beckmen et al. 1995). The adverse effects of DA on fetal survival in California sea lions in these recent events have led to questions about DA's role in episodic sea lion reproductive failure documented in the Channel Islands (F. Gulland pers. comm.). This reproductive problem was previously thought to be linked to DDT exposure or to be due to viral infection, with no conclusive findings. Studies are currently underway to characterize the maternal transfer of DA to fetuses and to assess the implications for pup survival.

Brevetoxins

Brevetoxin-producing red tides appear off the west coast of Florida in the late summer/early fall almost annually and may persist for several months. When offshore blooms are carried into coastal waters, they have their most significant impacts on coastal ecosystems, causing widespread mortality among marine organisms. Brevetoxins have long been suspected to be responsible for mortality of Florida manatees (*Trichechus manatus latirostris*) and bottlenose dolphins (*Tursiops truncatus*) in the Gulf of Mexico and along the southern U.S. Atlantic Coast. With improvements in detection methods over the past decade, the presence of PbTx in both manatees and dolphins during red tide–associated mortality events has now been confirmed.

Layne (1965) first reported a potential link between a Florida red tide and the deaths of seven manatees during March and April 1963 in the Fort Myers area of southwestern Florida. Further circumstantial evidence for the involvement of PbTxs in manatee deaths came from an epizootic that occurred between February and April 1982 when thirty-eight manatees died in the lower Caloosahatchee River and nearby bays in the vicinity of Fort Myers (O'Shea et al. 1991). The timing of the mortality event coincided with the presence of a persistent *K. brevis* bloom and associated fish kills and cormorant (*Phalacrocorax auritus*) deaths. Behavior of the affected individuals included disorientation, the inability to submerge or maintain a horizontal position, listlessness, flexing of the back, lip flaring, and labored breathing. In most cases, their stomachs were full, indicating recent feeding, with the contents consisting largely of seagrasses and filter-feeding tunicates (*Molgula* spp.), which are found in association with seagrasses. However, tunicates tested for PbTx using the standard mouse bioassay, the only test available at the time, did not contain measurable amounts of PbTx. Thus, although PbTx was certainly linked to this event by circumstantial evidence, it was not conclusively identified as the causative agent.

Since the 1982 epizootic, deaths of individual manatees are often observed in association with red tides, but mass mortalities of these animals are rare. Mass mortalities remarkably parallel to the 1982 event occurred in 1996, 2002, and 2003, all between the months of February and April, along approximately a 100-mile stretch of southwestern Florida centered around the mouth of the Caloosahatchee River. Although red tides have occurred almost every year along this coastline, high numbers of red tide–related manatee deaths are not observed every year. Rather, each mass die-off has occurred when *K. brevis* red tides, initiated in the fall of the previous year, persisted until, or reinitiated in, early spring along the coast and in the embayments behind the barrier islands that dot the southwestern coast of Florida.

The Caloosahatchee River is the epicenter of these events to some extent because a large portion of the southwestern Florida manatee population winters at a warm-water refuge at the intersection of the Caloosahatchee and Orange rivers and generally begins its dispersal out of the river during the early spring. Thus, it is the unfortunate coincidence of a persistent red tide with the time at which manatees are concentrated in these waters that precipitates mass mortality events. In the winter-spring periods of 1982, 1996, 2002, and 2003 red tide made unusual appearances inside the barrier islands of southwestern Florida. *Karenia brevis,* which typically develops 18–74 km offshore at low concentrations, usually comes inshore during the fall and early winter and then dissipates (Tester and Steidinger 1997). Red tides do not usually appear inshore during the late winter and spring months, when manatees are congregated in low-salinity or freshwater areas, in the warmer waters of the coastal power plants, at warm-water spring refugia, or in residential canals (Reynolds and Wilcox 1986). High-salinity areas (above 24 ppt) allow persistently high concentrations of *K. brevis* cells ($>1 \times 10^3$/ml) to be maintained. In the spring, as the water temperatures rise, manatees at the Fort Myers power plant usually disperse downstream into the inshore bays to forage. If a red tide has come inshore during this period, then the likelihood of manatees being exposed to it during their postwinter movements is fairly high, depending on which areas the manatees move to and their proximity to the red tide bloom (Landsberg and Steidinger 1998, Landsberg 2002).

In 1996, 149 manatees died in association with a red tide; in 2002, 34 died; and in 2003, 96 manatees died (E. Haubold pers. comm.). In all of these events, PbTx was found in the stomach contents of the manatees, which consisted largely of seagrass and sometimes (but not always) tunicates or attached epifauna that were suspected in the 1982 event. Improved detection methods developed largely during the 1990s, including receptor assay, enzyme-linked immunoabsorbent assay (ELISA), and high-performance liquid chromatography–mass spectrometry (HPLC-MS), enabled confirmation of PbTx as the causative agent in these events. In addition to being found in stomach contents, PbTx was also found in liver, kidney, and lung tissue. Based on extrapolation from the human symptomatic dose of 1.6 mg per 68-kg person (ca. 24 μg/kg), Baden (1996) estimated that a dose of less than16 mg would be sufficient to cause symptoms in a 700-kg manatee (ca. 22 μg/kg). Based on PbTx concentrations in stomach contents, it was estimated that by oral exposure, a manatee eating 7% of its body weight a day could reach this dosage of PbTx.

Histopathological analysis of tissues from the manatees affected in 1996 showed severe congestion in nasal tissues, bronchi, lungs, kidney, and brain (Bossart et al. 1998). Staining with an anti-PbTx antibody showed intense positive staining of lymphocytes and macrophages in these tissues (Bossart et al. 1998). These results suggest that the effects of PbTx in manatees may not have been from ingestion alone but may also have resulted in part from chronic inhalation. Indirect evidence suggests that, in addition to its neurotoxic effects, PbTx may be immunosuppressive by inhibiting proteases present in lymphocytes that function in antigen presentation (Baden 1996, Bossart et al. 1998). Retrospective immunohistochemical analysis of tissues from animals affected in the 1982 epizootic also showed positive staining for PbTx in tissue macrophages.

The 2002 event differed somewhat from the others in that manatee deaths began not during, but several weeks after, the demise of the red tide that appeared in the area in mid-February. This raised questions about the source of the toxin in that event. Toxicity of shellfish had been observed in the absence of an active bloom, and the lipophilic nature of PbTxs suggests that they may bioaccumulate in the benthos following a bloom. Seagrasses from the area had high levels of PbTx, both on the grass blades and in associated filter-feeding organisms several months after termination of the 2002 red tide (Flewelling et al. 2004). The duration of PbTx in the food web following the termination of a red tide has not yet been established, but current data indicate that chronic exposure to PbTx is possible for some time following bloom events. Because PbTxs are lipophilic, they are likely sequestered in fatty tissues of manatees, either as active compounds or metabolites (about which little is known at this time). Thus, important questions regarding the impacts of PbTxs on manatees have to do with their metabolism, clearance rates, and the body burdens of PbTxs in animals repeatedly exposed to red tides. The southwestern Florida population has the lowest adult survival rate of all Florida manatee populations. Recent manatee population modeling efforts by the Florida Fish and Wildlife Conservation Commission included catastrophic red tide–related mortality events in southwestern Florida. When existing population conditions were held constant but catastrophes including cold-related mortality and red tide–related mortality were considered, then the probability of a greater than 50% decline in the southwest population in the next 45 years is greater than 85%. The remaining three manatee subpopulations, which have not been affected by red tide mortality, had 0–3.1% probability of a 50% decline under similar conditions, but without red tide mortality factored into the model (E. Haubold pers. comm., Florida Fish and Wildlife Conservation Commission 2003).

As with manatees, die-offs of bottlenose dolphins have long been linked circumstantially to red tides. The earliest report of the co-occurrence of a red tide with a mass mortality of bottlenose dolphins was in southwestern Florida in 1946–1947 when an extensive red tide persisted for 8 months from November 1946 to August 1947 between Florida Bay and St. Petersburg (Gunter et al. 1948). At the time, the identity of *K. brevis* as the causative organism was tenuous, and the toxin was unidentified.

Brevetoxin was also proposed as the causative agent in an unprecedented die-off of more than 740 bottlenose dolphins from June 1987 to February 1988. Strandings began in New Jersey and expanded southward during the fall and winter. In the fall of 1987 a rare bloom of *K. brevis* was carried via the Loop Current from the Gulf of Mexico into the Atlantic, where it became entrained in the Gulf Stream and moved northward as far as North Carolina. A shoreward intrusion of warm Gulf Stream water onto the narrow continental shelf in North Carolina carried the bloom shoreward in late October and allowed it to persist in coastal waters until March 1988, resulting in toxic shellfish and human NSP intoxications (Tester et al. 1991). The unusual presence of a PbTx-producing bloom in U.S. mid-Atlantic coastal waters prompted the investigation of PbTx as a possible factor in the dolphin mortality event.

The evidence of PbTx involvement remains equivocal because the analytical methods available at the time were inadequate for confirmation. PbTx was identified in fish from the stomach contents of one dolphin that stranded in Florida in January 1988 and in the livers of seventeen beached dolphins (Baden 1989). However, five of the eight positive livers were from animals that had stranded in Virginia in August and September, prior to the *K. brevis* bloom that was documented in North Carolina beginning in late October. The other three positive animals were from Florida and stranded in January and February 1988 (Geraci 1989). For the animals from Virginia to have died from PbTx exposure would require either the existence of a bloom of *K. brevis* in the Gulf Stream prior to the bloom in North Carolina or a transfer of the toxin to the dolphins through the food chain, via a migratory prey species that acquired it in Florida waters. Based on the calculated transport rates of the Gulf Stream system, Tester et al. (1991) suggested that the origin of the October North Carolina bloom was likely a bloom that occurred in Charlotte Harbor in September 1987. Transport to North Carolina would have taken 22 to 54 days, placing it in North Carolina by 3–23 October, which makes the first scenario unlikely. The second scenario remains unconfirmed.

In the absence of more definitive analytical methods at the time, instrumental analysis suggested that the toxin actively present in livers was PbTx-2. However, more recent data suggest that PbTx-2 is rapidly metabolized to PbTx-3 and to polar metabolites by both shellfish and mammals (Poli et al. 2000, Plakas et al. 2002). Thus, it is unlikely that only PbTx-2 would have been present. This further calls into question the identity of PbTx-2 in the livers of dolphins from the 1987 mortality event. In a retrospective study, histologic analysis identified morbillivirus antigen in forty-two of the seventy-nine dolphins examined (Lipscomb et al. 1994). Morbillivirus infection in other mammals causes immunosuppression, which in turn commonly promotes opportunistic infections. The frequent presence of fungal and bacterial infections in these dolphins was consistent with an immunocompromised condition. Although this was the first report of morbillivirus in bottlenose dolphins, morbilliviruses have been suggested as major causative agents in pinniped and cetacean die-offs worldwide (Lipscomb et al. 1994). The example of the 1987–1988 dolphin die-off demonstrates the complexity of trying to isolate and identify the contributions of multiple etiological agents in mortality events. We currently do not have information on the PbTx body burdens of dolphins exposed intermittently to red tides; neither has the proposed immune suppression by PbTx been confirmed. Thus, the potential contribution of PbTx to the 1987–1988 mortality event, both as a chronic factor in immunosuppression and as an acute neurotoxin, can neither be confirmed nor ruled out.

Analysis of the involvement of PbTx in the most recent dolphin epizootic associated with red tide benefited from improved methods. During the period August 1999 to February 2000, more than 120 bottlenose dolphins stranded along the Florida panhandle, a fourfold increase over the historical (background) stranding incidence in this area. Two peaks of strandings coincided with a persistent *K. brevis* bloom that came ashore in two pulses, the first around St. Joseph Bay in September 1999 and the second around Choctawhatchee Bay from November 1999 through January 2000. The strandings were evenly distributed between the sexes and among age classes. Most animals were in good physical condition. Analysis for morbillivirus was negative in the animals tested. Histopathological examination showed significant upper respiratory tract lesions (Mase et al. 2000) and PbTx-specific antibody staining was found in the lungs and spleens of two fresh carcasses. In addition, PbTx was confirmed in 29% of the stranded animals, with the highest concentrations in stomach contents, followed by liver and kidney (Van Dolah and Leighfield unpubl. data). Brevetoxin was not found in either the spleen or the lungs, suggesting that the dolphins had obtained the toxin primarily via their food and not through inhalation. This was the first event in which the metabolism of PbTx could be addressed, as the HPLC-MS methods capable of these analyses have been developed only in the last 5 years. Stomach contents, when identifiable, consisted of fish, and PbTx-3 was the predominant form of PbTx present. As PbTx-2 is the principal toxin produced by *K. brevis*, this implies that PbTx-2 had been metabolized to PbTx-3 in the fish, as demonstrated in shellfish (Plakas et al. 2002), or that PbTx-3 had been selectively retained. The primary form identified in liver was also PbTx-3, although an unidentified peak believed to be a novel metabolite was also found. None of the polar metabolites of PbTx that have been identified in shellfish was present.

These data collectively provide evidence of PbTx involvement in the 1999–2000 dolphin mortality event. However, neither acute nor chronic adverse effect levels have been defined for PbTx exposure in *Tursiops*, and the lethal dose has yet to be determined. In this mortality event, cell counts of *K. brevis* were as high as 4×10^6 cells/liter in adjacent waters, with PbTx concentrations of up to 26 μg/liter (Leighfield and Van Dolah unpubl. data). This cell concentration is not higher than concentrations frequently encountered farther south on the west coast of Florida, where *Tursiops* are exposed to *K. brevis* red tides essentially annually. As in both the 1982 and 1996 manatee events, the *K. brevis* blooms associated with the Florida panhandle dolphin die-offs were trapped in semienclosed embayments, and judging by DNA analysis the affected individuals were from coastal, as opposed to offshore, populations (P. Rosel pers. comm.). Possible explanations for the unusually severe effect of the 1999–2000 bloom on bottlenose dolphins include the involvement of another yet-unidentified agent that increased their susceptibility to PbTx and the possibility that these animals were physiologically or behaviorally "naive" to *K. brevis* blooms, which are historically rare on this part of the Florida coast. However, there is no known basis for acquiring physiological tolerance to PbTx.

Ciguatera

Involvement of ciguatera in the morbidity or mortality of marine mammals is speculative. It has been proposed as one of several potential factors in the decline of Hawaiian monk seals (*Monachus schauislandi*). Other suggested factors include habitat degradation by human activities, starvation that is due to fishery competition for prey, mobbing behavior, predation by sharks, and a long-term (decadal) decline in productivity in the central North Pacific that is impairing their food supply (Craig and Ragen 1999, Ragen and Lavigne 1999). For the past several decades, the largest number of monk seals has hauled out at French Frigate Shoals, an atoll approximately 950 km northwest of Oahu. This number peaked in the 1980s, when the population was believed to be near carrying capacity.

The population decline over the past decade has been attributed primarily to the poor survival of juveniles, associated with reduced sizes of weaned pups, emaciation, and slower growth rates of juveniles (Craig and Ragen 1999). Numbers elsewhere in the atoll chain have also diminished, including those at Kure Atoll and Midway Atoll. The relocation of young female monk seals from French Frigate Shoals to Kure Atoll, as part of the Hawaiian monk seal recovery effort, has been successful in increasing the number of monk seals at Kure. Midway Atoll, the site of a former naval air station, experienced the most severe decline in monk seal numbers, probably as a result of human disturbances (Gilmartin and Antonelis 1998). With closure of the air station, a recovery strategy was implemented, and young females were relocated to Midway from French Frigate Shoals. However, only two of the eighteen translocated animals survived beyond 1 year (Gilmartin and Antonelis 1998).

The reasons for this poor survival rate are not clear as prey populations were believed to be adequate, but one hypothesis is that the reefs at Midway support a high incidence of ciguatera. This suspicion was based in part on the possible involvement of ciguatera in the deaths of fifty seals on Laysan Island in 1978 (Gilmartin et al. 1980) and a historical record of ciguatera outbreaks in humans on Midway (Wilson and Jokiel 1986). Preliminary surveys of fish species in Midway Lagoon known to be prey of monk seals were carried out in 1986 (Wilson and Jokiel 1986) and 1992 (Vanderlip and Sakumoto 1993) using the "stick test" immunoassay for ciguatera (Hokama 1985). These surveys yielded similar results: more than half of all fish tested were positive or borderline-positive for the presence of ciguatera. Unfortunately, this assay and its revisions are known to yield false positives, making the results ambiguous (Vanderlip and Sakumoto 1993, Dickey et al. 1994). Therefore, further evaluation of the occurrence of ciguateric fish in Midway Lagoon and the potential impact on monk seals is recommended (Gilmartin and Antonelis 1998).

FACTORS REGULATING THE OCCURRENCE OF HARMFUL ALGAL BLOOMS

Over the past several decades, the occurrence of harmful or toxic algal incidents has increased in many parts of the world, both in frequency and in geographic distribution (Fig. 6.2; Hallegraeff 1993, Anderson 1994, Van Dolah 2000). There has been much speculation about the causes and significance of the observed increase. What is not clear is the extent to which the reported increase can be attributed to increased awareness of the issues and the consequent establishment of research programs and surveillance systems that have helped identify problems not previously recognized. Certainly, increased study of the issue has led to the discovery of toxic species in places where no previous data were available, the recognition of toxicity in species not previously known to be toxic, the characterization of new toxins, and development of improved toxin testing methods. However, of particular concern is determining the extent to which the recent expansion in harmful algal blooms is a consequence of anthropogenic activities that might be modified in order to reverse the current trends. In some cases, human activities clearly do contribute to the problem through nutrient loading, the inadvertent transport of harmful species to novel places, and overfishing. Human activities may also indirectly contribute to the increased frequency or expansion of species ranges through large-scale effects of global climate change.

Phytoplankton productivity in coastal and estuarine waters is primarily limited by the availability of nitrogen and phosphorus, and anthropogenic nitrogen loading is correctly blamed for accelerated phytoplankton production in coastal waters (Pearl 1999). Anthropogenic nutrient loading of coastal waters comes from atmospheric deposition of industrial effluent, coastal development, agricultural runoff, and aquaculture. Consequently, coastal waters in developed countries have experienced a long-term increase in both nitrogen and phosphorus, by a factor of more than four, compared with preindustrial times (Nixon 1995). An example of blooms of a toxic species increasing in response to nutrient loading is the occurrence of *Pseudonitzschia* spp. in Louisiana waters around the plume of the Mississippi River. In the past four decades, the flux of nitrogen from the Mississippi River has increased more than fourfold (National Research Council 2000b). *Pseudonitzschia* blooms occur in the spring, when nutrient loading from the river is highest (Dortch et al. 1997, Pan et al. 2001). Both historical records and frustules preserved in sediment cores indicate a large increase in *Pseudonitzschia* spp. abundance since the 1950s, concurrent with increases in nitrogen loading (Parsons et al. 2002).

Harmful algal blooms resulting from nutrient loading can indeed be managed, and there are numerous examples in which phytoplankton biomass in general or the incidence of harmful algal blooms in particular has decreased following the implementation of nutrient controls (see Anderson et al. 2002). An often-cited example is the correlation between the eightfold increase in frequency of algal blooms in Tolo Harbour, Hong Kong, from 1976 to 1989 and a sixfold increase in human population during that period, which was accompanied by a 2.5-fold increase in nutrient loading (Lam and Ho 1989). In the late 1980s pollution loadings in Tolo Harbour were decreased by diversion of sewage efflu-

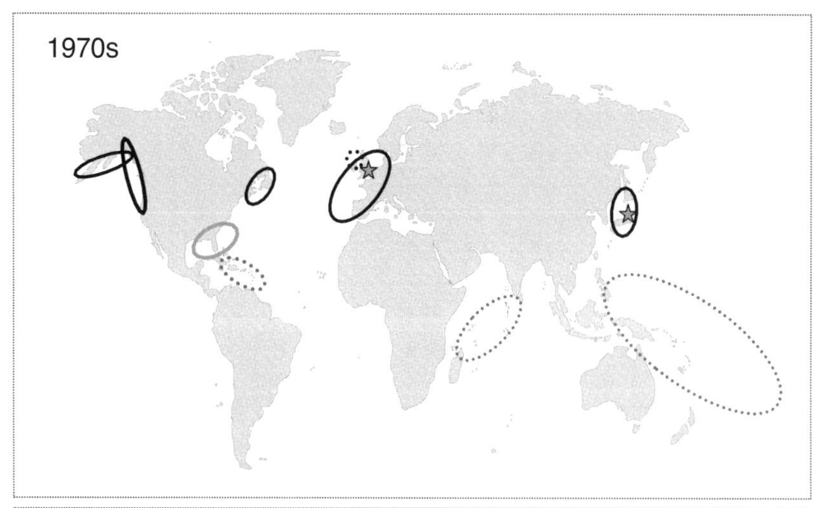

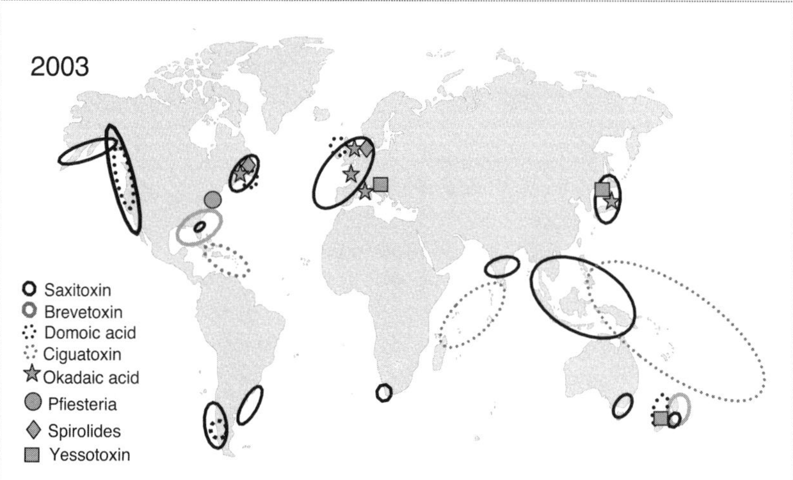

Fig. 6.2. Increase in the global distribution of the major classes of algal toxins with known (open circles) or potential (shapes) adverse effects on marine mammals.

ent to Victoria Harbour, which resulted in a resurgence of diatoms and a decrease in dinoflagellates and algal blooms (Yung et al. 1997). Not only is the absolute abundance of nutrients important, but input of these nutrients tends to alter nutrient ratios in coastal waters, particularly nitrogen/phosphorus or nitrogen/silicon, resulting in conditions that favor the growth of harmful species. The decrease in total nutrient loading in Tolo Harbour was also accompanied by a change in nutrient ratios, with the nitrogen/phosphorus ratio decreasing from 20:1 to 11:1 (Hodgkiss and Ho 1997), which favored the growth of a different suite of microalgal species.

A second direct impact of human activities on the occurrence of harmful algal blooms is the introduction of nonindigenous algal species to new regions. A principal mechanism for the transfer of nonindigenous species from one marine ecosystem to another is via ships' ballast water (water used to stabilize ships that are not carrying cargo; Ruiz et al. 1997). Water is pumped into ballast tanks in the port of origin and released when the ship enters its destination port prior to loading cargo. Once released, if conditions are conducive to growth, organisms carried in the ballast water may establish populations in the receiving port. A well-known example of ballast water–mediated introduction of a harmful algal species is that of STX-producing dinoflagellates from Japan to Australia, where paralytic shellfish poisoning was unknown prior to 1986 (McMinn et al. 1997). An estimated 10 billion tons of ballast water is transported each year, making it a major source of inadvertent transfer of exotic species (Ruiz et al. 1997). To address this issue, regulations have been established by the International Maritime Organization governing the open-ocean exchange of ballast water (Hallegraeff 1998).

Overfishing, a third mechanism by which human activity has been demonstrated to contribute to harmful algal blooms, can have dramatic top-down effects on community structure, including the phytoplankton community, at both local and regional levels. A dramatic example is the shift in Jamaican coral reefs from a coral-dominated to an algae-dominated community, caused in large part by the removal of herbivorous fishes and the subsequent overgrowth of corals by algae (Hughes 1994). Algae-dominated reefs favor the growth of the ciguatera dinoflagellate, *Gambierdiscus toxicus;* thus, reefs on the north side of Jamaica now suffer persistent ciguateric conditions.

However, it is important to recognize that the development of harmful algal blooms is, in large part, a natural phenomenon controlled primarily by processes such as circulation and upwelling. The toxic algal blooms with the greatest known impacts on marine mammals are among those driven by natural processes. For example, STX-producing dinoflagellate blooms have been recorded in eastern Canadian waters for well over 100 years. A natural expansion of the range of *Alexandrium tamarense* southward into New England occurred in 1972 in the aftermath of a September hurricane (Anderson 1994). Because this dinoflagellate produces an overwintering cyst, seed beds for subsequent blooms were established, and STX-producing blooms are now an annual occurrence as far south as Cape Cod. Likewise, blooms of the Florida red tide are recorded as far back as the 1800s, and annual records of red tides collected by the Florida Fish and Wildlife Conservation Commission indicate that the frequency of occurrence has not increased in the past 50 years. In fact, blooms of the Florida red tide dinoflagellate begin offshore in oligotrophic waters, and this species is well adapted to growth in low-nutrient conditions. Its role in the western Florida shelf ecosystem is unknown, but, as is true of wildfires in western North America, proposals to "annihilate" red tide are probably ill conceived.

The importance of large-scale circulation patterns and long-term climate cycles in controlling local events is becoming widely recognized. For example, the initiation of red tides in the Gulf of Mexico is now believed to depend on the input of iron from dust storms that originate in Africa (Walsh and Steidinger 2001). Changes in the intensity of iron deposition, and therefore the intensity of phytoplankton blooms, are thus affected by drought conditions on another continent. Such conditions may be under the influence of the North Atlantic Oscillation, which underwent a regime shift in the mid-1970s concurrent with the documented increase of marine diseases in the western North Atlantic (Hayes et al. 2001). Similarly, productivity of the North Pacific Ocean appears to vary with a 50-year cycle (Chavez et al. 2003), with warm periods favoring a "sardine regime" alternating with cool periods favoring an "anchovy regime" on the west coast of North America, each lasting approximately 25 years. The mid-1990s are believed to have brought a regime shift favoring cooler water, stronger upwelling with increased nutrient input expected to support greater phytoplankton productivity, and a larger anchovy population. This regime shift corresponds to the occurrence of large *Pseudonitzschia* blooms on the California coast, such as those that resulted in the DA marine mammal mortality events in 1998, 2000, and 2002. The predicted 25-year tenure of this phase in the cycle implies that such events may become routine on the west coast of North America (M. Silver pers. comm.).

On a shorter time scale, it is well established that many human disease outbreaks peak during unusual climatic events, including droughts, storm events that produce heavy rainfall, and El Niño phenomena (Colwell 1996, Harvell et al. 1999). The occurrence of diseases in marine species, including coral bleaching and shellfish diseases are linked with El Niño events (Epstein et al. 1998), and the geographic expansions of human intoxication by STX and ciguatera in the Indo-Pacific have been convincingly linked to the occurrence of El Niño conditions (McLean 1989, Hales et al. 1999). Historically, El Niño events have occurred at a frequency of one or two per decade. However, since the mid-1970s, they have become more frequent and have persisted longer. The increase in these events late in the century, at a time when sea surface temperatures rose by 0.8°C, has led to the speculation that global warming may be an underlying cause (Harvell et al. 1999). If human activities do indeed alter the nature of large-scale oceanic and climatic processes through the effects of greenhouse gases, then any hope of managing the trend toward increased frequency and extent of harmful algal blooms will depend on concerted international action.

PROSPECTS FOR MANAGING THE IMPACTS OF ALGAL TOXINS ON MARINE MAMMALS

Management options for harmful algal blooms have been considered in detail by Boesch et al. (1997) and include strategies for reducing their incidence (*prevention*), actions to terminate or contain existing blooms (*control*), and steps to reduce human and environmental health risks (*mitigation*). Management options for harmful algal blooms as they affect marine mammals should also be considered using the same framework for prevention, control, and mitigation.

Prevention

The algal blooms that have been documented to date as being responsible for marine mammal mortality events are largely natural phenomena that have occurred in the absence of (at least direct) anthropogenic influences. In these cases, preventing the occurrence of the blooms is neither feasible nor advisable. However, at least two notable examples where prevention could play a role are the potential involvement of STX in bottlenose dolphin die-offs in the Indian River Lagoon and of CTX in Hawaiian monk seal deaths on disturbed reefs. Our potential to prevent STX-producing algal blooms in the Indian River Lagoon can only be determined by better understanding the dynamics there and the conditions that initiate blooms of *Pyrodinium*. Studies directed toward these questions are currently under way by the Florida Fish and Wildlife Conservation Commission (J. Landsberg pers. comm.). Similarly, our potential to prevent ciguateric conditions in reef systems supporting the Hawaiian monk seal requires better understanding of the extent and dynamics of the conditions in these reefs. Our lack of insight into this seemingly basic question rests in the inadequacy of current detection methods for ciguatera toxins in the food web. Thus, development of tools with which to assess ciguateric conditions is a prerequisite to preventing their occurrence.

Control

Control strategies for protection of marine mammals are both species and location specific. In some cases, action may be undertaken by local regulators to avert bloom effects. For example, it was proposed following the 1996 manatee mortality event that red tides in the area of the Caloosahatchee River outfall might be prevented by increasing freshwater outflow and thus decreasing salinity in the embayments to levels below the lower limit of tolerance for the dinoflagellate. This strategy, although viewed positively by the South Florida Water Management District (J. Landsberg pers. comm.), has not yet been attempted and is likely to be successful in preventing manatee exposures only in an instance where a red tide occurs prior to the manatees' movement out of the river. The potential release of high concentrations of PbTxs into the ecosystem by water-management activities may cause unknown negative impacts on manatees. The fact that PbTxs are bound up in the food web means that we cannot preclude the possibility that manatees would be chronically exposed to them via ingestion of toxins released after a freshwater discharge (J. Landsberg pers. comm.).

Studies on the control of *K. brevis* blooms by the localized application of flocculants, such as clays, to remove algal cells from the water column are currently under way. This method is presently employed for other harmful algal species on a massive scale in fish-farming areas of the Korean Peninsula (Archembault et al. 2002). However, it is important to examine closely the potentially adverse effects of the use of flocculants, which may result in the transfer of toxins to the benthic community, as trophic transfer of both PbTx and DA to marine mammals through the benthic food web has already been demonstrated.

Mitigation

The early detection of harmful algal species and toxins is a critical component of mitigation of HAB effects. Many coastal states have algal monitoring programs in place that are designed to anticipate and give prior notice of impending toxic blooms. Of particular relevance are the programs in Maine, Florida, and California, which routinely do an excellent job of protecting public health. The existing monitoring programs depend on the time-consuming task of microscopic examination and enumeration of algal cells present in a water sample, followed by toxin testing in shellfish when a prescribed cell concentration has been detected. These programs are generally shore based and are therefore not adequate to protect marine animals that are exposed before the toxic blooms move into coastal sampling sites. In fact, the deaths of dolphins caused by DA early in 2003 in southern California served as a sentinel for the California Department of Health Services, enabling it to identify DA as a developing problem in an area for which it lacked bloom data (G. Langlois pers. comm.).

Currently significant research efforts are targeted at the development of remote-detection methods, capable of real-time monitoring and suitable for deployment on moored buoys or robotic submersible vehicles. Several different approaches have been taken, including the use of species-specific DNA probes, species- and toxin-specific antibody probes, in situ mass spectrometry, and in situ video-imaging systems that can be deployed on moored buoys or unmanned mobile submersibles (see Sellner et al. 2003). In addition, satellite imaging of ocean color and spectral analysis of pigment profiles have been used successfully to identify offshore blooms prior to their movement into coastal waters, giving coastal managers early warning of toxic conditions (Stumpf et al. 2002). Implementation of these real-time monitoring tools is a priority of the National Oceanic and Atmospheric Administration (NOAA) Monitoring and Event Response for Harmful Algal Blooms (MERHAB) program, which currently supports pilot projects in Florida, Washington, and the northeastern United States that employ the sensors described above. Efforts should be made to coordinate with ongoing programs to ensure that such monitoring tools are being used in problem areas where we currently lack forewarning of algal blooms known to adversely affect marine mammals (e.g., the Channel Islands, California).

However, establishment of early warning systems is useful only if further mitigation strategies for marine mammals are in place. In the case of endangered species whose populations were decimated by earlier human activities and which may now be threatened by repeated exposure to algal toxins as a consequence of long-term climate cycles (e.g., southern sea otters), the physical removal of the animals might be considered as a management option. The cost of such an activity might be tremendous, and any decision to attempt a temporary or permanent translocation is bound to be controversial.

Mitigation of the toxic effects on live-stranded marine mammals is a strategy that has been employed in the case of both DA and PbTx poisonings. Currently, treatments for live-stranded animals exposed to PbTx are limited to supportive care. In contrast, 3 years of experience with repeated DA events affecting California sea lions has yielded an effective course of treatment, including application of the antiseizure drugs diasepem, lorazepam, and phenobarbitone and the use of dexamethasone to reduce cerebral edema. Such treatment has resulted in the successful rehabilitation and release of numerous seizing animals (Gulland 2000).

Antidotes to toxin exposure that are effective for humans may also be applicable for the rehabilitation of intoxicated marine mammals. For example, because STX is classified as a chemical weapon, antidote development has been pursued by the U.S. Army Medical Research Institute. Both an STX antibody and nontoxic STX derivatives have demonstrated efficacy against the acute effects of STX in humans. Application of the STX antibody was considered during the Mediterranean monk seal mortality event, and it was made available by the U.S. Army (M. Poli pers. comm.); however, in the absence of any previous experimental treatments of marine mammals, administration of the antibody to monk seals in rookeries was deemed premature at the time. In recent laboratory studies of respiratory exposure to PbTx, both a synthetic derivative of PbTx and a naturally occurring inactive PbTx analogue have been reported to provide protection from PbTx-induced respiratory effects (Abraham et al. 2005). For PbTx exposure via ingestion, recent studies in mice have demonstrated protective effects of cholestyramine, a cholesterol-lowering drug that reduces bioavailability of fat-soluble substances (Gordon et al. 2002). Preliminary evidence suggests that cholestyramine is similarly effective in ameliorating the long-term neurological effects of CTX in human patients (R.C. Shoemaker pers. comm.). Consideration should be given to the administration of such therapeutic measures to stranded animals during Florida red tide events. However, the application of any of the above procedures during an acute marine mammal intoxication event requires investment in development of effective procedures for their administration.

In summary, because toxic algal blooms with known adverse effects on marine mammals are naturally occurring events, options for managing their impacts on marine mammals currently rest primarily in mitigation strategies. How heavily we as a society invest in such strategies for mitigating the effects of HABs on marine mammals will be predicated upon our better understanding of the long-term consequences of their effects. During the past decade, significant progress has been made in understanding both bloom dynamics and their toxic effects on a number of marine mammal species. However, our level of understanding is not balanced for each class of toxicity. From this, a number of information gaps, different for each class of toxins, have emerged that should be viewed as priority research agenda items in the next decade.

PRIORITY INFORMATION NEEDS FOR SPECIFIC ALGAL TOXINS

- *Domoic acid—a formal risk assessment.* Although a single DA-related mortality event is thought not to have a serious overall impact on the breeding populations of California sea lions, formal risk assessment should be undertaken to determine the long-term consequences of repeated exposures if, in fact, bloom events are expected to continue or increase over a 25-year cycle. Data on exposure levels, body burdens, toxin effect levels, clearance rates, maternal transfer of toxin, reproductive failure, and chronic effects of toxin exposure on the fitness of individuals and populations are beginning to accumulate for DA, making this toxin class a good candidate for formal risk assessment in marine mammals on the West Coast of the United States.
- *Brevetoxin—identification of effect levels.* The effect levels of PbTxs in manatees and bottlenose dolphins are not yet known. Information is lacking on the metabolism and detoxification of PbTx as well as the accumulated body burden of PbTx in marine mammals repeatedly exposed to red tides on the west coast of Florida. The analytical methods for carrying out studies to fill those gaps in knowledge are now available and should be pursued. In addition, insight should be sought into the apparent immunosuppression in PbTx-exposed animals and the apparent differential sensitivity of repeatedly exposed individuals as compared to individuals from naïve populations. Such information is needed if we are to understand the long-term consequences of PbTx exposure for marine mammal populations.
- *Ciguatera—improvement of toxin detection methods.* The role of ciguatera in the survival of the Hawaiian monk seal cannot be adequately addressed without the assessment of ciguateric fish in the reefs supporting the monk seals. Such assessment requires the development of adequate toxin detection methods, which has been difficult because of the exquisite potency and complex structures of CTXs and a lack (worldwide) of toxin standards. Thus, investment is needed in detection "infrastructure" before any kind of management plan can be developed and implemented. Once in hand, these methods should be employed to characterize the status of ciguatera on reefs supporting Hawaiian monk seals.
- *Saxitoxin in right whales—hazard identification.* Evidence for a role of STX in reproductive dysfunction in North Atlantic right whales remains circumstantial. Detailed study of the spatial/temporal fluctuations of PSP toxin levels in their primary prey species, *Calanus finmarchicus,* relative to the distribution of feeding right whales is needed for a critical assessment of the potential for exposure throughout the summer feeding season. Such a study should include testing right whale feces for PSP toxins, from the time of entry into northern feeding grounds until exit toward the southern calving ground. Because many of these animals can be identified using photographic records, the data could be used to assess the duration and possibly the extent of toxin exposure for individuals.

ACKNOWLEDGMENTS

I acknowledge the insightful contributions of the following people to this chapter and to the prioritization of research needs: Jan Landsberg, Frances Gulland, Elsa McDowell, Mary Silver, Greg Doucette, John Ramsdell, Randall Reeves, and William Perrin.

JOHN HILDEBRAND

7

Impacts of Anthropogenic Sound

THERE IS GROWING CONCERN THAT sound introduced into the sea by human activities has detrimental effects on marine mammals. For example, mounting evidence suggests that high-intensity anthropogenic sound from sonar and airguns leads to strandings and subsequent mortality of beaked whales. Although the mechanisms of injury in these events are unclear, the species affected and the implicated sound levels follow a consistent pattern. A more pervasive, yet more subtle, problem may be the effects of increases in background noise levels from commercial shipping. Higher levels of background noise may interfere with marine mammals' ability to detect sounds, whether calls from their own species, echoes from prey, or natural sounds that aid in navigation or foraging. Noise may affect developmental, reproductive, or immune functions and cause more generalized stress. The effects of other pollutants (e.g., chemicals) may be additive or synergistic with those of noise. As Read (this volume) and Plagányi and Butterworth (this volume) suggest, human activities may have both direct and indirect consequences. For instance, noise may have ecosystem-scale effects, including impacts on species that are marine mammal prey.

Sources of anthropogenic sound are becoming both more pervasive and more powerful, increasing both oceanic background noise levels and peak sound intensity levels. Anthropogenic activities in the ocean have increased over the past 50 years, resulting in more low-frequency (<1,000 Hz) and mid-frequency (1–20 kHz) noise. Sources of anthropogenic noise include commercial shipping, defense-related activities, hydrocarbon exploration and development, research, and recreation.

Anthropogenic sound is created in the ocean both purposefully and unintentionally. The result is noise pollution that is high intensity and acute, as well as lower level and chronic. Many sources of noise are located along well-traveled paths in the sea, particularly in coastal and continental shelf waters, areas that often include important marine mammal habitats.

There is sufficient evidence to conclude that some high-intensity sounds are harmful and, on occasion, fatal to marine mammals. Given the opportunity, the animals may avoid high-intensity sound, but in some extreme cases there has been documentation of injury from anthropogenic sound exposure. Multiple mass strandings of beaked whales following high-intensity sound exposure demonstrate a repeating pattern of events. Following exposure to high-intensity sonar or airguns, beaked whales have been known to strand on the shore, and if human intervention does not return them to the sea they die. Understanding the causes and consequences of beaked whale mass stranding should be a high research priority. What, then, are the mechanisms for damage or disturbance?

A major impediment to assessing the biological effects of ocean noise is the lack of knowledge concerning marine mammal responses to sound. Behavioral data from the wild are needed to examine those responses so that effects can be assessed. Significant effects may prove to be confined to a few individuals exposed at high sound pressure levels or they may be occurring at a population level as a result of widespread exposure. Discerning population-level effects is challenging as observations must be conducted over long distances and extended time periods.

Sound is an extremely efficient way of propagating energy through the ocean, and marine mammals have evolved to exploit its potential. Many marine mammals use sound as a primary means for underwater communication and sensing. Toothed whales have developed sophisticated echolocation systems to sense and track prey and engage in complex exchanges of vocalizations with members of their own species. Baleen whales have developed long-range acoustic communication systems to facilitate mating and social interaction. Some baleen whales produce intricately patterned songs that continue for hours or days. Marine mammals may use sound from natural sources as a guide for navigation, prey detection, and avoidance of predation. The sound environment of the ocean is an important aspect of marine mammal habitat and we can expect marine mammals to choose their locations and modify their behavior based, in part, on natural and anthropogenic sounds.

Human presence at sea is normally on the surface, and the sounds that we produce within the water are rarely given much consideration. The air-sea interface constitutes a substantial sound barrier. Sound waves in the water are reduced in intensity by a factor of more than a thousand when they cross the air-sea boundary, which means that we are effectively insulated from the noise produced by the rotating propellers that drive our ships or by the high-intensity sonar used to measure depths or probe the sea bottom. The conflict between human and marine mammal use of the sea is fundamentally a consequence of the fact that we do not inhabit the same sound environment. Marine mammals live with their ears in the water, and we live, even at sea, mainly with our ears in the air.

A notable exception is the military use of submarines, where stealth is required, so minimization of sound production then becomes crucial to human survival. Reduction in radiated sound has been achieved by placing rotating machinery on isolating mounts and by designing efficient propellers that thrust without unnecessary vibration and cavitation. Thus, when it has been important to keep the sea quiet, the necessary technology has been developed and made available.

NOISE LEVELS IN THE OCEAN

Sound is a vibration or acoustic wave that travels through some medium, such as air or water. Acoustic waves can be described either by the speed and direction at which a small piece of the medium vibrates, called the particle velocity, or by the corresponding pressure associated with the vibration. Frequency is the rate of vibration, given in hertz (Hz) or cycles per second; we perceive frequency as the pitch of the sound. A tone is a sound of a constant frequency that continues for a substantial time. A pulse is a sound of short duration and may include a broad range of frequencies.

In water, the pressure of sound waves is typically measured with a device called a hydrophone. When discussing background noise, the implicit assumption is that sound pressure fluctuations are being described, although it is not clear whether a particular marine organism is affected by particle velocity or pressure. Sound pressure is measured in pascals (Pa) in the international system of units (SI), although it is expressed in bars by the geophysical community (1 Pa = 10^{-5} bar). Because mammalian hearing and sound production cover a wide range of pressure values, the sound pressure level (SPL) is usually measured on a logarithmic scale called the decibel (dB), and compared against a 1 μPa reference (P_o) for underwater sound as follows:

$$\text{SPL dB re } 1\ \mu\text{Pa} = 10\log_{10}(P/P_o)^2 = 20\log_{10}(P/P_o)$$

Pressure amplitude can be measured either as a root-mean-squared (RMS) or peak value. (Note that in this chapter I use RMS values unless otherwise noted.) Pressure is squared in the above expression as a proxy for acoustic intensity, that is, the power flow per unit area in the sound wave, with units of watts/m^2. Sound intensity is the product of pressure (P) and particle velocity (v). Acousticians working in one medium (water or air) use the fact that for simple plane waves the pressure and particle velocity are related by the characteristic impedance (Z) of the medium:

$$Z = P/v$$

This allows the acoustic intensity (I) to be related to the pressure squared divided by the impedance:

$$I = 10\log_{10}[P^2/(ZI_o)]$$

Acoustic power is obtained by integrating intensity over some area, and acoustic energy is the power integrated over some time period. The same acoustic energy can be obtained from a high-intensity source lasting a short time (impulse) or a low-intensity source lasting a long time (continuous wave).

When sound propagates from water into air, there is a 30-dB (1,000×) decrease in acoustic intensity because the characteristic impedance of water (~1,500,000 kg/s-m^2) is much greater than that of air (~415 kg/s-m^2). This means that sounds made by a high-intensity underwater source (such as a sonar) are not transmitted into the air with the same intensity. In essence, sailors and seafaring passengers are protected from the sounds produced in the sea. Without the air-sea boundary for protection, there would be a strong incentive to protect human hearing from the noise of sonars and cavitating ships' propellers. (Note that for sound in air a different reference level is used, $P_o = 20\ \mu$Pa, and this may be a source of confusion when comparing sound under water and in air.)

Underwater sounds are classified according to whether they are transient or continuous. Transient sounds are of short duration and occur singly, irregularly, or as part of a repeating pattern. For instance, an explosion represents a single transient, whereas the periodic pulses from a ship's sonar are patterned transients. Broadband short-duration transients are called pulses and sound like clicks or bangs. Continuous sounds, which occur without pauses, are further classified as periodic, such as the sound from rotating machinery or pumps, or aperiodic, such as the sound of a ship breaking ice.

Pulsed sounds often are measured in terms of their peak-to-peak pressure, whereas continuous sounds are measured in terms of their RMS pressure. The method of converting between RMS and peak-to-peak pressures is well defined for continuous-wave signals (add 9 dB to the RMS pressure to get the peak-to-peak pressure). However, for pulsed sounds, the conversion is problematic because the duration of the signal included dramatically alters the result. For a brief pulse, peak-to-peak pressure is measured from the highest and lowest portions of the waveform, whereas RMS pressure is difficult to interpret because it depends on the duration over which the signal is measured. An alternative for pulsed signals is to estimate the total energy, rather than the peak-to-peak pressure or intensity. Energy is proportional to the time integral of the squared pressure, described in the units μPa^2-s. For brief pulses, energy in dB re 1 μPa^2-s is less than peak-squared pressure values in the same units. As others have warned, better standardization of measurement methods for pulsed underwater sounds is urgently needed to permit meaningful comparisons (Green and Moore 1995).

Ambient noise in the ocean is the background sound that incorporates the broad range of individual sources, some identified and others not. Ocean noise may come both from distant sources, such as ships, or from nearby, such as the waves breaking directly above the listener. Although ambient noise is always present, the individual sources that contribute to it do not necessarily create sound continuously.

Acoustic pressures are analyzed into their frequency components or spectrum using a Fourier transform (Bracewell 2000). One way to express the result is as a power spectral density with units of μPa^2/Hz. Note that the bandwidth of the power spectral density is explicitly part of the unit, and by convention noise measurements are presented in 1-Hz-wide frequency bins. Hearing and other auditory measures are often presented in one-octave or one-third-octave frequency bins as an approximation for the filtering characteristics involved in hearing. Source measurements are typically given for varying bandwidths with the following equation allowing for conversion to 1-Hz frequency bins:

$$\Delta\text{dB} = 10\log_{10}(\text{bandwidth in Hz})$$

The sound level received from a source depends on the distance between the source and receiver, as well as on the propagation characteristics of the environment. Therefore, the distance at which a source measurement was made must be specified, and the convention is to normalize the pressure to an approximation of what would be received at a range of 1 m from the source (dB re μPa at 1 m). When arrays of sources are used, this convention overestimates actual source levels in the near field (e.g., at a 1-m range) but provides a good way to predict source levels in the far field (e.g., at a 1-km range).

NATURAL OCEAN ACOUSTIC ENVIRONMENT

The ambient background noise of the ocean is highly variable. At a given time and place, a broad range of sources may be combined. In addition, conditions at a particular location may affect how well ambient sounds are received (e.g., sound propagation, water depth, and bathymetry). Natural phenomena known to contribute to oceanic ambient noise include: (a) wind, waves, and swell patterns, (b) bubbles, (c) currents and turbulence, (d) earthquakes, (e) precipitation, (f) ice cover and activity, and (g) marine life.

Wind, Waves, and Ice

Ocean surface motions that are due to wind, sea state, and swell patterns are the dominant physical mechanisms for natural background noise in the ocean. Noise is primarily associated with wind acting on the surface, causing wave activity. In the absence of anthropogenic and biological sound, ambient noise is wind dependent over an extremely broad frequency band from below 1 Hz to at least 100 kHz. At frequencies below 10 Hz, interactions of surface waves are the dominant mechanisms for sound generation. Across the remainder of the band from 10 Hz to100 kHz, oscillating bubbles in the water column are the primary noise source, both as individual bubbles and as bubble clouds.

In early descriptions, ocean noise was related to sea state (Knudsen et al. 1948). By this theory, noise levels increase with increasing sea state to the same degree across the entire frequency band from 1 to 100 kHz. More recent work suggests that noise is better correlated with wind speed than with sea state or wave height. The correlation between noise and wind speed allows for more accurate prediction, as sea states are more difficult to estimate than wind speeds. In the open ocean, the noise of breaking waves is correlated with wind speed. Spilling and plunging breakers raise underwater sound levels by more than 20 dB across the band from 10 Hz to 10 kHz (Wilson et al. 1985). Precipitation is another factor that can increase ambient noise levels by up to 35 dB across a broad band of frequencies from 100 Hz to more than 20 kHz (Nystuen and Farmer 1987).

Ice cover alters the ocean noise field depending on its type and degree—for instance, whether it is shore-fast pack ice, moving pack ice, or at the marginal ice zone (Milne 1967). Shore-fast pack ice isolates the water column from the effects of wind and results in a decrease in ambient noise of 10–20 dB. Sounds from ice cracking, however, may increase noise levels by as much as 30 dB. Ice cracking can generate broadband pulses up to 1 kHz lasting for 1 s or longer. Interaction of ocean waves with the marginal ice zone may raise noise levels by 4–12 dB (Diachok and Winikur 1974).

Earthquakes and Thunder/Lightning

Earthquakes and thunder/lightning are transient natural sound sources. Underwater recordings of thunder/lightning from storms 5–10 km distant show peak energy between 50 and 250 Hz, up to 15 dB above background levels (Dubrovsky and Kosterin 1993). Seismic energy from undersea earthquakes couples into the ocean and is called T-phase (tertiary) in addition to the usual P-phase (primary) and S-phase (secondary) seismic waves that are observed on land. At ranges of less than 100 km, T-phase energy can have frequencies greater than 100 Hz, with peak energy at 5–20 Hz. It can be as much as 30–40 dB above background noise, with a sharp onset, and can last from a few seconds to several minutes (Schreiner et al. 1995).

ANTHROPOGENIC SOUND

Human activity in the marine environment is now an important component of oceanic background noise. Sound is used both as a tool for probing the ocean and as a byproduct of other activities. Anthropogenic sound sources vary in space and time but may be grouped into general categories: (a) explosions, (b) large commercial ships, (c) airguns and other seismic exploration devices, (d) military sonars, (e) navigation and depth-finding sonars, (f) research sound sources, (g) acoustic harassment devices (AHDs) and pingers, (h) polar icebreakers, (i) offshore drilling and other industrial activity, and (j) small ships, boats, and personal watercraft.

Explosions

Two classes of man-made explosions create high sound levels in the ocean: nuclear and chemical. Nuclear devices have been tested underwater in the ocean, in the atmosphere above the ocean, and on oceanic islands. In 1963 all nuclear states signed the Limited Test Ban Treaty, pledging to stop testing nuclear weapons underwater. The Comprehensive Test Ban Treaty was adopted in 1996, whereby the major nuclear powers pledged to discontinue all nuclear testing. The most recent oceanic tests were conducted by France in 1995–1996 on the islands of Fangataufa and Mururoa in the South Pacific. There is currently a low probability of continued ocean testing of nuclear devices, although the situation could change with geopolitical developments over the coming years or decades.

Nuclear explosions are extremely strong sources of underwater sound. Their source levels are expressed as an equivalent weight of chemical explosives with fission devices yielding the equivalent of tens to hundreds of kilotons and fusion devices yielding the equivalent of tens of megatons. Past tests likely had significant impacts on marine mammals in the vicinity of the test sites, although no marine mammal monitoring or stranding data are available. To ensure compliance with the Comprehensive Test Ban Treaty, an international monitoring system is being implemented, including a series of marine hydrophones and terrestrial (island) seismic sensors to detect high-intensity sounds. This information is transmitted, in real time, to the International Data Centre, where analysts evaluate the data for indications of nuclear explosions. The physical character of the oceans allows the sounds of such explosions to travel for extremely long distances with little energy loss, and monitoring is conducted over a large fraction of the world's oceans with a small number of stations. The network designed for ocean monitoring currently includes eleven stations, located primarily in the Southern Hemisphere.

Chemical explosions are more portable and more easily conducted in an ocean setting and have been used for oceanic research, for construction, and for military testing. A surprisingly large number (300–4,000 per month) of underwater explosions were reported in the North Pacific during the 1960s (Spiess et al. 1968). At one time chemical explosions were commonly used for marine seismic exploration, but they have been replaced by airgun arrays, which provide a more reliable source signature. Chemical explosions continue to be used in the construction and removal of undersea structures, primarily by the oil industry, but the frequency of detonations presumably has decreased over the past few decades.

New classes of military vessels undergo tests, called ship-shock trials, to determine their ability to withstand explosions (Commander Naval Air Warfare Center 1994). During a ship-shock trial, a large chemical explosion (e.g., 10,000 lb) is detonated near the vessel's hull and measurements of hull stress are taken. Other Navy activities that involve underwater explosions include "Sinkex," in which torpedoes or other chemical explosives are used to sink retired ships; weapons being tested during development; and operational stores being test fired to monitor their military readiness. During the recent war in Iraq, Navy SEALS disposed of a dozen 500-lb sea mines confiscated from the Iraqi navy by detonating them simultaneously in the Persian Gulf, creating a blast that could be heard 50 miles away in Kuwait (Dao 2003).

The spectral and amplitude characteristics of chemical explosions vary with the weight of the charge and the depth of the detonation. The RMS source level of the initial shock wave, a large component of the energy, is given by

$$\text{SPL dB re } 1\mu\text{Pa at } 1 \text{ m} = 269 \text{ dB} + 7.53 \times \log_{10}(w)$$

where w is the charge weight in pounds (Urick 1975). For instance, 100 lb of TNT produces a shock wave SPL of 284 dB re 1μPa at 1 m with an almost constant frequency content from 10 to 1,000 Hz. The energy from the bubble pulse oscillations contribute approximately an additional 5 dB of source level, yielding a total SPL of 289 dB re 1μPa at 1 m. The signal duration can be obtained from the first few oscillations of the bubble pulse, in this case lasting one-third to one-half a second.

Commercial Shipping

Commercial shipping is the principal source of low-frequency (5–500 Hz) background noise in the world's oceans. Ships contribute to the noise level over large geographic areas, and the sounds of individual vessels are often spatially and temporally indistinguishable in distant vessel traffic. Noise from vessel traffic at high latitudes propagates particularly well over long distances because the oceanic sound channel (zone of most efficient sound propagation) in those regions reaches the surface.

Ships' noise is generated primarily from (a) propeller action, (b) propulsion machinery, and (c) hydraulic flow over the hull. Propeller noise is associated with cavitation (Ross 1987, 1993), the creation of voids from zones of pressure below the ambient water pressure. The collapse of these voids generates sound. Cavitation creates both broadband and tonal sounds, as it may be modulated by blade-passage frequencies and their harmonics, which are called the blade lines in a spectrum. The broadband and tonal components produced by cavitation account for 80–85% of ship-radiated noise power (Ross 1987). Propeller noise also may be generated by unsteady propeller blade-passage forces, and there is additional ship noise from propulsion machinery.

Particular vessels produce unique acoustic signatures described by their source levels and frequency bands. Sharp tonal peaks produced by rotating and reciprocating machinery such as diesel engines, diesel generators, pumps, fans, blowers, hydraulic power plants, and other auxiliaries often are seen in these acoustic signatures. Hydrodynamic flow over the ship's hull and hull appendages is an important broadband sound-generating mechanism, especially with increased ship speed. At relatively short ranges and in isolated environments, the spectral characteristics of individual ships can be discerned. At distant ranges in the open ocean, multiple ships contribute to the background noise, and the sum of many distant sources creates broad spectral peaks of noise in the 5- to 500-Hz band.

Models for representative sound spectra for different classes of ships have been developed by the U.S. Navy. The research ambient noise directionality (RANDI) model (Wagstaff 1973, Schreiner 1990, Breeding 1993) uses ship length and speed as well as an empirically derived power law to determine the broadband (5–500 Hz) spectral level for various classes of vessels. Peak spectral densities for individual commercial ships range from 195 dB re $\mu Pa^2/Hz$ at 1 m for fast-moving (20 knots) supertankers to 140 dB re $\mu Pa^2/Hz$ at 1 m for small fishing vessels. Source-level models also have been developed for the propeller tonal blade lines, occurring at 6–10 Hz and their harmonics for most of the world's large merchant fleet (Gray and Greeley 1980).

Shipping vessel traffic is not uniformly distributed. The major commercial shipping lanes follow great circle routes or coastlines to minimize the distance traveled. Dozens of major ports and "megaports" handle the majority of the traffic, but hundreds of small harbors and ports host smaller volumes of traffic. The U.S. Navy lists 521 ports and 3,762 traffic lanes in its catalog of commercial and transportation marine traffic (Emery et al. 2001). Vessels found in areas outside major shipping lanes include fishing vessels, military ships, scientific research ships, and recreational craft—the last typically found nearshore.

Lloyd's Register of the world's commercial fleet for the year 2001 listed 92,817 vessels (National Research Council 2003b). The principal types (their numbers in parentheses) are cargo/passenger transport (34,704), fishing (23,841), towing/dredging (13,835), oil tankers (10,941), bulk dry transport (6,357), and offshore supply (3,139), but gross tonnage may be a more important index of sound production than vessel numbers. From that perspective, oil tankers and bulk dry transport vessels represent nearly 50% of the total tonnage but less than 19% of the total number of vessels.

Vessel operation statistics indicate steady growth in vessel traffic over the past few decades (Mazzuca 2001). There has been an increase in both the number of vessels and in the tonnage of goods shipped. For example, there has been a 30% increase in the volume of goods shipped by the U.S. fleet (by flag and ownership) over the past 20 years (U.S. Maritime Administration 2003). Oceanic shipping is an efficient means of transporting large quantities of goods and materials globally. The globalization of economic infrastructure means that more raw materials, as well as finished goods, require long-distance transport. The economic incentives for oceanic shipping are strong and, in the near term, there is no viable alternative for transporting large-tonnage materials to distant locations.

The bulk of U.S. waterborne trade is conducted through a few ports (Table 7.1). For instance, the combined California ports of Los Angeles and Long Beach carry 37% of the total trade as measured by 20-ft-equivalent container traffic

Table 7.1 U.S. foreign waterborne trade, calendar year 2002

Rank	U.S. port	Total	Export	Import
1	Los Angeles	4,060	866	3,194
2	Long Beach	3,184	717	2,467
3	New York	2,627	747	1,879
4	Charleston	1,197	521	676
5	Savannah	1,014	453	561
6	Norfolk	982	431	551
7	Oakland	979	469	482
8	Houston	851	430	420
9	Seattle	850	338	512
10	Tacoma	769	278	491
Total		19,729	6,814	12,916

Source: U.S. Maritime Administration (2003).

Note: Units are thousands of twenty-foot-equivalent (TEU) containers.

(U.S. Maritime Administration 2003). Within the U.S. Exclusive Economic Zone, this concentrates the noise from shipping into the regions adjacent to these ports and their approaches. Significant concentrations of shipping traffic also occur in New York (13%) and the Puget Sound (Washington) area (8%).

Short sea shipping is commercial waterborne transportation that does not transit an ocean. It is an alternative form of commercial transportation that uses inland and coastal waterways to move commercial freight from major domestic ports to its destination. The U.S. Maritime Administration and the European Commission are playing active roles in the development of short sea shipping to help reduce freight congestion on national rail and highway systems. Short sea shipping already accounts for 41% of the total goods transport market within Europe, compared to 44% by road and 8% by rail (European Commission 2001). Short sea shipping is of particular concern with respect to shipping noise and marine mammals because of its coastal setting.

Seismic Exploration

Seismic reflection profiling uses high-intensity sound to image the Earth's crust. It is the primary technique for finding and monitoring reserves of oil and natural gas and is used extensively by the fossil-fuel extraction industries. It is also used by academic and government researchers to gather information for studies on Earth's origin and tectonic history.

Arrays of airguns are the sound-producing elements in seismic reflection profiling (Dragoset 1984, 2000). Airguns release a specified volume of air under high pressure, creating a sound pressure wave from the expansion and contraction of the released air bubble. To yield high intensities, multiple airguns are fired with precise timing to produce a coherent pulse of sound. Oil industry airgun arrays typically involve twelve to forty-eight individual guns, towed about 200 m behind a vessel, which operate at pressures of 2,000 psi and are distributed over a region that measures 20 × 20 m.

The pressure output of an airgun array is proportional to its operating pressure, the number of airguns, and the cube root of the total gun volume. For consistency with the underwater acoustic literature, airgun-array source levels are back-calculated to an equivalent source concentrated into a 1-m-radius volume, yielding source levels as high as 259 dB peak re 1 μPa at 1 m output pressure (Greene and Moore 1995). This effective source level predicts pressures in the far field of the array, but in the near field the maximum pressure levels encountered are limited to 220–230 dB peak re 1 μPa. The far-field pressure from an airgun array is focused vertically, being about 6 dB stronger in the vertical direction than in the horizontal direction for typical arrays. The peak pressure levels for industry arrays are in the 5- to 300-Hz range.

Airguns are towed at speeds of about 5 knots and are typically fired every 10 s. A seagoing seismic-reflection operation includes a series of parallel passes through an area by a vessel towing an airgun array as well as six to ten seismic receiving streamers (hydrophone arrays). A recent practice is the use of repeated seismic reflection surveys for "time-lapse" monitoring of producing oil fields, called "4-D" surveys. More than ninety seismic vessels are available worldwide (Schmidt 2004), and about 20% of them are conducting field operations at any given time (Tolstoy et al. 2004).

Offshore oil and gas exploration and construction activities occur along continental margins. Areas of current activity include northern Alaska and northwestern Canada, eastern Canada, the U.S. and Mexican Gulf of Mexico, Venezuela, Brazil, West Africa, South Africa, North Sea, Middle East, northwestern Australia, New Zealand, southern China, Vietnam, Malaysia, Indonesia, and the Sea of Okhotsk. New areas of exploration include the deepwater U.S. Gulf of Mexico and deepwater West Africa, both of which have seen increasing activity in the past 5–10 years.

A recent study of ambient noise in the North Atlantic suggests that airgun activity along the continental margins propagates into the deep ocean and is a significant component of low-frequency noise (Nieukirk et al. 2004). Sounds from airguns were recorded almost continuously during the summer, originating at locations over 3,000 km from the receiving hydrophones.

Sonar

Sonar systems intentionally create acoustic energy to probe the ocean. They seek information about objects within the water column, at the sea bottom, or within the sediment. Active sonar emits high-intensity acoustic energy and receives reflected and/or scattered energy. A wide range of sonar systems is in use for both civilian and military applications. For purposes of discussion, sonar systems can be categorized as low-frequency (<1 kHz), mid-frequency (1–20 kHz), and high-frequency (>20 kHz).

Military sonars are used for target detection, localization, and classification. They generally cover a broader frequency range with higher source levels than civilian sonars and are operated during both training exercises and combat. Because far more time is spent in training than in combat, training exercises may be the primary context in which military sonar is used. Low-frequency active (LFA) sonars are used for broadscale surveillance; they are designed to allow submarine tracking over scales of many hundreds to thousands of kilometers. Specialized support ships are used to deploy LFA sonars, which consist of arrays of source elements suspended vertically below the ship. The U.S. Navy's surveillance towed array sensor system (SURTASS) LFA sonar uses an array of eighteen projectors operating in the frequency range from 100 to 500 Hz, with a 215 dB re 1 μPa at 1 m source level for each projector (Johnson 2002). These systems are designed to project beams of energy in a horizontal direction, and the effective source level of an LFA array, when viewed horizontally, can be 235 dB re 1 μPa at 1 m or higher. The signal includes both constant-frequency (CF) and frequency-modulated (FM) components with a bandwidth of approximately 30 Hz. A ping sequence can last from 6 to 100 s, with a time between pings of 6–15 min and a typical duty cycle of 10–15%. Signal transmissions are emitted in patterned sequences that may last for days or weeks.

Mid-frequency tactical antisubmarine warfare (ASW) sonars are designed to detect submarines over several tens of kilometers. They are incorporated into the hulls of submarine-hunting surface vessels such as destroyers, cruisers, and frigates (see Table 7.2). There are 117 of these sonars on U.S. Navy ships currently in active service and equivalent systems in foreign navies (e.g., British, Canadian, French) bring the worldwide count to about 300 (Watts 2003). The AN/SQS-53C is the most advanced surface ship ASW sonar in use by the U.S. Navy, and it generates FM pulses of 1–2 s duration in the 1- to 5-kHz band, at source levels of 235 dB re 1 μPa at 1 m or higher (Evans and England 2001). This sonar has a nominal 40° vertical beam width (dependent on frequency), directed 3° down from the horizontal. The AN/SQS-53C is designed to perform direct-path ASW search, detection, localization, and tracking from a hull-mounted transducer array of 576 elements housed in a bulbous dome located below the waterline of the ship's bow. These systems are used to track both surface and submerged vessels, often detecting surface ships at greater range than many radar systems.

Table 7.2 U.S. Navy surface ships with mid-frequency antisubmarine warfare sonars

Type of ship	Class	Type of sonar	Number in use
Cruiser	Ticonderoga	SQS-53	27
Destroyer	Spruance	SQS-53	11
	Arleigh Burke	SQS-53	49
Frigate	Oliver Hazard Perry	SQS-56	30
Total			117

Other mid-frequency sonars are used by the Navy for depth sounding, communication between platforms, and device activation. High-frequency sonars are incorporated either into weapons (torpedoes and mines) or weapon countermeasures (mine countermeasures or antitorpedo devices). They are designed to operate over ranges of a few hundred meters to a few kilometers. Mine-hunting sonars operate at tens of kilohertz for mine detection and above 100 kHz for mine localization. These sonars are highly directional and use pulsed signals. Other high-frequency military sonars include side-scan sonar for seafloor mapping, generally operated at frequencies near 100 kHz.

Over the past decade, there has been a trend in the U.S. Navy to emphasize training operations in coastal and shallow-water settings. There are now plans to construct shallow-water training ranges on both the West and East coasts of the United States.

Commercial sonars are designed for fish finding, depth sounding, and sub-bottom profiling. They typically generate sound at frequencies of 3–200 kHz, with only a narrow frequency band generated by an individual sonar system. Source levels range from 150–235 dB re 1 μPa at 1 m. Commercial depth sounders and fish finders are typically designed to focus sound into a downward beam. Depth sounders and sub-bottom profilers are designed, respectively, to locate the sea bottom and to probe beneath it. They are operated primarily in nearshore and shallow environments. Fish finders are used in both deep and shallow areas.

The acoustic characteristics of small-scale commercial sonars are unlikely to change significantly in the future since they are limited by several key physical properties. At the low-frequency end (about 3 kHz), they are limited by the physical dimensions of the transducers. At the high-frequency end (200 kHz), they are limited by severe attenuation of sound. Likewise, the maximum power level that can be emitted by a single transducer (200 dB re 1 μPa at 1 m) is limited by cavitation at shallow depths of operation. Higher power levels can be achieved by constructing arrays of sensors on the hull of the vessel. For example, multibeam echo-sounding systems (e.g., SeaBEAM or Hydrosweep) form narrow directional beams (e.g., 1° beam width) of sound and are used for precise depth sounding. Using hull-mounted arrays of transducers, these systems can achieve 235 dB re 1 μPa at 1 m source levels; they are typically operated at 12–15 kHz in deep water and at higher frequencies (up to 100 kHz) in shallow water. They may ensonify a swath of a few tens of kilometers along the track of the ship.

Sonar is an extremely efficient means for fish finding and depth sounding/sub-bottom profiling. Nearly all of the 80,000 vessels in the world's commercial fleet and many of the 17 million small boats owned in the United States are equipped with some form of commercial sonar, and new applications may lead to even greater proliferation of these systems. It is possible that the impact of the large number of these systems in use may be offset to some degree by their limited range.

Research Sound Sources

Research in underwater acoustic propagation and acoustical oceanography often involves the use of sound. Almost all of the programs in the United States are sponsored by the Office of Naval Research, and the information obtained is of value for improving military sonar systems. The sound sources used for these studies are either commercially available transducers or systems specially designed to meet specific research requirements. A wide variety of signals, bandwidths, source levels, and duty cycles are transmitted during these projects. The spatial extent of most experiments is tens of kilometers, but basin-scale projects such as the Acoustic Thermometry of Ocean Climate (ATOC) program have also been undertaken.

The ATOC (later the North Pacific Acoustic Laboratory [NPAL]) project was initiated in the early 1990s to study ocean warming and received much attention from regulatory agencies, the public, and the scientific community because of concerns regarding the potential impact of its sound source on marine mammals (Baggeroer et al. 1998). This program was extensively discussed in two National Research Council (NRC) reports (National Research Council 1994, 2000a). The ATOC source has a 195 dB re 1 μPa at 1 m level and is deployed at 939 m, near the axis of the deep sound channel (Howe 1996). It is designed to study the entire North Pacific basin, with the sounds being received by the U.S. Navy's fixed hydrophone arrays. The transmitted signal is centered at 75 Hz with a bandwidth of 37.5 Hz. It broadcasts at 4-h intervals with a "ramp-up" period of 5 min and a full-power signal duration of 20 min. The long time frame for operation of this experiment was a key aspect that led to questions regarding its potential impact on marine mammals (Potter 1994).

Another basin-scale sonar research project uses drifting sources (Rossby et al. 1986), called SOFAR floats. These devices drift at depth and periodically emit a high-intensity tone (195 dB re 1 μPa at 1 m) that is frequency swept at 200–300 Hz or a continuous signal at 185–310 Hz with a duration of 120 s or more. The sounds are detected by distant receivers and their timing is used to determine the float location and therefore its drift, as a proxy for deep currents.

Acoustic Deterrent Devices and Pingers

Acoustic deterrent devices (ADD) use sound in an effort to repel marine mammals from fishing activities. The idea behind these devices is that they keep the animals away by introducing a local acoustic annoyance or alerting signal. Pingers are used in some fisheries to reduce the bycatch of marine mammals by alerting them to the presence of, or driving them away from, a net or other entangling object. These are typically low-power ADDs with source levels of 130–150 dB re 1 μPa at 1 m. Acoustic harassment devices (AHDs) are used to reduce depredation by marine mammals on caught or cultured fish. These are high-powered devices with source levels of 185–195 dB re 1 μPa at 1 m. Both pingers and AHDs have frequencies in the 5- to 160-kHz band, and generate pulses lasting from 2 to 2,000 ms. To reduce habituation, a single device may transmit with a variety of waveforms and time intervals.

Pingers have been shown to be effective in reducing bycatch, at least for some marine mammal species in some settings (Kraus et al. 1997, Culik et al. 2001, Bordino et al. 2002). A trial of pinger use in the California drift gillnet fishery for swordfish and sharks showed that for both cetaceans and pinnipeds, the entanglement rate in nets with pingers was only one-third of what it was in nets without these devices (Barlow and Cameron 2003). Likewise, a large-scale trial of pingers in Danish gillnet fisheries showed a reduction in bycatch of harbor porpoises (Larsen 1997, Vinther 1999).

Concerns have arisen that the use of AHDs in aquaculture facilities leads to unintended displacement of marine mammals, for example, killer whales (Morton and Symonds 2002) and harbor porpoises (Olesiuk et al. 2002) in the vicinity of salmon farms off British Columbia. Likewise, there are concerns that widespread use of AHDs may lead to the exclusion of porpoises from important feeding habitat (Johnston 2002). AHDs have sufficiently high source levels that they could result in hearing damage to marine mammals exposed at close range.

Polar Icebreakers

Ice-breaking ships are a source of noise in the polar regions (Erbe and Farmer 2000). Two types of noise have been identified in association with ice breaking: bubbler system noise and propeller cavitation noise. Some ice-breaking ships are equipped with a bubbler system that blows high-pressure air into the water around the ship to push floating ice away. While the bubbler system is operating, the noise is continuous, with a broadband spectrum below 5 kHz. A source level of 192 dB re 1 μPa at 1 m in one-twelfth-octave bands has been reported for bubbler system noise. Icebreaker propeller cavitation noise is associated with the ship's ramming the ice with its propeller turning at high speed. The spectrum of propeller cavitation noise is broadband up to at least 20 kHz, and has a source level of 197 dB re 1 μPa at 1 m.

Industrial Activities, Offshore Drilling, and Construction

Industrial activities and construction both in the ocean and along the shoreline can contribute to underwater noise. Examples include coastal power plants, pile driving, dredging, tunnel boring, power-generating wind mills, and canal lock operations (Greene and Moore 1995). The coupling of these sounds into the marine environment is poorly understood, but it is generally more efficient at low frequencies.

Marine dredging is commonly conducted in coastal waters to deepen channels and harbors, reclaim land, and mine seabed resources. Reported source levels for dredging operations range from 160 to 180 dB re 1 μPa at 1 m for one-third-octave bands with peak intensity between 50 and 500 Hz (Greene and Moore 1995).

Oil and gas production activities that generate marine noise include drilling, offshore structure emplacement and removal, and related transportation. Sound pressure levels associated with drilling are the highest with maximum broadband (10 Hz–10 kHz) energy of about 190 dB re 1 μPa at 1 m. Drill-ship noise comes from both the drilling machinery and the propellers and thrusters used for station keeping. Jack-up rigs are the most commonly used offshore drilling devices, followed by platform drill rigs. Drilling generates ancillary noise from the movements of supply boats and support helicopters. Emplacement of offshore structures creates localized noise for brief time periods, and powerful support vessels are used to transport these large structures from the point of fabrication to the point of emplacement. This activity may last for a few weeks and may occur eight to ten times a year worldwide. Additional noise is generated during oil production activities, which include borehole casing, cementing, perforating, pumping, pipe laying, pile driving, and ship and helicopter support. Production activities can generate received levels as high as 135 dB re 1 μPa at 1 km from the source (Greene and Moore 1995), which suggests source levels as high as 195 dB re 1 μPa at 1 m with peak frequencies at 40–100 Hz.

Oil and gas production is moving from shallow-water settings into water depths of up to 3,000 m. Deepwater drilling and production have the potential to generate higher levels of noise than shallow-water production, owing to the use of drill ships and floating production facilities. In addition, noise generated in deep water may be more easily coupled into the deep sound channel for long-range propagation. The worldwide count of offshore mobile drill rigs in use fluctuates with business conditions, but there is a growing number of drill rigs available, with an increase of approximately 10% over the past 5 years.

Small Ships, Boats, and Personal Watercraft

Small vessels do not contribute significantly to the global ocean sound environment, but may be important local sound sources, particularly in coastal settings. Examples of sound levels for whale-watching boats range from 115–127 dB re 1 μPa at 1 m for one-third-octave bands (Au and Green 2000) and 145–169 dB re 1 μPa at 1 m for one-twelfth-octave bands, with increased sound levels for high-speed operation (Erbe 2002). A recent study of noise levels from small powerboats suggests peak spectral levels in the 350- to 1200-Hz band of 145–150 dB re 1 μPa^2/Hz at 1 m (Bartlett and Wilson 2002). The total number of recreational craft in operation is poorly documented although about 17 million small boats are owned in the United States (National Marine Manufacturers Association 2003). The vessel categories are outboard (8.4 million), inboard (1.7 million), stern drive (1.8 million), personal watercraft (1.4 million), sailboats (1.6 million), and miscellaneous (2.5 million). In the inshore waters of Florida, there are nearly 1 million registered recreational boaters (U.S. Fish and Wildlife Service 2001), and the number of boats in operation is raised seasonally by an influx of boats from out of state.

Comparison of Anthropogenic Sound Sources

The anthropogenic sound sources discussed previously are summarized by source level and other parameters in Table 7.3, ordered by their intensity. For sources constructed from arrays of elements (e.g., military sonars and airguns), the individual source elements can be widely distributed. In this case, the source level is given for a range of 1 m to standardize the calculation, but in practice the actual levels experienced near the source never reach the stated levels. Instead, these levels are used to calculate accurately what the source level is at longer ranges, where the distance to the source is much greater than the source dimensions. Table 7.3 is designed to approximate the potential for these sources to impact acutely or injure marine mammals. In practice, the sensitivity of marine mammals to various kinds of sound is another important consideration, as discussed later in the chapter.

Underwater nuclear tests and ship-shock trials produce the highest overall sound pressure levels, yet these are rare events and so may be assumed to have limited impact overall. Military SURTASS-LFA sonars and large-volume airgun arrays both have high SPLs. The long ping lengths and high duty cycles of LFA sonars increase their total energy levels; both the SURTASS-LFA and airgun arrays have dominant energy at low frequencies, where long-range propagation is likely. Mid-frequency military sonars (such as the SQS-53C) have shorter ping durations and more moderate duty cycles; because they operate at middle frequencies, propagation effects also limit their range. Concern for the impact of these sonars is for local settings (as discussed later in this chapter).

Commercial supertankers are arguably the most nearly ubiquitous sources of high-intensity, with more than 10,000 vessels operating worldwide. Concern with these noise sources is concentrated near major ports and along the most heavily traveled shipping lanes. The moored research sound source for the ATOC project is a source level equivalent to a supertanker, although it operates on a low duty cycle. AHDs have high source levels, whereas ADDs have relatively moderate source levels. Multibeam hull-mounted echo sounders have high source levels, but their narrow beam widths and medium frequencies limit their range and the area that they ensonify. Research acoustic floats (RAFOS) produce a moderately high source level but are operated at a very low duty cycle. Fishing vessels have moderate source levels and may represent at least local acoustic annoyances.

Anthropogenic Noise Energy Budget per Year

An annual energy budget is one approach to comparing the contribution of each anthropogenic noise source. The approach taken here is to consider the acoustic energy output at the source itself, rather than the sum of many sources af-

Table 7.3 Comparison of anthropogenic underwater sound sources ordered by their short-term (~ 1 s) energy output, approximating their potential for acute or injurious effects

Sound source	SPL (dB re 1 μPa at1m)	Ping energy (dB re 1 μPa^2-s)	Ping duration	Duty cycle (%)	Peak frequency (Hz)	Bandwidth (Hz)	Directionality
Underwater nuclear device (30 kiloton)	328	337	8 s	Intermittent	Low	Broad	Omnidirectional
Ship shock trial (10,000 lb TNT)	299	302	2 s	Intermittent	Low	Broad	Omnidirectional
Military sonar (SURTASS/LFA)	235	243	6–100 s	10	250	30	Horizontal
Airgun array (2000 psi, 8000 in.^{3})	256	241	30 ms	0.3	50	150	Vertical
Military sonar mid-frequency (SQS-53C)	235	232	0.5–2 s	6	2,600–3,300	Narrow	Horizontal
Supertanker (337 m length, 18 knots)	185		Continuous	100	23	5–100	Omnidirectional
Research sonar (ATOC source)	195	226	1200 s	8	75	37.5	Omnidirectional
Acoustic harassment device	185	185	0.5–2 s	50	10,000	600	Omnidirectional
Multibeam echosounder (hull-mounted)	235	218	20 ms	0.4	12,000	Narrow	Vertical
Research sonar (RAFOS float)	195	216	120 s	Small	250	100	Omnidirectional
Fishing Vessel (12 m length, 7 knots)	151		Continuous	100	300	250–1000	Omnidirectional
Acoustic deterrent device (AquaMark300)	132	127	300 ms	8	10,000	2000	Omnidirectional

ter propagation within the ocean, as would be experienced by a receiver at a particular location. Ambient noise distributions at a given location result from a complex distribution of worldwide sources and variable acoustic propagation. The question considered here is a simpler one: what is the total energy output from each source type at the location of the source. That is, all sources are assumed to be at a compact location, at a range of 1 m, and the total annual energy output of each source type is estimated. This is clearly not the most desirable form of energy budget but is amenable to a manageable tabulation. Table 7.3 shows the approximate potential of these sources to produce chronic effects on marine mammal populations. However, many other factors have to be considered before these data can be used to help understand the impacts of sound on marine mammals, including the distribution of sources in space and time and the varying sensitivities of marine mammals to sound.

Starting with the source pressure levels given in Table 7.3, the additional information needed to go from sound pressure to total energy includes the source directionality, duration, rate of usage, and total number of sources. The first step is to convert sound pressure level (p) to acoustic intensity (I), obtained by dividing the squared pressure by the acoustic impedance (Z):

$$I = |p\vec{v}| = |p|^2/Z \quad [\text{watts/m}^2]$$

The next step is to account for the directionality of the source. For omnidirectional sources, the acoustic power (P) is given by the solid angle (A) emitted by the source (for an omnidirectional source this is 4π, the area of a sphere of 1 m radius) multiplied by the acoustic intensity (I):

$$P = AI \quad [\text{watts} = \text{joules/sec}]$$

The energy per source transmission or ping (E_{ping}) is given by the acoustic power multiplied by the duration of the transmission:

$$E_{ping} = PT_{ping} \quad [\text{joules}]$$

The number of source pings per year per source and the total number of sources in operation yield the annual energy output for each source type:

$$E_{total} = E_{ping} N_{pings/year} N_{sources} \quad \text{[joules]}$$

For continuous sources, the energy of 1 s of transmission is used for E_{ping}, and the number of seconds the source is in operation per year is used for $N_{pings/year}$.

A proposed annual anthropogenic energy budget is presented in Table 7.4, starting with the sources and sound pressure levels from Table 7.3. Underwater nuclear explosions, even assuming a 20-year recurrence rate, top the annual energy budget with 2.1×10^{15} J. This is comparable to a small power plant of 100 MW with an annual energy output of 3.2×10^{15} J. The highest-energy regularly operated sound sources are the airgun arrays from 90 vessels operating for 80 days/year to produce 3.9×10^{13} J. Military sonars for antisubmarine warfare (SQS-53C) used on 300 vessels for 30 days/year produce 2.6×10^{13} J. The contribution from shipping comes mostly from the largest vessel classes, with 11,000 supertankers, operating 300 days/year, to yield 3.7×10^{12} J. Lesser contributions are made by other vessel classes (e.g., merchant and fishing) and by navigation and research sonars. For comparison at the low-energy end, a symphony orchestra produces about 10 W of sound energy, and would emit 3.2×10^{8} J over the course of a year.

LONG-TERM TRENDS IN OCEAN NOISE

Overall trends for the level of sound in the sea can be broken down into anthropogenic and nonanthropogenic components. For instance, there is evidence that global climate change may have resulted in higher sea states (Bacon and Carter 1993, Graham and Diaz 2001), which would have the effect of increasing background noise levels. Over the past few decades, however, it is likely that increases in anthropogenic noise have been more prominent. In order of importance, the anthropogenic sources most likely to have contributed to increased noise are commercial shipping, offshore oil and gas exploration and drilling, and naval and other uses of sonar.

Waters surrounding Australia, which are remote from most commercial shipping, allow the effects of anthropogenic and natural noise to be separated. At low frequency (50 Hz), Australian data suggest that ocean noise levels may be as low as 50 dB re 1 μPa^2 /Hz, which is about 30–40 dB below levels in North American and European waters (Cato and McCauley 2002). These data further suggest that wind/wave noise increases at low frequencies, in contrast to the predictions of the deepwater curves developed from Northern Hemisphere data (Wenz 1962). Cato (1998) pointed to the difficulty of separating wind/wave-generated noise from shipping noise in North American datasets.

Trends in background noise and anthropogenic activity levels suggest that ocean noise levels increased by 10 dB or more between 1950 and 1975 (Ross 1987, 1993). These trends are most apparent in the eastern Pacific and eastern and western Atlantic, where they are attributed to increases in commercial shipping. A doubling of the number of ships explains 3–5 dB, and greater average ship speeds, propulsion power, and propeller tip speeds may explain an additional 6 dB.

Other data on long-term noise trends come from a comparison of historical U.S. Navy acoustic array data (Wenz 1969) with modern recordings along the west coast of North America (Andrew et al. 2002). A low-frequency noise increase of 10 dB over 33 years was observed at a site off the central California coast. The explanation for a noise increase in this band is the growth in commercial shipping in terms of both the number of ships and the gross tonnage. From 1972 to 1999 the total number of ships in the world's fleet in-

Table 7.4 Comparison of anthropogenic underwater sound sources ordered by their total annual energy output

Sound source	Intensity (dB re 1 W/m²)	Directionality (angle)	Power (dB re 1 W)	Number of sources	Operate (days/year)	Repetition (pings/day)	Total energy (J)
Underwater nuclear explosions	146	4π	157	1	0.05	1	2.1×10^{15}
Airgun arrays	61	π	66	90	80	4320	3.9×10^{13}
Military sonar (mid-frequency)	53	π/2	55	300	30	4,320	2.6×10^{13}
Supertankers	3.2	2π	11	11,000	300	86,400	3.7×10^{12}
Ship-shock trials	117	4π	128	1	0.5	1	3.3×10^{12}
Military sonar (SURTASS/LFA)	53	π	58	1	30	175	1.7×10^{11}
Merchant vessels	−17	2π	−8.8	40,000	300	86,400	1.4×10^{11}
Navigation sonar	−1.8	π	3.2	100,000	100	86,400	3.6×10^{10}
Fishing vessels	−31	2π	−23	25,000	150	86,400	1.7×10^{9}
Research sonar	13	4π	24	10	4	86,400	9.1×10^{8}

Note: Although this table is designed to approximate the potential of these sources to produce chronic effects, many other factors must be considered, including the distribution of sources in space and time and the sensitivities of marine mammals to sound.

creased from approximately 57,000 to 87,000, and the total gross tonnage increased from 268 to 543 million gross tons.

Mazzuca (2001) compared the results of Wenz (1969), Ross (1987), and Andrew et al. (2002) to derive an overall increase of 16 dB in low-frequency noise from 1950 to 2000. This corresponds to a doubling of noise power (3 dB) every decade for the past 50 years, equivalent to a 7% annual increase in noise. During this period the number of ships in the world fleet tripled (from 30,000 to 87,000) and the gross tonnage increased by a factor of 6.5 (from 85 to 550 million gross tons) (National Research Council 2003b; from McCarthy and Miller 2002).

OCEAN NOISE RESEARCH PRIORITIES

Ocean noise is an important component of the marine environment. Data on ocean noise trends are scarce, despite substantial investments by the U.S. government in the collection of underwater sound data for military purposes (e.g., SOSUS and other ASW monitoring systems). Expanding use of the sea for commercial shipping, oil and gas development, and advanced warfare has resulted in noise levels that are at least ten times higher today than they were a few decades ago. Without some effort to monitor, reduce, or at least cap them, these noise levels are likely to increase and further degrade the marine acoustic environment. Recommendations for tracking and improving our understanding of ocean noise sources are presented below.

Priority 1: Initiate long-term global ocean noise monitoring. A long-term monitoring program is needed to track future changes in ocean noise (National Research Council 2003b:90). Acoustic data should be included in global ocean observing systems now being developed by U.S. and international research agencies. Data from these monitoring systems should be openly available and presented in a manner accessible to decision makers in industry, the military, and regulatory agencies.

Priority 2: Analyze historical marine anthropogenic noise data. In tandem with the effort to monitor present-day ocean noise, a program should be developed to collect, organize, and standardize data on ocean noise and related anthropogenic activities (National Research Council 2003b:89). Infrastructure appropriate for maintaining an archive of these data already exists (i.e., the National Oceanographic Data Center, www.nodc.noaa.gov). Currently, data regarding shipping, seismic exploration, oil and gas production, and other marine activities are either not collected or are difficult to obtain and analyze because they are maintained by separate organizations. International cooperation in this effort should be encouraged.

Priority 3: Develop global models for ocean noise. Marine noise measurements and source data should be used to develop a global model of ocean noise (National Research Council 2003b:92) that incorporates both transient and continuous noise sources. The development of an accurate global model depends on access to ocean noise data and anthropogenic activity data collected by long-term monitoring, as described previously.

Priority 4: Report signal characteristics for anthropogenic noise sources. An important component of model development is better understanding of the signal characteristics for representative anthropogenic sound sources. The description of each source should include enough information to allow reconstruction of its character (e.g., frequency content, pressure and/or particle-velocity time series, duration, repetition rate).

Priority 5: Determine the relationship between anthropogenic activity level and noise level. Research should be conducted relating the overall levels of anthropogenic activity (such as the types and numbers of vessels) with the resulting noise (National Research Council 2003b:90). These correlations will help to extend noise modeling to areas without direct long-term monitoring, but where anthropogenic noise sources are present.

HOW SOUND AFFECTS MARINE MAMMALS

The responses of marine mammals to sound depend on a range of factors, including (a) the sound pressure level and other properties, for example, frequency, duration, novelty, and habituation; (b) the physical and behavioral state of the animals; and (c) the ambient acoustic and ecological features of the environment. Richardson et al. (1995) reviewed marine mammal responses to specific sound sources, but our present understanding of how marine mammals respond to sound is insufficient to allow reliable predictions of behavioral responses either to intense sounds or to long-term increases in ambient background noise.

In humans, the perceived loudness of a sound involves not only hearing sensitivity but also psychological and physiological factors (Beranek and Ver 1992). A loudness-level scale has been developed from detailed testing, where the human subject judges the relative loudness of two sounds; for instance, the phon (in dB) compares the loudness level of tones of varying frequency to a 1-kHz reference tone. In practice, the annoyance level of a sound depends on a range of factors apart from loudness, such as the sound's fluctuation; intermittent sounds are more annoying than continuous ones. The degree to which human and terrestrial animal studies can be reliably extrapolated to marine mammals is uncertain because there are vast differences in the role of sound in sensing the marine and terrestrial environments and the ambient and biologically significant sounds, such as those of predators, differ in each setting.

MARINE MAMMAL SOUND PRODUCTION

The frequency band of sounds that are important to marine mammals matches, or extends beyond, the range of the sounds that they produce. Marine mammal call frequencies generally show an inverse correlation with body size (Watkins and Wartzok 1985), with mysticetes having larger bodies and lower frequency calls than odontocetes.

For mysticetes, most sound production is in the low-frequency range of 10–2,000 Hz (Edds-Walton 1997). Mysticete sounds can be broadly characterized as (a) tonal calls, (b) FM sweeps, (c) pulsed tonals, and (d) broadband grunt-like sounds and are generated either as individual calls or combined into patterned sequences or songs. For odontocetes, most sound production is in the mid-frequency and high-frequency range of 1–200 kHz (Matthews et al. 1999). Odontocetes produce (a) broadband clicks with peak energy between 5 and 150 kHz, varying by species, (b) burst-pulse click trains, and (c) tonal or FM whistles that range from 1 to 25 kHz. Pinnipeds that breed on land produce a limited array of barks and clicks ranging from less than 1–4 kHz. Those that mate in the water produce complex vocalizations during the breeding season. All pinnipeds, the sea otter, and manatees use sound to establish and maintain the mother-young bond, especially when attempting to reunite after separation (Sandegren et al. 1973, Hartman 1979).

The ability to use self-generated sounds to obtain information about objects and features of the environment, called echolocation, has been demonstrated for at least thirteen odontocete species (Richardson et al. 1995). No odontocete has been shown to be incapable of echolocation, and echolocation clicks have been observed in all recorded species. These echolocation sounds are produced in forward-directed beams, using specialized fats in the forehead (melon) as acoustic lenses. Some species of odontocetes produce few or no whistles and very-high-frequency clicks with peak spectra above 100 kHz. Examples include the Amazon river dolphin (*Inia geoffrensis;* Norris et al. 1972), and the harbor porpoise (*Phocoena phocoena;* Kamminga 1988).

Other odontocetes produce clicks with peak spectra below 80 kHz and use whistles regularly. Examples include bottlenose dolphins (*Tursiops* spp.), which are coastal, and the pantropical spotted dolphin (*Stenella attenuata*), which often occurs in offshore waters. Deep-diving odontocetes, such as sperm whales (*Physeter macrocephalus*) and beaked whales (Ziphiidae), are only known to produce clicks (Mohl et al. 2000, Hooker and Whitehead 2002, Johnson et al. 2004). Some odontocete whistles have been described as "signature" calls that identify individuals (Caldwell and Caldwell 1965). Sounds produced by killer whales are known to be group specific (Ford 1991, Tyack 2000), and the patterned "coda" click sequences made by sperm whales show geographic variation (Rendell and Whitehead 2003).

Source levels for odontocete clicks have been reported to be as high as 228 dB re 1 μPa at 1 m for false killer whales (*Pseudorca crassidens;* Thomas and Turl 1990) and for bottlenose dolphins echolocating in the presence of noise (Au 1993), and 232 dB re 1 μPa at 1 m for male sperm whales (Mohl et al. 2000). The short duration of such echolocation clicks (50–200 μs) means that their total energy is low (197 dB re 1 μPa^2-s) although their source levels are high. Odontocete whistles have lower source levels than their clicks, ranging from less than 110 dB re 1 μPa at 1 m for the spinner dolphin (*Stenella longirostris;* Watkins and Schevill 1974) to 169 dB re 1 μPa at 1 m for bottlenose dolphins (Janik 2000). The detection range for odontocete clicks and whistles is about 5 km, although greater detection ranges also have been reported (Leaper et al. 1992, Barlow and Taylor 1998, Gordon et al. 2000).

Mysticete calls can be detected over long ranges (Payne and Webb 1971). For instance, blue whales (*Balaenoptera musculus*) produce low-frequency (10–100 Hz) calls with estimated source levels of 185 dB re 1 μPa at 1 m (McDonald et al. 2001), which are detectable at ranges of 100 km or more, depending on the acoustic propagation. Most large mysticetes (blue, fin, bowhead, right, humpback, Bryde's, minke, and gray whales) are known to vocalize at frequencies below 1 kHz, with estimated source levels as high as 185 dB re 1 μPa at 1 m (Richardson et al. 1995).

Source levels and frequencies have been estimated for the underwater calls of several species of pinnipeds. Examples are the Weddell seal (*Leptonychotes weddellii*), which produces calls from 148 to 193 dB re 1 μPa at 1 m at frequencies of 0.2–12.8 kHz (Thomas and Kuechle 1982), and the Ross seal (*Ommatophoca rossii*), which produces calls at 1–4 kHz (Watkins and Ray 1985). These calls can be detected at ranges of several kilometers both in the open ocean and under ice (Wartzok et al. 1982).

MARINE MAMMAL HEARING

Sound propagates efficiently underwater, and one reflection of its importance to marine mammals is their development of broader hearing frequency ranges than is typical for terrestrial mammals. Audiograms have been produced for eleven species of odontocetes and nine species of pinnipeds, out of a total of approximately 127 marine mammal species (Wartzok and Ketten 1999, Nachtigall et al. 2000). All hearing data are from species that are small enough to be held in captivity. Direct hearing data are not available for species that are not readily tested by conventional audiometric methods. For the latter, audiograms must be estimated from mathematical models based on ear anatomy or inferred from the sounds they produce and field-exposure experiments (Wartzok and Ketten 1999).

Most delphinids are thought to have functional hearing from 200 Hz to 100 kHz, and some smaller species may hear frequencies as high as 200 kHz. Delphinid audiograms measured to date show peak sensitivity between 20 and 80 kHz, along with moderate sensitivity at 1–20 kHz. Since ambient noise decreases at high frequencies, odontocetes may find it advantageous to hear and echolocate at high frequencies to avoid low-frequency background noise. Another factor favoring high frequencies for some odontocetes is the acoustic background noise of inshore and riverine environments. Higher frequencies also give better spatial resolution, but at the expense of diminished propagation distance.

Based on modeling but with no measured audiograms, mysticete hearing probably ranges between 20 Hz and 20–30 kHz. Several of the larger species, such as blue and fin whales, are thought to hear at infrasonic (down to ~10 Hz) frequencies. Pinniped audiogram data suggest that their best hearing is between 1 and 20 kHz. True seals (phocids) tend to hear higher frequencies underwater than fur seals and sea lions (otariids). Some pinnipeds hear moderately well in both water and air, whereas others are better adapted for underwater than for in-air hearing. Sea lions have the most terrestrially adapted hearing, and elephant seals have the most marine-adapted hearing, with good sensitivity below 1 kHz (Kastak and Schusterman 1998).

Marine mammal hearing is adapted to an aquatic environment, yet their hearing anatomy evolved from terrestrial ancestors. The divergence in hearing physiology between land and marine mammals is most pronounced in the external ears, which are absent in most marine mammal species, and in the middle ears, which are extensively modified in marine mammals. Because levels of ambient noise in the sea can vary by many orders of magnitude as a result of storms and other natural phenomena, marine mammals may have developed resilient mechanisms to guard against hearing loss. Existing studies do not allow for prediction of the impacts of high-intensity sounds on marine mammal hearing except in general terms.

Hearing Losses

Hearing thresholds may be degraded by exposure to high-intensity sound. Hearing losses are classified as either temporary threshold shifts (TTS) or permanent threshold shifts (PTS), where threshold shift refers to the raising of the minimum sound level needed for audibility. Repeated TTS is thought to lead to PTS. The extent of hearing loss is related to the sound power spectrum, the hearing sensitivity, and the duration of exposure. High-intensity, impulsive blasts can damage cetacean ears (Ketten et al. 1993). Hearing losses reduce the range for communication, interfere with foraging capacity, increase vulnerability to predators, and may cause erratic behavior with respect to migration, mating, and stranding. For cetaceans, which are highly dependent on their acoustic sense, both TTS and PTS should be considered serious cause for concern.

Relatively few data are available on hearing loss in marine mammals. Experiments on captive bottlenose dolphins suggest that TTS are observed at levels of 193–196 dB re 1 μPa for exposure to 1-s tones at 20 kHz (Ridgway et al. 1997). Work with impulsive sources (seismic waterguns) suggests that exposure to sound pressure levels of 217 dB re 1 μPa and total energy fluxes of 186 dB re 1 μPa^2-s produces TTS in beluga whales (*Delphinapterus leucas;* Finneran et al. 2002). One hypothesis is that animals are most vulnerable to TTS at or near the frequencies of their greatest hearing sensitivity. For baleen whales, this suggests low-frequency sensitivity and for smaller cetaceans, mid-frequency and high-frequency sensitivity. It also raises the question of why marine mammals (apparently) do not damage their hearing by their own sound production, as both the tonal and impulsive sounds that they produce can be comparable in sound level to those found to induce TTS in the controlled experiments mentioned previously. It is thought that internal mechanisms may protect an animal from its own vocalizations.

Masking

Acoustic signals are detected against the ambient background noise. When background noise increases, it may reduce an animal's ability to detect relevant sound; this is called masking. Noise is effective for masking when it is within a critical band (CB) of frequency around the desired signal. The amount by which a pure tone must exceed the noise spectral level to be audible is called the critical ratio (CR). The CR is related to the bandwidth (CB) within which background noise affects the animal's ability to detect a sound. Estimates of marine mammal CBs and CRs come from captive odontocetes and pinnipeds (Richardson et al. 1995). For all species, the CB expressed as a percentage is broader at low frequencies (25–75% at 100 Hz), and narrower (1–10%) at middle and high frequencies (1–100 kHz). This suggests that band-limited noise is more effective at masking low frequencies than middle and high frequencies. An animal's directional hearing capabilities may help it avoid masking by resolving the different directions of propagation between the signal and the noise. A directivity index of as much as 20 dB has been measured for bottlenose dolphins (Au and Moore 1984). Directional hearing is less acute in pinnipeds.

Erbe and associates have studied masking of beluga whale sounds by icebreaker noise (Erbe and Farmer 1998, 2000, Erbe 2000), including construction of software to model this process (Erbe et al. 1999). Icebreaker noise from ramming, ice cracking, and bubbler systems produced masking at noise-to-signal ratios of 15–29 dB. The predicted zone of masking for beluga calls from ramming noise was 40 km (Erbe and Farmer 2000). Beluga whales' vocal output changes when they are moved to locations with higher background noise (Au et al. 1985). With noise at low frequencies, an animal increases both the sound pressure level and the frequency of its vocalizations, perhaps in an attempt to avoid or overcome masking. Beluga whales also have been observed to increase call rates and shift to higher call frequencies in response to boat noise (Lesage et al. 1999). Likewise, it has been suggested that killer whales shift their call frequencies in response to the presence of whale-watching boats (Foote et al. 2004).

Hearing Development

Does increased noise in the oceans cause developmental problems for young animals? High-noise environments affect auditory development in very young rats (Chang and Merzenich 2003). Brain circuits that receive and interpret sound did not develop at the same rate in animals living in an environment of high continuous background noise as in animals that were raised in a quiet environment. It took three to four times as long for rats raised in a noisy environment to reach the basic benchmarks of auditory development. For marine mammals, comparable data may be difficult to obtain, but the potential for developmental impairment should be an important consideration when assessing the impacts of ocean noise.

NONAUDITORY SOUND IMPACTS

Nonauditory effects involve the interaction of sound with marine mammal physiology. Sound is known to have direct and indirect physiological effects on mammals apart from its effects on hearing discussed previously. The symptoms of these physiological effects range from subtle disturbance to stress, injury, and death.

The physiology of marine mammals is uniquely adapted to life underwater. For example, deep-diving species have specialized cardiovascular and pulmonary systems that allow breath holding and accommodation to changes in pressure. The same physiology that allows marine mammals to spend extended periods underwater and make deep dives may also create vulnerabilities to sound exposure, and their physiological responses to such exposure may differ from those of humans and other terrestrial mammals.

Research on human divers, laboratory terrestrial animals, and captive marine mammals suggests that exposure to underwater sound can produce nonauditory physiological effects. The range of potential impacts may include physiological stress, neurosensory effects, effects on balance (vestibular response), tissue damage from acoustic resonance, gas bubble formation and/or growth in tissues and blood, and blast-trauma injury.

The term stress is used to describe physiological changes that occur in the immune and neuroendocrine systems following exposure to a stressor. Physiological stress responses are not fully understood; however, indicators of stress that is due to noise have been measured in marine mammals. For instance, dolphins experience heart rate changes in response to sound exposure (Miksis et al. 2001). A beluga whale showed increased stress hormone levels (norepinephrine, epinephrine, and dopamine) with increased sound exposure level (Romano et al. 2004). Prolonged noise-induced stress can lead to debilitation such as infertility, pathological changes in digestive and reproductive organs, and reduced growth, as documented for some fish and invertebrates (Banner and Hyatt 1973, Lagardere 1982).

Cases of neurologic disturbance have been described for human divers exposed to intense underwater sound (160–180 dB re 1 μPa for 15 min). Symptoms during exposure included head vibrations, lightheadedness, somnolence, and an inability to concentrate. These divers reported recurrent symptoms days to weeks after exposure, including, in one case, a partial seizure 16 months after the initial exposure (Stevens et al. 1999). Effects of this type have yet to be studied in marine mammals.

Sound exposure in humans may elicit a vestibular response or dizziness (called the Tullio phenomenon) at thresholds as low as 101–136 dB re 1 μPa (Erlich and Lawson 1980). When human diver vestibular function was assessed before and after underwater sound exposure, transient effects were detected immediately postexposure to 160 dB re 1 μPa for 15 min (Clark et al. 1996). Likewise, rats exposed to 180 dB re 1 μPa for 5 min exhibited mild transient impairment in vestibulomotor function (Laurer et al. 2002), and vestibular effects have been detected in guinea pigs immediately following underwater sound exposure of 160 dB re 1 μPa for 5 min (Jackson and Kopke 1998).

Acoustic resonance can lead to an amplification of pressure within mammalian air cavities in response to sound. Lung and other air cavity resonance is important for establishing thresholds for injury because at any given level of excitation, the vibration amplitude is greatest at resonance. In vivo and theoretical studies related to tissue damage support a damage threshold of the order of 180–190 dB re 1 μPa (Cudahy et al. 1999, Cudahy and Ellison n.d.). These studies also provide a relationship between resonance and body mass, based on underwater measurements of terrestrial mammals, including humans, as well as extrapolation from in-air results. Finneran (2003) measured resonance frequencies for beluga whale and bottlenose dolphin lungs directly and found them to be at low frequencies (30 and 36 Hz, respectively). An important issue for resonance effects is the tuning or amplification effect of the resonance. The degree of tuning (defined as Q, with high Q indicating sharper tuning) that has been measured in vivo in the lungs of pigs and humans is from 3 to 5 (Martin et al. 2000), and for beluga whales and bottlenose dolphins is 2.5 and 3.1, respectively (Finneran 2003). This suggests that a moderate level of amplification (a factor of 3) occurs at resonance frequencies.

Sound can increase gas bubble presence in mammalian tissues, especially when dissolved gases are abundant as a result of repeated dives. Human divers are obliged to decompress slowly following dives to prevent bubble formation, whereas deep-diving marine mammals have evolved a means to avoid decompression sickness during their routine diving activity. Intense sound generates bubbles—in vivo cavitation (ter Harr et al. 1982); it also leads to bubble growth—rectified diffusion (Crum and Mao 1996, Houser et al. 2001). The growth of bubbles increases the potential for blocked arteries.

Intense pressures from sources such as explosions can damage air-filled cavities, such as lungs, sinuses, ears, and intestines (Cudahy et al. 1999). A dramatic pressure drop, such as occurs from blast waves, may cause air-filled organs to rupture. Research on blast damage in animals suggests that the mechanical impact of a short-duration pressure pulse (positive acoustic impulse) is best correlated with organ damage (Greene and Moore 1995). Peak pressures of 222 dB re 1 μPa result in perforation and hemorrhage of air-filled intestines in rats (Bauman et al. 1997). Lethal peak pressures of 237 dB re 1 μPa cause pulmonary contusion, hemorrhage, barotraumas, and arterial gas embolisms in sheep (Fletcher et al. 1976). Two humpback whales were found dead following a nearby 5,000-kg explosion, and examination of the temporal bones in their ears revealed significant blast trauma (Ketten et al. 1993).

EFFECTS OF NOISE ON MARINE MAMMAL BEHAVIOR

The behavioral responses of marine mammals to noise are complex and poorly understood (Richardson et al. 1995). Responses may depend on hearing sensitivity, behavioral state, habituation or desensitization, age, sex, presence of offspring, location of exposure, and proximity to a shoreline. They may range from subtle changes in surfacing and breathing patterns to cessation of vocalization to active avoidance or escape from the region of highest sound levels. For instance, several studies suggest that bowhead whales follow a pattern of shorter surfacings, shorter dives, fewer blows per surfacing, and longer intervals between blows when exposed to anthropogenic noise (Richardson et al. 1995), even at moderate received levels (114 dB re 1 μPa). Another common response pattern is a reduction or cessation of vocalization, such as for right whales in response to boat noise (Watkins 1986), bowheads in response to playback of industrial sounds (Wartzok et al. 1989), sperm whales in response to pulses from acoustic pingers (Watkins and Schevill 1975) and in the presence of military sonar (Watkins et al. 1985), and sperm and pilot whales (*Globicephala* spp.) in response to an acoustic source for oceanographic research (Bowles et al. 1994). Moreover, humpback whales lengthen their song cycles when exposed to the LFA source (Miller et al. 2000), move away from mid-frequency sonar (Maybaum 1993), and tend to cease vocalizations when near boats (Watkins 1986). Beluga whales adjust their echolocation clicks to higher frequencies and higher source levels in the presence of increased background noise (Au et al. 1985). Gray whales (*Eschrichtius robustus*) exhibited an avoidance response when exposed to airgun noise, and their response became stronger as the source level increased from 164 to 180 dB re 1 μPa (Malme et al. 1984). They also preferentially avoided LFA transmissions conducted in a landward direction (Tyack and Clark 1998).

Marine mammals have been observed to have little or no reaction to some anthropogenic sounds. For example, sperm whales continued calling when they encountered echosounders (Watkins 1977) and when they were exposed to received sound levels of 180 dB re 1μPa from a detonator (Madsen and Mohl 2000). A fin whale (*Balaenoptera physalus*) continued to call with no change in rate, level, or frequency in the presence of noise from a container ship (Edds 1988).

Age and sex are important factors in noise sensitivity. For instance, juvenile and pregnant Steller sea lions (*Eumetopias jubatus*) are more likely to leave a haul-out site in response to aircraft overflights than are territory-holding males and females with young (Calkins 1979). Walruses (*Odobenus rosmarus*) may stampede and crush calves (Loughrey 1959) or temporarily abandon them (Fay et al. 1984) when exposed to sounds from aircraft or vessels. In gray whales, cow-calf pairs are considered more sensitive than other age or sex classes to disturbance by whale-watching boats (Tilt 1985), and humpback groups containing at least one calf appear to be more sensitive to vessel traffic than are groups without calves (Bauer et al. 1993).

Marine mammal responses also appear to be affected by the context of the exposure, for example, by the location of the source relative to that of the animal, by the motion of the source, and by the onset of the source and its repetition (random versus periodic and predictable). Fin whales are more tolerant of a stationary than a moving source (Watkins 1986). Humpback whales are less likely to react to a continuous source than to one with a sudden onset (Malme et al. 1985). California sea lions (*Zalophus californianus*) and harbor seals (*Phoca vitulina*) react at greater range from a ship when they are hauled out, and this is also true of walruses (Fay et al. 1984). Bowheads are more responsive to overflights of aircraft when they are in shallow water (Richardson and Malme 1993). In the St. Lawrence River, beluga whales are less likely to change their swimming and diving patterns in the presence of vessels moving at low speed than in the presence of fast-moving boats (Blane and Jaakson 1994). In Alaska, beluga whales feeding on river salmon may stop and move downstream in response to noise from small boats, whereas they are relatively unresponsive to noise from fishing boats (Stewart et al. 1982). In Bristol Bay, beluga whales continue to feed when surrounded by fishing vessels, and they may resist dispersal even when deliberately harassed (Fish and Vania 1971). In Sarasota Bay, bottlenose dolphins had longer interbreath intervals during approaches by small boats (Nowacek et al. 2001). In Kings Bay, Florida, manatees' use of boat-free sanctuaries increased as the number of boats in the bay increased (Buckingham et al. 1999).

Only a few studies document long-term marine mammal responses to anthropogenic sound, suggesting abandonment of habitat in some cases. At Guerrero Negro Lagoon in Baja California, shipping and dredging associated with a salt works may have induced gray whales to abandon the area through most of the 1960s (Bryant et al. 1984). After these activities stopped, the lagoon was reoccupied, first by single whales and later by cow-calf pairs. Killer whales (*Orcinus orca*) in the British Columbia region were displaced from Broughton Archipelago in 1993–1999, a period when acoustic harassment devices were in use on salmon farms (Morton and Symonds 2002).

HABITUATION AND TOLERANCE OF NOISE

Habituation is the loss of responsiveness to noise over time. A diminution of responsiveness over time may be due to the animals' becoming accustomed to, and no longer threatened by, the signal. Alternatively, animals may return to the noisy area because of its importance, despite the annoying nature of the sound. The best evidence for habituation of marine mammals to intense sound comes from attempts to use acoustic harassment devices (AHDs) to keep marine mammals away from aquaculture facilities or fishing equipment (Jefferson and Curry 1994). For instance, there is evidence that harbor seals habituate to AHDs, partly because they modify their swimming behavior to keep their heads out of the water when they are in high-intensity sound fields (Mate and Harvey 1987). Likewise, harbor porpoises have been shown to habituate to gillnet pingers over a span of 10 or 11 days (Cox et al. 2002).

Observations of responses to whale watching and other vessels also suggest some level of habituation to noise. Near Cape Cod, common minke whales (*Balaenoptera acutorostrata*) changed from being attracted to vessels to appearing generally uninterested, fin whales from flight reactions to disinterest; and humpback whales from mixed, but usually strongly negative, to strongly positive reactions (Watkins 1986). At San Ignacio Lagoon, Baja California, gray whales become less likely to flee from whale-watching boats as the season progresses (Jones and Swartz 1984).

Habituation does not signify that hearing loss or injury from high-intensity sounds has not occurred. Humpback whales in Newfoundland remained in a feeding area near where there was seafloor blasting (Todd et al. 1996). Received sound pressure levels at 1 mi from the explosions were typically 145–150 dB re 1 μPa at 240–450 Hz, with presumed source levels of 295–300 dB re 1 μPa at 1 m based on the size of the explosive charges. The whales showed no clear reaction to the blasting in terms of behavior, movement, or residence time. However, increased incidental entrapment in nets followed the blast exposure (Todd et al. 1996). In addition, two whales were found dead after a 5,000-kg explosion, and examination of the temporal bones of their inner ears revealed significant blast trauma (Ketten et al. 1993). This incident highlights the difficulty of using overt behavioral reactions to monitor the effects of noise or high-intensity sound on marine mammals.

INCIDENTS OF MASS STRANDING ASSOCIATED WITH HIGH-INTENSITY SOUND

Multiple-animal strandings ("mass strandings") have been associated with the use of high-intensity sonar during naval operations and with the use of airguns during seismic reflection profiling. A key characteristic of these incidents is that they predominantly involved beaked whales, particularly Cuvier's beaked whales (*Ziphius cavirostris*). In many of the areas where such events occurred, Cuvier's beaked whale was not thought to be the most abundant cetacean species present.

Odontocetes are known to mass strand, that is, to come ashore in groups of two or more animals (Walsh, M., et al. 2001). Mass strandings of beaked whales, however, are relatively rare. The National Museum of Natural History, Smithsonian Institution (J. Mead pers. comm.) has compiled a global list of Cuvier's beaked whale strandings involving two or more animals (Table 7.5). Except for a stranding of two individuals in 1914, there are no records of multiple-animal strandings until 1963. Between 1963 and 2004, three to ten mass strandings of Cuvier's beaked whales were reported per decade (Fig. 7.1) although improved reporting may be a factor in explaining the increasing number of mass-stranding events detected in recent decades.

The first published suggestion of a connection between beaked whale strandings and naval activity was by Simmonds and Lopez-Jurado (1991). They described a set of three multianimal strandings associated with naval activity in the Canary Islands in 1985, 1988, and 1989, and additional incidents of beaked whale mass strandings in the Canary Islands were noted in 1986 and 1987. These authors did not posit a connection between beaked whale mass strandings and the use of ASW sonar, but rather related them to the nearby presence of naval operations.

The increased incidence of multianimal beaked whale stranding events can be correlated with the advent of mid-frequency ASW sonar. Prototypes of hull-mounted ASW sonars (e.g., SQS-23 and 26) were first tested in 1957 (Gerken 1986) and were deployed on a broad range of naval ships (frigates, cruisers, and destroyers) belonging to the United States and other nations beginning in the early 1960s. This timing coincides with increased reports of mass strandings of Cuvier's beaked whales (Fig. 7.1). Eleven out of thirty-two of the documented mass strandings of these whales have been associated with concurrent naval activities. Efforts to record marine mammal strandings worldwide have been intensified during the past few decades, so, again, greater efficiency of reporting may be a factor in the increased numbers recorded.

An examination of the circumstances surrounding these mass strandings may help to define the association with the occurrence of high-intensity sound. Two such strandings have been documented by detailed investigative reports: the Kyparissiakos Gulf, Greece, incident of May 1996 (D'Amico and Verboom 1998) and the Bahamas incident of March 2000 (Evans and England 2001). Examination of other beaked whale mass strandings provides additional perspective on the diversity of sound sources, environment, and conditions associated with these events.

Kyparissiakos Gulf, Greece, May 1996

Frantzis (1998, 2004) first drew attention to a mass stranding of Cuvier's beaked whales in the Ionian Sea that coin-

Table 7.5 Strandings of at least two Cuvier's beaked whales, from Smithsonian records

Year	Location	Species (numbers)	Correlated activity
1914	United States (New York)	Zc (2)	
1963	Italy	Zc (15+)	
1965	Puerto Rico	Zc (5)	
1968	Bahamas	Zc (4)	
1974	Corsica	Zc (3), *Stenella coeruleoalba* (1)	Naval patrol
1974	Lesser Antilles	Zc (4)	Naval explosion
1975	Lesser Antilles	Zc (3)	
1980	Bahamas	Zc (3)	
1981	Bermuda	Zc (4)	
1981	United States (Alaska)	Zc (2)	
1983	Galapagos	Zc (6)	
1985	Canary Islands	Zc (12+), Me (1)	Naval maneuvers
1986	Canary Islands	Zc (5), Me (1)	
1987	Canary Islands	Zc (group), Me (2)	
1987	Italy	Zc (2)	
1988	Canary Islands	Zc (3), Me (1), *Hyperoodon ampullatus* (1), *Kogia breviceps* (2)	Naval maneuvers
1989	Canary Islands	Zc (19+), Me (2), Md (3)	Naval maneuvers
1991	Canary Islands	Zc (2)	Naval maneuvers
1991	Lesser Antilles	Zc (4)	
1993	Taiwan	Zc (2)	
1994	Taiwan	Zc (2)	
1996	Greece	Zc (14)	Navy LFAS trials
1997	Greece	Zc (3)	
1997	Greece	Zc (8)	
1998	Puerto Rico	Zc (5)	
2000	Bahamas	Zc (9), Md (3), unidentified ziphiids (2), *Balaenoptera acutorostrata* (2), *Stenella frontalis* (1)	Naval maneuvers
2000	Galápagos	Zc (3)	Seismic airgun
2000	Madeira	Zc (3)	Naval maneuvers
2001	Solomon Islands	Zc (2)	
2002	Canary Islands	Zc (7), Me (2), Md (1), unidentified ziphiids (9)	Naval maneuvers
2002	Baja California	Zc (2)	Seismic airgun
2004	Canary Islands	Zc (2)	Naval maneuvers

Source: J. Mead (pers. comm.), with updates by the author.

Note: Zc, Cuvier's beaked whale. Other beaked whales that stranded during these events included Gervais' beaked whale, *Mesoplodon europaeus* (Me), and Blainville's beaked whale, *Mesoplodon densirostris* (Md).

cided with tests of ASW sonar by the North Atlantic Treaty Organization (NATO). Twelve of these animals stranded along 38 km of coastline on 12–13 May 1996; another stranded along the same coastline about 20 km to the north on 16 May and was driven back out to sea. Two weeks later, one more animal was found decomposing on a remote beach on the neighboring Zákinthos Island, located northeast of the strandings on the mainland. Twelve of these fourteen animals stranded alive, with no apparent disease or pathogenic cause. These strandings coincided with a 4-day period (12–16 May) when the vessel NRV *Alliance* was towing an acoustic source in the vicinity, primarily at depths between 70 and 85 m. The source generated both low- and mid-frequency sound at source levels of 226 dB re 1 μPa at 1m. The transmitted low-frequency signal included a 2-s upsweep at 450–650 Hz, and a 2-s continuous tone at 700 Hz. The mid-frequency signal included a 2-s upsweep at 2.8–3.2 kHz and a 2-s tone at 3.3 kHz. Both frequencies were projected as horizontally directed beams with vertical beamwidths of about 20°. Three source tows of about 2 h duration were conducted each day, and the beaked whale strandings occurred nearest in time to the first two source runs of 12 May (runs 9 and 10; D'Amico and Verboom 1998) and the last two source runs of 13 May (runs 13 and 14; D'Amico and Verboom 1998). Sound propagation modeling suggests that sound pressure levels only exceeded 190 dB re 1 μPa at ranges of less than 100 m. The sound levels present broadly throughout the Kyparissiakos Gulf are thought to have been in the range of 140–160 dB re 1 μPa (D'Amico and Verboom 1998).

The association in space and time between stranding locations and acoustic source tracks suggests that the animals were affected by the ASW sonar (D'Amico and Verboom 1998). Figure 7.2 shows the acoustic source tracks and stranding locations for 12 and 13 May. There is a general correlation between the offshore source track locations and the inshore stranding locations. The 13 May source track is shifted northward from the 12 May track, and likewise some

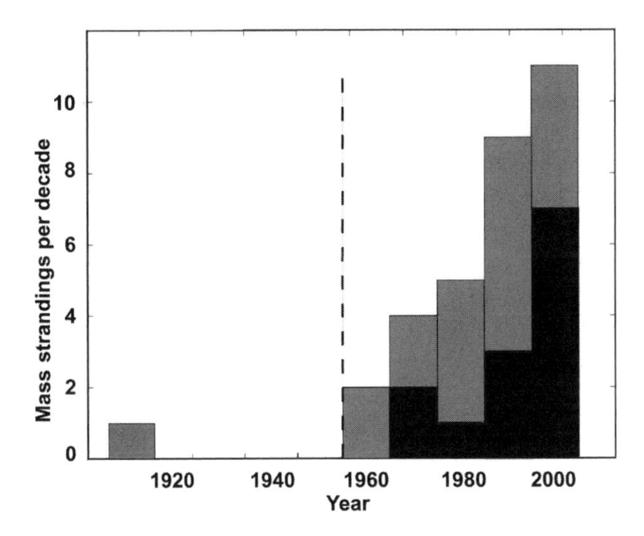

Fig. 7.1. Reported Cuvier's beaked whale mass stranding events tabulated by decade. Total numbers of reported events per decade (gray) are shown, along with the events documented in association with naval activities or airgun usage (black). Incidence of mass-stranding events increased following the deployment of mid-frequency ASW sonar beginning in the 1960s (dashed line), although the level of effort dedicated to reporting has also increased during this period.

of the 13 May stranding locations are located farther north. Correlation of stranding times and source track locations for 12 May suggests that at least three of the six animals with known stranding times were affected by the 0600–0800 h source tow as their stranding times precede the 1100–1300 h source tow. Assuming that they were near the source when they were exposed to the sound (and therefore exposed at levels above 190 dB re 1 μPa), their swimming distances to reach the shore would be approximately 30 nmi, covered at speeds of approximately 10 knots. Alternatively, they might have been exposed at locations inshore relative to the source, which would suggest lower sound exposure levels (140–160 dB re 1 μPa) but shorter swimming distances and speeds. The two 12 May afternoon stranding locations with known times likewise occurred at swimming distances of 20–30 nmi from the source track locations.

Bahamas, 15–16 March 2000

Sixteen cetaceans were found stranded along the Providence Channel in the Bahama Islands during a 2-day period in March 2000, and the episode was correlated with a U.S. Navy training exercise using ASW sonar (Evans and England 2001). The stranded animals were predominantly beaked whales (seven *Ziphius cavirostris,* three *Mesoplodon densirostris,* and two ziphiid sp.), although two minke whales were among the animals that live-stranded. One Atlantic spotted dolphin (*Stenella frontalis*) stranded at a somewhat distant location and is thought to have died of unrelated causes. Eight of the beaked whales died and the remaining animals were refloated, their fates remaining unknown (Balcomb and Claridge 2001). Tissue samples were collected from five of the dead beaked whales. Gross necropsy results suggested that all five were in good body condition; none showed evidence of debilitating disease. Hemorrhages were found in the acoustic fats of the head, the inner ears, and spaces around the brain, with no evidence of external blunt-force trauma. The pattern of injury in the freshest specimens suggested that the ears were structurally intact and that the animals were alive at the time of injury (Ketten et al. 2004).

Five Navy ships were operating hull-mounted ASW sonars in the area, four of which were described in a preliminary report (Evans and England 2001). Of the four ships described, two operated SQS-53C hull-mounted sonars and two used SQS-56 hull-mounted sonars (Watts 2003). The former were operated at 2.6 and 3.3 kHz with a source level of 235 dB re 1 μPa at 1 m or higher and ping lengths of 0.5–2 s alternating between tones and FM sweeps. The latter were operated at 6.8, 7.5, and 8.2 kHz at 223 dB re 1 μPa at 1 m. Integrated sound exposures of 160–165 dB re 1 μPa for 50–150 s would have been experienced in near-surface waters (15 m depth) throughout much of the Providence Channel on 15 March 2000 (Evans and England 2001). Peak sound pressure levels above 185 dB re 1 μPa would have been experienced only within a few hundred meters of the ship tracklines along the central portion of the channel.

The association in space and time between the stranding locations and acoustic source tracks shown in Figure 7.3 is

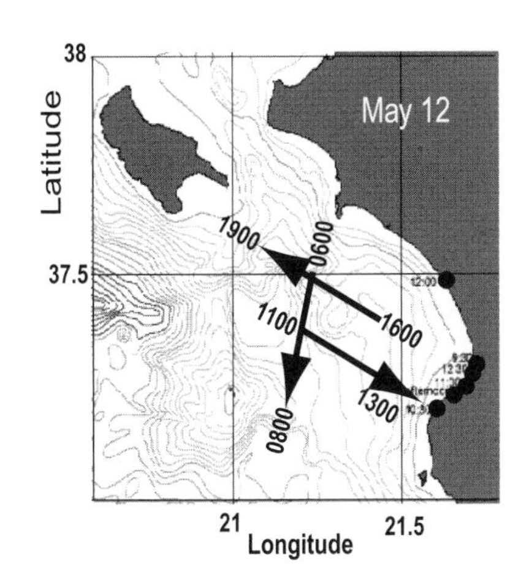

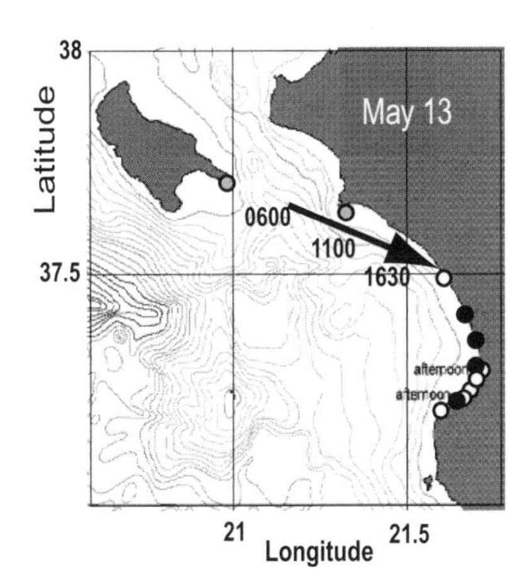

Fig. 7.2. Kyparissiakos Gulf, Greece, Cuvier's beaked whale (*Ziphius cavirostris*) mass stranding locations and acoustic source tracks for 12 May and 13 May 1996. Times for the source tracks and strandings are indicated in GMT. The stranding locations for each day are indicated by black dots on the respective maps. For 13 May, strandings occurring on the previous day are indicated by white dots; strandings occurring after 13 May are shown by gray dots (D'Amico and Verboom 1998, Frantzis 2004).

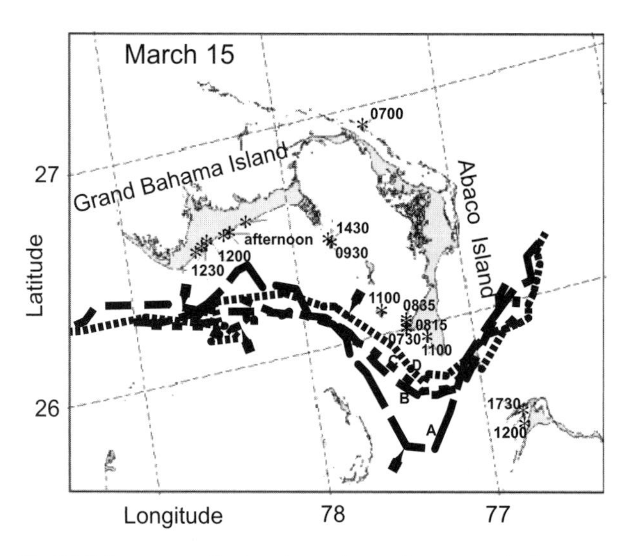

Fig. 7.3. Bahamas stranding locations and acoustic source tracks for 15 March 2000. Stranding locations (asterisks) are shown, along with known stranding times (all times are given as local). Tracks for four U.S. Navy ships are shown moving from east to west through the Providence Channel. Ship tracks are shown as broken lines (ships are given designations A, B, C, and D following Evans and England 2001). Arrows along each ship track designates 0730 h, the time of the first recorded stranding, located at the south end of Abaco Island. Ships B and C entered the channel near Abaco before midnight on 15 March, whereas ships A and D entered the channel at about 0400 h (Evans and England 2001).

compelling evidence that these animals were affected by the high-intensity sound sources (Evans and England 2001). A few hours before midnight on 15 March, two source ships (designated B and C) entered the Providence Channel off the southwest end of Abaco Island and moved through the channel toward the west. Both ships were using active sonar as they entered the channel (ship B 53C sonar, and ship C 56 sonar) with pulse repetition rates of about 24 s. The second pair of source ships (A with 53C sonar, and D with 56 sonar) entered the channel at about 0400 h on 15 March, again from the east and moving west.

A cluster of strandings occurred at the south end of Abaco Island during the morning of 15 March, with the first recorded stranding at 0730 h. These strandings appear to have occurred at the same time or soon after the second pair of source ships passed through the channel south of Abaco, and about 8 h after the first pair of source ships passed this point in the channel. These standings occurred at minimum ranges of 10–30 nmi from the ships' closest points of approach. During the late morning, the source ships moved northwestward, approaching Grand Bahama Island, and a cluster of noon and afternoon strandings occurred on the south coast of Grand Bahama Island, again with minimum source-to-shore ranges of 20–30 nmi. Balcomb and Claridge (2001) noted that individual Cuvier's beaked whales that had been identified photographically in this region previously have not been sighted subsequent to the stranding event, suggesting that the beaked whale mortality was higher than simply the number of whales that were known to have mass stranded.

The highest sound exposures would have been experienced by animals distributed at locations along the source tracks, suggesting that following exposure they might have swum toward the stranding sites 10–30 nmi distant. However, data on beaked whale distribution (K. Balcomb and D. Claridge pers. comm.) suggest that *Mesoplodon densirostris* is found predominantly along the margins of the channel, in waters about 500 m deep, and that *Z. cavirostris* is also found along the channel margins, but in deeper waters. Combining the source modeling and the animal distribution data suggests that sound at moderate exposure levels (e.g., 150–160 dB re 1 μPa for 50–150 s) would have been received at the most likely locations of beaked whales in Providence Channel (Evans and England 2001).

Madeira, May 2000

Three Cuvier's beaked whales stranded in May 2000 on the Madeira Archipelago in the northeastern Atlantic (Freitas 2004). The area south of Madeira Island, specifically the deep channel between Madeira and Porto Santo islands, is a known location for *Ziphius cavirostris* sightings. The animals stranded on 9, 13, and 14 May: two subadults (one male, one female) and a female of unknown age. The two subadults were examined and found to have eye hemorrhages, pleural hemorrhages, and lesions of the lungs (Freitas 2004). It was concluded that they had stranded while still alive. The third animal was found in an advanced state of decomposition and was not examined in detail. A NATO exercise was signaled by the presence of naval vessels and aircraft in the deepwater channel between the islands during the period 9–14 May. The exercise was reported to have involved one aircraft carrier, three submarines, and more than forty surface vessels. Details of the acoustic sources in use during the exercise have not been made available.

Canary Islands, 24 September 2002

A mass stranding of fourteen to nineteen beaked whales occurred on the Canary Islands of Fuerteventura and Lanzarote, associated with naval maneuvers by Spain and other NATO countries on 24–25 September 2002. The stranded animals included seven *Ziphius cavirostris,* two *Mesoplodon europaeus,* and one *M. densirostris.* On 24 September a total of fourteen animals were found stranded; five were dead, three were alive and subsequently died, and six were pushed back to sea. Five more animals were found dead and in a decomposed state between 25 and 28 September. It is possible that these included animals that had been pushed out to sea and then had stranded again. Preliminary necropsy results for six of the beaked whales suggest that they were healthy. The strandings occurred at dawn or in the early morning, and the animals that were found alive all appeared disoriented. Those that were found dead had been feeding shortly prior to stranding (Martín et al. 2004).

Necropsies and dissections (Fernández 2004) revealed no visible signs of traumatic lesions physically caused by ship strikes, fishing activities, or blunt trauma generally. The stomach contents, and their freshness and digestive status, indicated that there was only a short period between the onset of illness and death.

Examination of these animals' heads and bodies revealed hemorrhages and bubbles (Fernández 2004). Hemorrhages were observed along acoustic paths in the head and in the brain and spinal cord. The hemorrhagic areas observed macroscopically in the acoustic fat were also demonstrated histologically. All of the animals were bleeding profusely from the eyes and there was evidence of multifocal petechial (pinpoint) hemorrhages. Fat embolisms were observed, which could have been responsible for hemorrhages in the macrovascular system. Focal hemorrhages were found in the dura mater, and there was a large quantity of blood in the subarachnoid space around the cranial spinal cord. A generalized congestion of the blood vessels in the brain was seen in all the fresh animals, and multifocal subarachnoid hemorrhages were detected. Additionally, in the tissues that were fresh, empty spaces and bubbles were seen inside the blood vessels. In sections of the brain, multifocal petechial hemorrhages were located mainly in the white matter. All the lungs presented general diffuse congestion, some subpleural hemorrhages, and alveolar edema. The kidneys were enlarged, with marked vascular congestion and hemorrhages in the capsular and interstitial areas. Degeneration (in vivo) of vestibucochlear portions of the ear was noted. Although the exact physical mechanism for these injuries is not known, several hypotheses currently under investigation are focused on nonauditory acoustic effects (Jepson et al. 2003).

The strandings occurred along the southeastern coast of the islands of Fuerteventura and Lanzarote. At the time of the 24–25 September strandings, ten NATO countries (Germany, Belgium, Canada, France, Greece, Norway, Portugal, Turkey, the United Kingdom, and the United States) were conducting a multinational naval exercise (known as NEOTAPON 2002); however, the acoustic sources employed during the exercise are not known at this time. The participating countries include ASW sonar in their capabilities although the details of their systems vary (Watts 2003). Common features of the sonars that may have been used in this exercise include high-amplitude source levels (SPL > 223 dB RMS re 1 μPa at 1 m), periodic (15–60 s) repetition, pulses (up to ~4 s), with significant energy at middle frequencies (3–10 kHz), and formed into horizontally directive beams (Watts 2003). Eight mass strandings of *Z. cavirostris* have been documented in the Canary Islands since 1985, and naval exercises have been recorded as associated with five of them (Table 7.5; Martín et al. 2004).

Gulf of California, 24 September 2002

A stranding of two *Ziphius cavirostris* occurred on 24 September 2002 on Isla San Jose in the Gulf of California, Mexico, coincident with seismic reflection profiling by the R/V *Maurice Ewing* operated by Columbia University, Lamont-Doherty Earth Observatory (Malakoff 2002, Taylor et al. 2004). On 24 September at about 1400–1600 h local time (2100–2300 h GMT), fishermen discovered two live-stranded whales and unsuccessfully attempted to push them back out to sea (J. Urbán-Ramírez pers. comm.). A group of marine biologists found the whales dead on 25 September (B. Taylor pers. comm.). By 27 September, when one carcass was necropsied, the advanced state of decomposition precluded determination of the cause of death.

On 24 September the R/V *Ewing* had been firing an array of twenty airguns with a total volume of 8,500 in.^{3} Such an array is expected to have an effective broadband source level of 256 dB peak re 1 μPa at 1m, with maximum energy at frequencies of 40–90 Hz. A later attempt to directly measure the array source level (Tolstoy et al. 2004) was unsuccessful owing to equipment malfunction and ambiguities in converting the pulsed signals into RMS pressure values. Source levels of airgun arrays at middle frequencies (1–5 kHz) are thought to be diminished from levels at low frequencies by 20–40 dB (Goold and Fish 1998, Tolstoy et al. 2004). The *Ewing* airguns were fired with a repetition rate of approximately 20 s (50 m distance between shots). Figure 7.4 shows the *Ewing* track for 24–25 September; the ship was on a transect line directed toward the stranding site and reached the closest point of approach (18 nmi) at 1430 h local time (2130 h GMT). These animals would have received the highest sound pressure levels if they were exposed at locations near the source tracks. Then, following exposure, they might have swum toward the stranding site 20–30 nmi distant. Alternatively, they might have been exposed at lower source levels at locations nearer the stranding site.

Summary of Beaked Whale Stranding Events

The mass strandings of beaked whales following exposure to sound from sonar or airguns present a consistent pattern of events. Cuvier's beaked whales are, by far, the most commonly involved species, making up 81% of the total number of stranded animals. Other beaked whales (including *Mesoplodon europaeus, M. densirostris,* and *Hyperoodon ampullatus*) account for 14% of the total, and other cetacean species (*Stenella coeruleoalba, Kogia breviceps,* and *Balaenoptera acutorostrata*) are sparsely represented. It is not clear whether: (a) *Ziphius cavirostris* are more prone to injury from high-intensity sound than other species, (b) their behavioral response to sound makes them more likely to strand, or (c) they are substantially more abundant than the other affected species in the areas and times of the exposures leading to the mass strandings. One, two, or three of these possibilities could apply. In any event, *Z. cavirostris* has proven to be the "miner's canary" for high-intensity sound impacts. The deployment of naval ASW sonars in the 1960s and the coincident increase in *Z. cavirostris* mass strandings suggest that lethal impacts of anthropogenic sound on cetaceans have been occurring for at least several decades.

The settings for these strandings are strikingly consistent: an island or archipelago with deep water nearby, appropri-

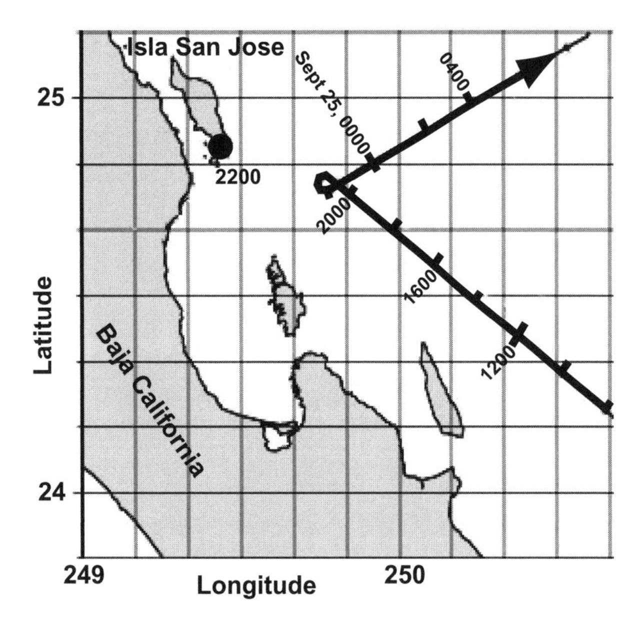

Fig. 7.4. Gulf of California *Ziphius cavirostris* stranding location (black dot) and best estimate for the time of stranding (2200 h GMT), along with the R/V *Ewing* track for 24–25 September 2002. Times along the ship's track are indicated in GMT. The ship's closest point of approach to the stranding site was 18 nmi.

ate for beaked whale foraging habitat. The conditions for mass stranding may be optimized when the sound source transits a deep channel between two islands, such as in the Bahamas, and apparently in the Madeira incident. When exposed to these sounds, some beaked whales swim to the nearest beach. The animals appear on the beach not as a tight cluster of individuals but rather distributed over miles of coastline. Such scatter in the distribution of stranding locations is an important characteristic, which has resulted in these events being called "atypical" mass strandings (Frantzis 1998, 2004, Brownell et al. 2004). The stranded animals die if they are not returned to the sea by human intervention, and the fate of the animals that are returned to the sea is unknown. Necropsies of stranded animals suggest internal bleeding in the eyes, ears, and brain, as well as fat embolisms.

The implicated sounds involve pulses with high-intensity source levels (235 dB re 1 μPa at 1 m) from sonar or airgun arrays. Middle frequencies (1–6 kHz) are clearly implicated in the sonar-induced stranding incidents. It is unclear whether low-frequency sound also has the potential of causing injury to beaked whales. Although airguns create predominantly low-frequency energy, they may also have ample mid-frequency energy. The actual sound exposure levels received by animals that later strand are unknown although in the best-documented events these levels may be bounded by careful sound propagation modeling and by knowledge of where the animals are most likely to be found. Source levels high enough to create permanent or temporary hearing loss would be experienced only at ranges close to the source (< 1 km). The sound exposures calculated for sites of most likely animal presence appear to be significantly lower. For instance, in the Bahamas, the most likely exposure levels appear to have been 150–160 dB re 1 μPa for 50–150 s, or less, well below the level expected to create hearing loss in odontocetes. Given that damage to hearing appears unlikely, other mechanisms are needed to explain the connection between sound exposure and stranding in beaked whales.

RESEARCH PRIORITIES FOR SOUND EFFECTS ON MARINE MAMMALS

A decade has passed since the National Research Council (1994) outlined a set of research priorities for understanding the effects of noise on marine mammals. In most of the areas outlined for study, a basic understanding is still lacking. Many of the same research priorities were reiterated by two subsequent National Research Council (2000a, 2003b) reports. The need to study the impacts of noise in the field rather than in captive settings means that a clear understanding may not become available for many years. There is also the need to differentiate the effects that are significant for individual animals from those that are significant on a population level. Addressing population-level impacts requires observations that are distributed in space and time and large numbers of observations to provide statistical power.

Priority 1: Understand, in detail, the causes of mass strandings of beaked whales. When exposed to high-intensity sound, some beaked whales strand and die. Understanding the causes and consequences of beaked whale mass stranding represents the highest research priority for marine mammalogists studying the conservation implications of exposure to sound. The sound levels implicated in these events is probably not sufficient to cause permanent or temporary hearing threshold shifts. What is the mechanism for damage or disturbance? The behavioral reaction is swift and vigorous on an individual level. The lack of close animal clustering on the beach suggests little or no social component to these strandings, yet the potential for large numbers of animals to strand suggests significance at a population level. What are the source characteristics that lead to damage? Is low-frequency sound (the primary energy component of airguns) as damaging as mid-frequency sound (used by SQS-53 ASW tactical sonars)? What sound pressure exposure level creates damage or disturbance? Beaked whale mass stranding events make it clear that high-intensity anthropogenic sound is a threat to at least some marine mammals, yet key parameters about beaked whale strandings must be understood before we can predict the impacts of high-intensity sound on other species in other settings.

Priority 2: Determine behavioral responses to anthropogenic sound. A key impediment to assessing the biological effects of ocean noise is the paucity of knowledge about marine mammal behavior and specifically the lack of understanding of their behavioral responses to anthropogenic sound. Behavioral data must be collected in the wild to provide a basis for understanding potential effects. Significant effects of ocean noise may be confined to a few individuals exposed to high sound pressure levels, or they may extend to entire populations as a result of widespread exposure. Controlled exposure experiments might be helpful in defining obvious or short-term effects on individuals but may not reveal long-term impacts. Discerning population-level effects is challenging as the observations must be conducted over long distances and extended periods of time, and there are many confounding influences. Relating migration and movement to noise level is one potential approach. Do marine mammals systematically avoid habitat areas with high noise levels? More subtle behavioral changes may be associated with exposure to high ambient ocean noise. Is natural sound (e.g., snapping shrimp) useful for prey location? A better understanding of how and why marine mammals make and use sound would greatly aid our ability to predict how ocean noise might be disruptive to marine mammal behavior. The sound avoidance response has been exploited to exclude marine mammals from fish pens and fishing operation areas using acoustic harassment devices. However, does their behavioral response take place before some hearing loss occurs? Our ignorance about marine mammal behavioral responses to sound is abysmal, and knowledge of this subject must be improved in the face of rising ocean noise levels.

Priority 3: Improve tools for assessing and measuring the behavior of marine mammals. Better research tools are needed to observe marine mammal behavior in the wild. Such tools are needed both to characterize normal behavior and to detect changes in behavior associated with anthropogenic sound. Acoustic recording tags are important for detailed behavioral studies in the presence of sound. Technical improvements are needed over current tags to increase their duration of attachment, to expand the volume of stored data, and to enhance the suite of available sensors. Improvements to the tag attachment system are particularly needed for large cetaceans, which cannot be captured for tagging. Current noninvasive attachments have limited duration, whereas invasive attachments may involve both disturbance and injury to the animal. Technology for passive acoustic tracking is another important component of behavioral study with the potential for improvement.

Priority 4: Develop tools to study marine mammal physiology in the wild. For many species of marine mammal, large numbers of individuals will probably never be maintained in captivity for study of their physiology, so tools are needed to study marine mammal physiology in the wild. For instance, indicators of stress may be used to assess the impact of anthropogenic noise. If stress factors can be measured from blubber or blood samples, perhaps biopsy or other tissue samples collected in the wild could reveal regional or population-wide stress levels associated with noise. Moreover, without a field method that can be rapidly deployed to determine hearing capabilities, it is difficult to collect audiometric data on all marine mammal species and under the full range of conditions where chronic noise may have degraded hearing capabilities. The ability to collect audiometric data on a beached or net-entangled animal is a first step and will be especially useful when high-intensity sound is suspected of having played a role in the animal's stranding.

Priority 5: Characterize and monitor marine mammal populations in areas of high-intensity sound generation. High-intensity anthropogenic ocean sound sources are primarily concentrated in well-defined zones: (a) at major commercial ports and along shipping lanes, (b) within military test and training sites, and (c) within regions of oil exploration and development. The marine mammal populations that inhabit or move through those zones should be characterized and monitored. A combination of visual and acoustic monitoring may be necessary for efficient assessment of marine mammal distributions, ambient noise, and anthropogenic sound. Such monitoring data will help determine whether noise is a factor in discouraging habitat use by marine mammals.

ACKNOWLEDGMENTS

I thank Randy Reeves, Suzanne Montgomery, Tim Ragen, David Cottingham, John Reynolds, and others at the Marine Mammal Commission for encouraging me to write this chapter and for their tireless editorial efforts. I also thank Dave Mellinger, Doug Nowacek, Bill Perrin, Ted Cranford, Sue Moore, Mark McDonald, and Sean Wiggins for helpful reviews and comments. I thank Gerald D'Spain for help with the ocean noise budget calculation. My understanding of this topic was aided by long discussions with Jack Caldwell, Tara Cox, Gerald D'Spain, Roger Gentry, Bob Gisiner, Mardi Hastings, Lee Langstaff, Rodger Melton, Paul Nachtigall, Naomi Rose, Peter Tyack, and Erin Vos.

TIMOTHY J. RAGEN

Assessing and Managing Marine Mammal Habitat in the United States

SIMPLY PUT, EVEN IF PEOPLE do not kill them outright, organisms cannot survive and populations cannot persist if we destroy their habitat. Human activities that threaten marine mammal habitat include coastal development (Fish and Wildlife Service 2001), destruction of bottom communities by fishing gear (Auster and Langton 1999), loss of prey to fisheries (Dayton et al. 2002, Plagányi and Butterworth this volume), creation of anoxic conditions (e.g., dead zones) that can result from disturbance of sediments or eutrophication (Rabalais et al. 1999), and loss of sea-ice habitat caused by global warming (Derocher et al. 2004, Moore this volume). Other human activities may threaten marine mammals by making their habitat less hospitable: for example, introduction of contaminants (O'Hara and O'Shea this volume) and pathogens (Miller et al. 2002, Gulland and Hall this volume) or stimulation of harmful algal blooms (Van Dolah this volume) through agricultural or urban runoff; oil spills (Loughlin 1994); and increased human-generated noise associated with shipping, use of sonar systems, and seismic research and exploration (Hildebrand this volume). One needs merely to contemplate the potential physical, biological, and ecological consequences of global warming, in the polar regions and elsewhere, to appreciate the potential for habitat-related decline of marine mammals and marine ecosystems.

These and other threats arising from human activities raise the fundamental question of whether managers are achieving our conservation goals for the marine environment generally and marine mammal habitat specifically. In the United States, society's aspirations with regard to the conservation of species, habitat, and ecosystems are expressed in a suite of laws, many of which were passed in the late 1960s and early

1970s (see Baur et al. 1999 for an overview). The National Environmental Policy Act (NEPA; 42 U.S.C. 4321 et seq.) seeks to "prevent or eliminate damage to the environment and biosphere." The Marine Mammal Protection Act of 1972 (MMPA; 16 U.S.C. 1361 et seq.) asserts that "efforts should be made to protect essential habitats, including the rookeries, mating grounds, and areas of similar significance for each species of marine mammal from the adverse effect of man's actions." The Endangered Species Act of 1973 (ESA; 16 U.S.C. 1531 et seq.) generally requires the identification of habitat critical to the survival and recovery of threatened and endangered species and attempts to prevent the destruction or adverse modification of that habitat through consultations required by Section 7 of the Act. The Magnuson-Stevens Fishery Conservation and Management Act of 1976 (MSA; 16 U.S.C. 1801 et seq.) defines "conservation and management" in a manner that requires rebuilding, restoring, or maintaining "the marine environment." These and other federal and state laws (e.g., the Coastal Zone Management Act of 1972 and the Marine Pollution Act of 1974) reflect society's intention to prevent human activities from degrading natural habitat to the extent that it can no longer support healthy natural ecosystems.

Precisely what constitutes ecosystem "health" and the tolerable level of degradation in that context are not clear. This lack of clarity is a significant shortcoming of the current approach to habitat conservation in the United States, and a more specific, science-based management framework for habitat conservation is needed. The purpose of this chapter is to promote discussion of a revised, more rigorous framework and to explore the role of science within it. I begin by describing scientific activities essential to support such a framework, including a description of species' distributions, identification of essential habitat features within those distributions, a description of the natural character or condition of those features (i.e., baseline data), an assessment of changes caused by human activities, and an evaluation of our ability to detect significant effects when they occur. I then propose some elements for the revised framework, which also would require considerable scientific input because much of the information and many of the functions needed fall within the scientific domain.

DEFINITIONS

Habitat

The term "habitat" is often used to refer to the physical and chemical (abiotic) components of an organism's environment. Here habitat is intended to include both the abiotic and biotic environment to emphasize that an organism must integrate and adapt to all the elements of its surroundings, including those that are living and those that are not. Excluding the biological components seems a denial of the interactions among the biotic and abiotic components and is therefore inconsistent with efforts to move toward a more holistic, ecosystem-based approach to research and management.

Habitat has a temporal dimension, with a history and a future. It changes diurnally, seasonally, annually, and so on up to an evolutionary scale. It is an integrated collection of selective factors that have driven the evolution of species. To persist, a species must adapt at a rate that keeps pace with changes in those factors. The rate at which human activities alter habitat is, therefore, an important consideration.

Defining habitat broadly leads to overlap in the content of this chapter and others in this volume. The accompanying chapters on anthropogenic noise, indirect fishery interactions, contaminants, disease, harmful algal blooms, and factors related to long-term environmental change all review topics that pertain to habitat quantity and quality. As such, these factors may adversely affect the environment in which marine mammals must find the resources to survive and reproduce. In this chapter, I attempt to address the issue of habitat degradation more generally by identifying concerns and making recommendations that apply to all or most of those other more specific topics.

Natural

Conservationists and managers often use the term "natural" to mean without human effect or influence. Regardless of whether one believes that humans should be considered as part of natural ecosystems, making this distinction is necessary for the purpose of assessing human effects on ecosystems. Congress made such a distinction when it established many U.S. conservation laws (see, e.g., Section 2 of the MMPA and Section 2 of the ESA). To make informed decisions about acceptable levels of human impact, it is essential that we understand and acknowledge human capacity to degrade the ecosystems that support us and other forms of life. Human abundance on Earth exceeded 6.4 billion in late 2004 (U.S. Census Bureau, www.census.gov/cgi-bin/ipc/popclockw) and will continue to increase in the foreseeable future. The combination of human population size and consumption patterns results in an enormous capacity to eradicate other forms of life and degrade their habitat. People are capable of managing and moderating their impact, and, to that end, the "natural" character of habitat or ecosystems provides a comparative baseline or reference for assessing changes caused by human activities.

State

Efforts to assess human impacts are confounded by the fact that ecosystems change, even under natural conditions. The term "state," as in the natural state of the environment, tends to imply a constancy that may be misleading. I use the term "state" to include nature's tendency to change on its own. How such changes are incorporated into environmental assessment and management has become a key issue in habitat conservation.

A MODEL OF HABITAT CONSERVATION

Society's values are the driving force behind conservation of marine mammals and their habitat. Congress passes laws to uphold those values. Managers develop regulations to implement the laws. The regulations are used to constrain human activities that could have significant undesirable effects on marine mammals and their habitat. Scientists play a number of key roles in habitat conservation (Fig. 8.1). The following section focuses on the more familiar of those roles.

ASSESSMENT OF MARINE MAMMAL HABITAT AND HUMAN IMPACTS UPON IT

Identifying Habitats of Concern

Section 7 consultations under the ESA and environmental assessments or impact statements under NEPA are two of the more important mechanisms for predicting and analyzing effects of human activities on the environment. In both analyses, one of the first steps is to identify the species that may be affected by a proposed action. This is accomplished by identifying those species whose distributions overlap the area affected by the action. For that purpose, adequate descriptions of species' distributions and the "action area" are needed, and both may require scientific information or input.

DESCRIBING MARINE MAMMAL DISTRIBUTIONS. In spite of extensive research during the past half-century, the lack of information about the geographic distributions of marine mammals often impedes analyses of potential effects. This is especially problematic for those species most in need of protection. Recent sightings of North Pacific right whales (*Eubalaena japonica*) have provided important (albeit limited) information on the summer distribution of a few whales (LeDuc et al. 2001) but, in general, the foraging and reproductive distributions of this highly endangered species are largely unknown (Shelden et al. 2005). Reproductive areas for the western gray whale (*Eschrichtius robustus*) population in the North Pacific are also unknown (Weller et al. 2002). Although scientists may speculate about threats to these whales within a broad geographic range, such threats cannot be described or quantified with any confidence. The distribution of vaquitas (*Phocoena sinus*) in the Gulf of California is only now being described at a spatial scale needed for recovery actions (International Committee for the Recovery of the Vaquita 1999), but the description is clouded by inconsistencies between survey sightings and the locations where animals have been caught incidentally in fisheries. The locations of foraging areas for the western population of Steller sea lions (*Eumetopias jubatus*) in the Gulf of Alaska and Bering Sea / Aleutian Island region have been debated intensely in recent years (National Marine Fisheries Service 2000). The at-sea distribution of Hawaiian monk seals (*Monachus schauinslandi*) was poorly known and based primarily on anecdotal observations until recently (Parrish et al. 2000). Similarly, the distributions of other marine mammals around the Northwestern Hawaiian Islands were essentially unstudied prior to a survey by the National Marine Fisheries Service in 2002. Even the distribution of the Florida manatee (*Trichechus manatus*) is incompletely described (i.e., the areas where calves are born and nursed during their first months of life) despite a history of intensive

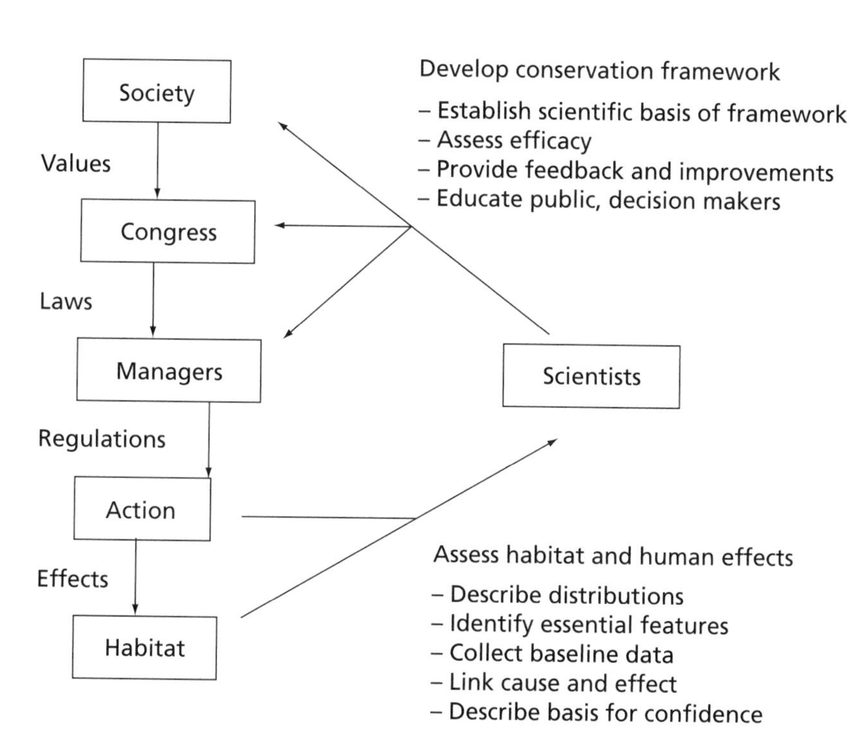

Fig. 8.1. A simple conceptual model of marine mammal habitat conservation. The hypothesis is as follows: Human activities will not have irreversible effects on marine mammal habitat. The model emphasizes the role of scientists in describing habitat, assessing human effects upon it, and developing a management framework to assess the efficacy of conservation efforts and provide feedback to society, Congress, and managers.

study and extensive conflict in a nearshore habitat used by manatees and boaters (Rathbun et al. 1995).

Most of these examples involve species or populations that are difficult to study because of their endangered status, but for the same reason they often are the subject of more than average attention (and resources) for research and management. Critical habitat has been formally designated for some of them, but this does not necessarily mean that their distributions are, in fact, well described. The distributions of species listed as endangered or threatened under the ESA or as depleted under the MMPA are often described from observations of reduced populations. As a consequence, the habitat that is essential or critical to their full recovery may not be recognized and suitably protected (hence, Section 4(a)(3) of the ESA). Finally, the distributions of many species are described largely from snapshot observations, with limited information on temporal variation or variation resulting from changes in oceanographic conditions, forage base, or other factors. This insufficiency of basic information impedes assessment of human effects.

DESCRIBING THE ACTION AREA. Better descriptions are needed for areas that have been or may be affected by human actions. Description of the area affected by some actions may be relatively straightforward, such as the pollution of enclosed or partially enclosed bodies of water (e.g., Puget Sound or San Francisco Bay). In other instances, the action area may be more difficult to describe because the effects of the action extend outside the area or beyond the time frame of the action itself. In ESA terminology, the "action area" is defined to include such temporal and spatial "downstream" effects (i.e., effects outside the area and beyond the time period in which an action occurs; 50 CFR Section 402.02). The so-called dead zone in the Gulf of Mexico (Rabalais et al. 1999) is literally a downstream effect of agricultural and other pollutants transported to sea via the Mississippi River and its tributaries. The input of nitrogen and phosphorus results in increased phytoplankton production that sinks to the bottom either on its own or in zooplankton fecal pellets. This additional organic matter is decomposed by oxygen-consuming bacteria, which can lead to hypoxia or anoxia. Pathogen pollution of waters off central California (Miller et al. 2002) is a downstream effect of sewage and waste disposal in urban areas. Construction of coastal dock facilities in Florida provides another example that may lead to downstream effects if the availability of new facilities provides boater access to manatee habitat that otherwise would be inaccessible.

Describing the action area fully and accurately can be difficult and can lead to controversy. Debate about the use of military sonar systems (SURTASS LFA) has been confounded by uncertainty regarding the distributions of potentially affected marine mammals as well as the range over which significant effects can occur or the amount of habitat significantly affected (*National Resources Defense Council* v. *Evans,* U.S. District Court for the Northern District of California, Case No. C-02-3805 EDL; see Hildebrand this volume). Similarly, analyses of the impact of the Alaska groundfish fisheries have been questioned on the grounds that they have failed to account for effects occurring outside the fishing area or season and resulting from removal of long-lived, mobile animals (*Greenpeace* v. *National Marine Fisheries Service,* U.S. District Court for the Western District of Washington at Seattle, Case No. C98-492Z). These and other examples indicate that rigorous assessment of potential human effects on habitat requires thorough description of the action area and must include spatial and temporal downstream effects.

More basic research is needed regarding the natural distributions of marine mammals and how they vary over time or as a function of other natural factors. Periodic systematic reviews of existing information concerning these distributions and movement patterns would identify important gaps in knowledge and guide needed research. Better descriptions also are needed of human action areas, or areas that are affected by human activities, including indirect or downstream effects.

Describing the Essential Features of Habitat

Identifying the spatial and temporal boundaries of an animal's distribution is one thing, but describing the essential features of its habitat and their significance is quite another. For marine mammals, the challenge is especially great because their oceanic environment is habitat "in motion." Important features of the habitat must be proximally related to basic needs such as prey; refuge from predators; suitable conditions for reproduction including mating and rearing of young, resting, and molting (for pinnipeds); and safety from extreme environmental events such as storms or high surf. Ultimately, the features must be determined by variation in physical and chemical properties (e.g., temperature, light, currents and fronts, depth and bottom characteristics, noise levels, ice, terrestrial haul-out habitat, salinity, and nutrients for biological productivity).

Human activities may alter the essential features of marine mammal habitat either directly or indirectly. Reductions in prey availability (e.g., salmon), water quality (e.g., contaminants), and areas relatively free of disturbance (e.g., whale watching) have been identified as factors implicated in the decline of the southern resident killer whale (*Orcinus orca*) population off the North American west coast (Krahn et al. 2002). The only known foraging habitat of the critically endangered western North Pacific population of gray whales may be effectively lost to the population during seismic testing associated with oil and gas exploration and drilling (Weller et al. 2002). Reynolds (1999) identified abundance of seagrasses, access to fresh water, and refuge from disturbance by human activities as important features of manatee habitat and suggested that the "loss of suitable habitat constitutes the greatest threat to survival of most manatee populations." Manatees in Florida waters have come to depend on warm-water refuges created by thermal discharge from power plants, and the future closing of those plants will result in a marked reduction in habitat available to manatees. The loss of Arctic and sub-Arctic sea-ice habitat from global warming will likely have enormous effects on some seal species and on the polar bears (*Ursus maritimus*) that depend on ice seals for food (Stirling 2002, Derocher et al. 2004). These and other examples illustrate the need to go beyond simply describing marine mammal distributions and instead to identify and evaluate habitat features that are essential to sustain populations.

Investigations of habitat (and human effects upon it) are often "reductionist" in their approach; that is, they reduce habitat to its component parts to study the individual significance of each. Research may usefully focus on single features of the habitat (e.g., prey), but the utility of this approach is limited because marine environments are complex and because the distribution and behavior of marine mammals almost certainly reflect the integration of multiple biotic and abiotic environmental features.

The alternative is a multivariate approach that includes multiple features of the habitat and is carried out over a long time period to determine how those features are integrated over time and space. The ongoing controversy about the respective roles of prey, predators, and ocean conditions in the decline of North Pacific Ocean/Bering Sea pinnipeds is a clear example of the need for such an approach. The California Cooperative Oceanic Fisheries Investigation (CalCOFI) exemplifies the type of multivariate, multidisciplinary research needed to investigate the marine environment generally and marine mammal habitat specifically. CalCOFI conducts systematic surveys with concurrent and cooperative studies of fisheries, ecology, and oceanography of the California Current ecosystem. The research is supported by a joint effort involving the California State Department of Fish and Game, the National Oceanic and Atmospheric Administration, the Southwest Fisheries Science Center of the National Marine Fisheries Service, the University of California, and the Marine Life Research Group at the Scripps Institution of Oceanography.

Another alternative would be to manage marine mammal habitat based on one or more indices of habitat quality assumed to integrate the various salient features (e.g., Fowler 2003). The "environmental carrying capacity" (K) is the best-known index of habitat quality and quantity; that is, the environment's capacity to support a population (e.g., Wilson and Bossart 1971). However, its value as a standard for assessing human impacts depends on whether natural changes in K can be distinguished from changes that are due to human influence. The above-mentioned reductions in pinniped abundance in the North Pacific Ocean and Bering Sea have been attributed to changes in K that are presumably due to natural regime shifts (National Marine Fisheries Service 2000). Such conclusions may be misleading if fisheries have artificially reduced the quality and quantity of marine mammal habitat and such human-related effects cannot be or have not been separated from presumed natural changes in K. Similarly, the optimum sustainable population level of a marine mammal population (Gerrodette and DeMaster 1990), which is a standard and goal of the Marine Mammal Protection Act, may be artificially reduced by human activities that have affected habitat quality and quantity. Thus the use of an index of habitat quality and quantity to assess human impact on the status of a marine mammal population and its habitat requires that managers be able to distinguish the effects of human activities from changes that have occurred naturally.

More research is needed to identify the essential features of marine mammal habitats, to investigate their nature and variability, and to relate them to the life histories of marine mammals. Long-term, large-scale, multivariate investigations of marine mammal habitat seem essential for a holistic, ecosystem-based approach to habitat conservation and management.

Describing the Natural (Baseline) Character of Essential Features over Space and Time

The ability to avoid unacceptable levels of human impact on marine ecosystems (however we define "unacceptable") depends heavily on whether scientists can distinguish between the unaffected and affected character of such systems. This requires that scientists be able to describe or reliably predict the natural character of a habitat or an environment. The inability to do so often stems from the absence of baseline data, which may be lacking for a number of reasons.

Baseline data often have not been collected before an action has been initiated. Fishery management, for example, often depends on data collected from the fisheries themselves, and sufficient information to assess even a targeted fish stock may not be available for years after a fishery has been initiated. Thus, the habitat (including the target stock) may have been altered significantly before any effort is made to assess baseline conditions. This problematic approach often results from a questionable sense of economic urgency, misunderstanding or underestimation of the time and effort needed to adequately characterize baseline conditions, and a reactive rather than proactive approach to management that assumes no significant effect until one is demonstrated.

The perception of baseline conditions may shift over time (Pauly 1995, Dayton et al. 1998). That is, the perception of what constitutes the pristine or unaffected state, which is needed as a basis for assessing impact, may be sequentially replaced according to the perspective of each new generation of scientists and managers. This shifting-baseline phenomenon may be reinforced by the initial failure to collect sufficient baseline data and is particularly problematic for species that have declined extensively (e.g., the vaquita). In the absence of such data, each new generation of scientists and managers not only has the option, but also may feel compelled, to rely on its own perspective. If suitable data had been collected and were available, they would have provided a long-term reference that could not be ignored.

Baseline conditions vary naturally and that variation may not have been adequately characterized. In the past several decades, the "balance-of-nature" paradigm seems to have been replaced by a paradigm of nearly unpredictable change. The evidence for environmental change is growing and irrefutable, but it does not necessarily follow that such change is always natural or that it is haphazard or unconstrained. (A normally distributed variable may be "random" but still have a defined, recognizable distribution, with defined mode, mean, and standard deviation, and defined tail areas that are occupied only very rarely.) If scientists cannot distinguish between natural and human-related effects, then managers and decision makers may rationalize the effects of human activities by wrongly attributing them to natural environmental variation. The growing evidence of environmental variation indicates that collection of adequate baseline information is critical because the natural character of the environment is difficult to describe or predict. This evidence also indicates that accurate characterization of baseline conditions or patterns requires patience. Relatively long-term comprehensive investigations of baseline conditions are needed if the results are to provide an appropriate basis for prediction and comparison. The responsibility for this greater effort should be borne by those whose activities (a) promise economic gain and (b) pose risks to the habitat, and the associated burden should be viewed as a necessary cost of doing business.

Finally, much of the world's marine mammal habitat has been altered to the extent that it is now nearly impossible to describe truly natural baseline conditions. In such cases, a different approach for assessing human-related habitat degradation is needed. One alternative is to acknowledge that the natural baseline cannot be determined and adopt current conditions as a comparative basis for determining future human effects. Doing so is tantamount to completing another iteration of the shifting-baseline phenomenon, but careful documentation of the adopted baseline should help deter further shifting. A second alternative is to allow the habitat (or some representative portion of it) to return to a more natural state by removing the factors that have degraded it. Marine protected areas may usefully serve this approach. A third alternative is to undertake active restoration efforts based on educated speculation about the nature of the pristine habitat.

If society truly intends to maintain healthy marine ecosystems by making informed decisions regarding the effects of human activities, then a fundamental shift in thinking is needed to ensure that scientists have collected adequate baseline information prior to initiation of further activity that may have a significant impact on the natural marine environment. Unless and until we are willing to shift our thinking, the marine environment, including marine mammal habitat, will remain at risk of continued degradation.

Detecting Significant Effects and Linking Them to Causes

Investigations of human effects on marine mammal habitat are hindered by the fact that multiple actions may be affecting the habitat simultaneously and acting together synergistically against a backdrop of complex ecology and natural environmental change. In addition, such investigations are subject to a number of biases.

First, scientists and managers are often preoccupied with single causes. Despite the fact that environmental assessments, impact statements, and ESA Section 7 consultations explicitly require analyses of cumulative effects, such analyses are usually limited to separate consideration of the possible effects of single factors. The individual factors are then weighed against each other, and those factors whose effects cannot be described or appear less significant are rejected or given little additional attention. The analyses often fail to consider all the factors together, including their interactions. Beginning in 1998, the debate regarding the effects of fisheries on Steller sea lions first pitted environmental regime shifts against the fisheries and then shifted to pit killer whale predation against the fisheries. Little attention and discussion has been devoted to the question of whether environmental conditions, fishery effects, and predation are interacting. Such interactions are plausible, if not likely: a regime shift leading to reduced availability of prey could effectively increase the sensitivity of sea lions to competition from fisheries, and the reduction in prey from the combined effects of a regime shift and fishing could force sea lions to spend more time at sea foraging, thereby increasing their vulnerability to killer whale predation. Evaluation of single factors in isolation is tantamount to parsing or fragmenting total effects into smaller increments, each, by itself, less likely to be detected or considered significant.

Second, scientific investigations are often constrained to short time frames and small geographic areas, leading researchers and managers to ignore or overlook the importance of temporal and spatial scaling. The time frame for research may be limited by a sense of economic urgency and the annual federal funding cycle. Spatial scales may be limited by the costs, including those for vessels, aircraft, and personnel. The result is that the long-term, large-scale research projects that are essential for investigating human impacts are often not undertaken or not conducted at the appropriate scale. Without such studies, the true nature and extent of human effects are less likely to be identified and decision makers are more likely to make Type II errors by incorrectly assuming no significant effects. The expectation of quick scientific resolution of a potential conflict is often unrealistic, particularly when an issue involves *K*-selected marine mammals with life histories and flexible behavioral patterns that tend to buffer against short-term adverse effects of environmental variation, including human impacts. Resolution of significant conservation issues often occurs on a scale of decades rather than years, and thoughtful long-term planning and realistic expectations seem essential if scientific investigations are to detect effects when they occur.

Third, in spite of professed concern regarding the effects of human activities on the marine environment and our limited ability to detect significant effects when they occur, managers and decision makers have failed to set aside sufficient "control" areas, such as marine protected areas, for studying natural conditions and assessing human effects. Where such areas have been established, they may be inadequate either because such impacts are already apparent and recovery is needed before the area can be legitimately considered "undisturbed," or the area was designated out of concern for a single species, failing to take into account the ecological relationships among the various components of the ecosystem (Witherell et al. 2000). In addition, marine mammals often have large home ranges and undertake long migrations. Areas established to protect them rarely encompass sufficient habitat and impacts outside the protected areas can have important effects, essentially undermining the benefits conferred by the protected areas.

Fourth, scientific efforts are constrained by economic considerations. Scientific studies that could resolve debate over human effects are sometimes precluded when the economic consequences of such studies are deemed unacceptable. This is tantamount to placing the burden of proof on those who are responsible for or concerned about conservation of the resources rather than on the proponents of an economic activity. For example, the debate over the effects of Alaska groundfish fisheries on Steller sea lions might have been resolved more efficiently and conclusively if certain research designs (e.g., ones involving the suspension of fishing over large areas) had not been rejected outright on the basis of economic considerations. With regard to extractive resource use, the burden of proof must be shifted onto the proponents of such use to ensure an effective management and conservation strategy.

Fifth, with regard to federally linked actions, the burden of proof is often placed on the agency responsible for providing habitat protection rather than on the agency proposing the action. This places a large and unreasonable burden on the conservation agency and reduces the chance of identifying significant effects. The Office of Protected Resources of the National Marine Fisheries Service, for example, has limited resources but at the same time is required to assume the burden of proof regarding a wide range of federal actions. At any given time, officials from that office may be consulting with the Navy regarding SURTASS LFA, the Minerals Management Service regarding oil and gas exploration and drilling, the Office of Sustainable Fisheries (also in the National Marine Fisheries Service) regarding federally managed fisheries, the Army Corps of Engineers regarding widening and dredging of ports, various agencies regarding discharge of pollutants into coastal waters, and so on. Lacking sufficient resources, the Office of Protected Resources may be overwhelmed, and significant effects may go undetected.

With regard to federal actions, the burden of proof belongs with the action agency. Although Congress explicitly intended as much (H.R. Conf. Report No. 697, 96th Congress, 2d Session 12; 1979) and the courts have interpreted the ESA accordingly (*TVA* v. *Hill 437,* U.S. 153; U.S. Supreme Court), the mere fact that the burden of proof has become an issue for judicial interpretation indicates the lack of clear understanding on the part of managers and decision makers. In practice, the burden of proof is often wrongly assigned until the given issue is brought into court for resolution.

The biases described previously reduce the likelihood that adverse effects of human activities on marine mammal habitat will be adequately detected, described, and mitigated. Decision makers must be informed about the limits imposed on science by inadequate support and unrealistic expectations. In view of the difficulty of assessing human effects under these constraints, precautionary measures (e.g., establishment of marine protected areas) and adaptive management approaches are essential. Holt and Talbot (1978) recommended such approaches 25 years ago; their recommendations were reiterated a decade ago by Mangel et al. (1996), and the same recommendations still apply.

Revealing Scientific Power

The single thing that distinguishes science from other human endeavors seeking knowledge and truth is its demand for, and dependence upon, measures of confidence in its methods, results, and conclusions. Virtually all elements of descriptive and experimental science (e.g., careful problem/issue definition, generation of alternative hypotheses, experimental design, statistical analysis of data, meticulous description and reporting, peer review, and repeatability) are intended to provide a basis for confidence in the outcome and conclusions. Even α, the probability level beyond which a null hypothesis is normally rejected, is an expression of confidence, for by setting α at a particular level (usually 0.05), we explicitly accept a certain probability of wrongly rejecting that hypothesis.

Equal attention is not usually paid to β, the probability of wrongly accepting the null hypothesis. Nor is sufficient attention paid to $1 - \beta$, the probability of detecting a significant effect when it occurs (i.e., statistical power). The concept of statistical power is essential to investigations of human impacts because it addresses questions of whether, and to what extent, significant human-caused changes can be detected.

The issue of statistical power is particularly important because of the extensive uncertainty surrounding some kinds of effects (e.g., fishing, climate change). As noted earlier, placement of the burden of proof on conservation advocates means that decision makers are more likely to make Type II than Type I errors, that is, to conclude incorrectly that there are no significant effects (Dayton 1998). As a result, habitat is more likely to be adversely altered than "overprotected" (see Taylor this volume).

The emphasis on statistical power may seem objectionable to scientists because it is equivalent to asking whether the quality of the existing science is good enough to identify effects when they occur. All too often, it is not. In addition, the issue of statistical power may be overlooked because the demand that regulatory decisions be based on "the best available scientific data" may be misinterpreted. "Data" are the records of measurements. "Good data" are measurements with known quality assurance (precision, accuracy, and representativeness). Unfortunately, the demand for "best data" has become confused with a demand for "best estimates," which is contrary to the need to consider the confidence interval for an estimate in making a decision (Taylor 1993). The term "best estimate" refers to a statistical operation for identifying a single value, the "point estimate," when there is uncertainty. We now understand that decision making should also take account of the probability that the true value is some value other than the point estimate. For decision makers and the public to be adequately informed about the potential for adverse effects on marine mammal habitat or the environment, it is essential that they be aware not only of the best available scientific data but also of the reliability and power of those data. In other words, we cannot adequately inform decision makers and the public without describing the level of confidence that we have in the best available science.

Scientists also may fail to address the issue of statistical power because describing power is not a trivial exercise. Thorough descriptions require thoughtful consideration of alternative hypotheses and our ability to discriminate among them. Such descriptions also may require difficult conceptual models and calculations. Nonetheless, if legislators are expected to support science aimed at addressing difficult issues and to make thoughtful and informed decisions about those issues, the information they are given must include an assessment of scientists' ability to detect significant effects and credibly identify their cause or causes when they occur.

EXPANSION OF THE MANAGEMENT AND CONSERVATION FRAMEWORK FOR HABITAT

The familiar tasks thus described are essential but not sufficient to ensure conservation of marine mammal habitat. To that end, and based on the shortcomings of the existing framework described above, the framework must be strengthened by the following initiatives:

1. *Translating society's general intent into specific, measurable goals and objectives for habitat conservation.* Although terms like "significant adverse effects" and "irreversible effects" are expressive of general intent, they are too vague or ill-defined to be useful for distinguishing effects that are acceptable from those that are not. The task of translation must be accomplished in conjunction with managers and decision makers, but it depends heavily upon scientific input (see Goodman this volume). The development of the potential biological removal concept exemplifies the development of a conservation and management framework with such specific, measurable goals and objectives (Wade and Angliss 1997).

2. *Developing an explicit, comprehensive strategy for achieving those goals and objectives.* The strategy should include marine protected areas that are sufficiently large and appropriately positioned, designed, and managed to serve as controls for studying human effects and to provide some insurance against mistakes in judgment (i.e., when human effects prove greater, or different, than anticipated). It should take into account spatial and temporal scaling of habitat features to ensure that measures of effects are appropriately scaled and lead to reliable conclusions about effects. It should include a precautionary and adaptive management approach that ensures that sufficient time and resources are made available to describe the potential effects of a proposed activity before it is allowed to gain such economic momentum that it becomes difficult to stop or modify. This approach should incorporate assessments of environmental effects during the design and planning stage of proposed actions before commitments are made and while it is feasible to modify the activity in response to identified concerns.

3. *Establishing and promoting a stronger intellectual foundation based on ecosystem science and community ecology.* The intellectual foundation for habitat management must be expanded to emphasize community ecology as well as population biology and natural history (Goodman et al. 2002). As a first step, studies of individual species (e.g., target species of fisheries) should be expanded to include information about their ecological interactions with predators, competitors, and prey.

4. *Defining habitat types to use as management units and determining their quantity and quality.* Examples of such units might include estuaries, bays, regions dominated by kelp forests or coral reefs, nearshore zones with different benthic structure or substrate (e.g., sand, rock), areas surrounding islands and atolls, areas delimited by bathymetry such as seamounts or banks, and various features of the continental shelf and slope. Such habitat identification/zoning is a common element of forest and land management and should serve the same purposes in the marine environment. Among other things, the establishment of management units should facilitate assessment of threats from human activities, identification of priority areas of research, and evaluation of the overall efficacy of the conservation and management framework.

5. *Emphasizing and requiring the collection of baseline information for each habitat or habitat type, including variation in the habitat and its components under natural conditions.* Collecting such information lends detail and specificity and, therefore, gives direction to well-intentioned but vague aspirations related to healthy ecosystems. Where environmental conditions within a habitat unit are so disturbed that natural baseline information cannot be collected and used as a comparative basis for determining human-related effects, alternative approaches (as described previously) have to be developed.

6. *Identifying key human-caused threats and requiring that they be comprehensively described.* These descriptions should take into account the dynamic nature of the environment and the flow and movements of its constituent elements (e.g., marine, coastal, atmospheric, and riverine transport processes). The descriptions also should include effects beyond the area and time period in which human activities occur. Assessment of human effects on marine mammal habitat should always be accompanied by an assessment of the statistical power to detect significant effects when they occur. This assessment should include a thorough discussion of alternative hypotheses and scientists' ability to evaluate those hypotheses and discriminate among them.

7. *Emphasizing comprehensive, multivariate, multidisciplinary research programs.* As habitat consists of multiple features that interact and change over time, multivariate, long-term studies are needed to reveal its ecological nature or character. Multidisciplinary teams (e.g., biologists, ecologists, oceanographers) should conduct the studies to ensure robust assessment.

8. *Providing comprehensive summary statistics that characterize the present condition of marine mammal habitat and how it has changed over time.* These statistics are essential for an overall determination of whether the conservation strategy is working and would include, for example, what portion of existing habitat is in relatively pristine condition; is being fished or overfished; is being polluted by contaminants; is being developed or is adjacent to development; or is experiencing dead zones, toxic algal blooms, or increased noise from commercial shipping or other human activities. They should be presented in terms understandable to scientists, managers, decision makers, and the public, and the presentation should convey the basis for confidence in the methodology and conclusions.

9. *Providing practical and specific guidance for restoration efforts where human impacts have adversely modified habitats beyond levels consistent with the goals and objectives of the framework.* Restoration, whether pursued actively (i.e., by taking restorative actions) or passively (i.e., by allowing natural processes to restore habitat), is essential in view of the fact that we have had, and likely will continue to have, effects inconsistent with our environmental goals and objectives. In such cases, the goal should be to restore habitat rather then simply accept its degraded condition as unavoidable or part of a new baseline.

10. *Incorporating feedback and adaptive mechanisms for assessing management efficacy and modifying the framework as needed.* Even when protection goals and objectives remain relatively fixed, the means for achieving them may require adjustment as more is learned about habitat, the threats to it, and the measures needed to preserve or restore it. If the public, managers, and decision makers are to make informed decisions about habitat conservation, they must have feedback on how well the overall strategy is working and the ability to make adjustments based on that feedback.

11. *Facilitating international cooperation in managing human activities that affect marine mammal habitat.* The geographic distribution of most marine mammals includes areas under the jurisdiction of different countries. Even for marine mammals found within the jurisdiction of a single country (e.g., the vaquita), the quantity and quality of their habitat may be adversely modified by activities occurring in other countries or conducted by citizens of other countries (e.g., fishing, discharge of pollutants into the atmosphere or oceans, burning of fossil fuels leading to climate change, reduced freshwater supply to estuaries). Ultimately, the protection of marine mammal habitat depends on the collective efforts of all the countries involved in activities affecting, or potentially affecting, that habitat. National laws such as the Marine Mammal Protection Act and the Endangered Species Act in the United States are not sufficient by themselves to provide such protection because they do not apply to all the relevant activities of other nations. For that reason, international agreements and cooperation are essential. The Convention on the Conservation of Antarctic Marine Living Resources (CCAMLR) and the Convention on Migratory Species are examples of treaties intended to promote and facilitate international cooperation.

12. *Securing the resources needed to develop and implement the framework.* Perhaps the most challenging aspect of developing and implementing an effective habitat conservation framework is convincing the American public and Congress that it is necessary, thereby ensuring that the requisite resources are made available. Research and monitoring require substantial investment in vessels, facilities, technology and equipment, and personnel from a variety of disciplines. However, the costs need not be borne only by taxpayers but rather should be borne largely by the proponents of activities likely to destroy or degrade habitat. Such programs should be viewed as investments that, in the long run, are likely to be far less costly than dealing with the consequences of inadequate management, including the loss of marine resources and the costs of restoration, if it is even possible. A convincing argument for such a proactive agenda requires careful descriptions of the associated risks and benefits, as well as a sophisticated means of communicating the argument to the right audience.

CONCLUSIONS

Is such an expanded framework realistic? I think so. Similar frameworks have been developed for other conservation purposes. Section 117 of the MMPA and its implementing regulations are based on a clear framework for assessing marine mammal stocks. Similarly, the implementation of Section 118 of the MMPA (Wade 1998, Taylor et al. 2000) and its implementing regulations are grounded in a science-based framework for limiting incidental mortality and serious injury of marine mammals in commercial fisheries. These frameworks are not without their limitations, but they provide a specific basis for defining marine mammal stocks, assessing their status and trends, identifying the nature and significance of some threats, identifying needed management actions, and tracking progress with respect to conservation goals.

Clearly, expansion of the current framework requires interaction among decision makers, managers, the public, and scientists. Scientists must play a leading role because so much of the information needed to develop and implement the expanded framework falls within the scientific domain. Only scientists are in a position to explore, characterize, and quantify aspects of the marine environment that have to be understood to achieve management goals. Society may hold certain values or seek certain goals, but over time these are most likely to be upheld or achieved if they are buttressed by scientific understanding.

To a large extent, habitat conservation consists of setting and abiding by limits on human activities. Although existing conservation laws express society's general wish and intent to protect habitat, the implementation of relevant laws is often controversial. Moreover, effectiveness is uncertain because, to date, clear limits have not been established and suitably specific measures of our effects relative to those limits have not been developed. The expanded framework described in this chapter is intended to address such shortcomings and facilitate scientific and management efforts toward the larger goal of preserving marine mammal habitat.

ACKNOWLEDGMENTS

I gratefully acknowledge and thank Daniel Goodman, Randall Reeves, John Reynolds, Brian Smith, and Brent Stewart for their helpful comments.

SUE E. MOORE

Long-Term Environmental Change and Marine Mammals

THERE IS NO DOUBT THAT Earth's environment is changing, possibly at an unprecedented rate (National Research Council 2002, Jensen 2003, Root et al. 2003). Overall, the planet is warming, glaciers and sea ice are retreating, sea levels are rising, and pollutants are accumulating both in the environment and within organisms (e.g., Overpeck et al. 1997, Rothrock et al. 1999, Comiso 2002, Albers and Loughlin 2003). Some of these changes are now interpreted in light of the "regime shift" concept, whereby decadal or multidecadal changes in oceanographic pattern and productivity are driven by oscillations in the position and intensity of large atmospheric pressure cells (e.g., Frances et al. 1998, Aanes et al. 2002, Chavez et al. 2003). Anthropogenic activities, such as the burning of fossil fuels, destruction of forests, and use of pesticides and fertilizers, also contribute to environmental changes (Allen et al. 2000, Intergovernmental Panel on Climate Change 2001). Further, it appears that abrupt environmental change has occurred repeatedly in the past (National Research Council 2002, Alley et al. 2003, Dokken and Nisancioglu 2004), and some scientists suggest that we may be on the verge of a similar event now if, for example, formation of deep water in the North Atlantic or Southern Ocean ceases (Broecker et al. 1999, Bard 2002, Hansen et al. 2004). In short, ecosystem changes are occurring that may prove challenging to extant species, particularly long-lived, highly specialized flora and fauna that are incapable of rapid adaptation.

Marine mammals fall into this category, with the added challenge that they tend to be cryptic. We understand relatively little about their behavioral ecology (e.g., how they find prey or select mates, when and why they move among ocean domains) and are, therefore, often stymied when asked to formulate proactive conservation and management agendas. Another major stumbling block for conservation biologists is the scale at which marine mammals live their lives (e.g., Bowen and Siniff 1999, Mangel and Hofman 1999, Reeves et al. 2002a,b). Most living marine mammal species have been extant for roughly 10 million years (Fordyce et al. 1994, Rice 1998), and individuals can live from about 10 to more than 100 years (e.g., George et al. 1999). Spatially, marine mammals can roam across ocean basins or live productive lives within small estuarine or coastal home ranges (Reeves et al. 2002a,b). How then do we predict or assess the effects of long-term environmental change on marine mammals?

The framework used here is one of ecological scale, as described by the interface between population biology and ecosystem science (S. A. Levin 1992, Norris 2003). Ecological scale for marine mammals extends from days to decades, and sometimes centuries, with spatial scale and associated ocean processes defined by species' natural history (Fig. 9.1). This chapter focuses on the question: Over what temporal and spatial scales is environmental change anticipated to occur and how might various marine mammal species respond given their natural history? In the context of this chapter, "long-term" means decades to centuries. The goal of framing the subject this way is to focus attention on immediate and long-term environmental threats to marine mammals that might be mediated though conservation agendas based upon proactive research.

OCEANOGRAPHIC DOMAINS AND MARINE MAMMAL BIOGEOGRAPHY

All marine mammals, even those that must return to land to breed, are adapted for living in the water. As most marine mammals live in the ocean, some understanding of oceanographic domains will facilitate discussion of their response to environmental change. Oceanographers frequently divide the marine environment into domains based upon latitude (polar, temperate, and tropical), bathymetry (coastal, shelf,

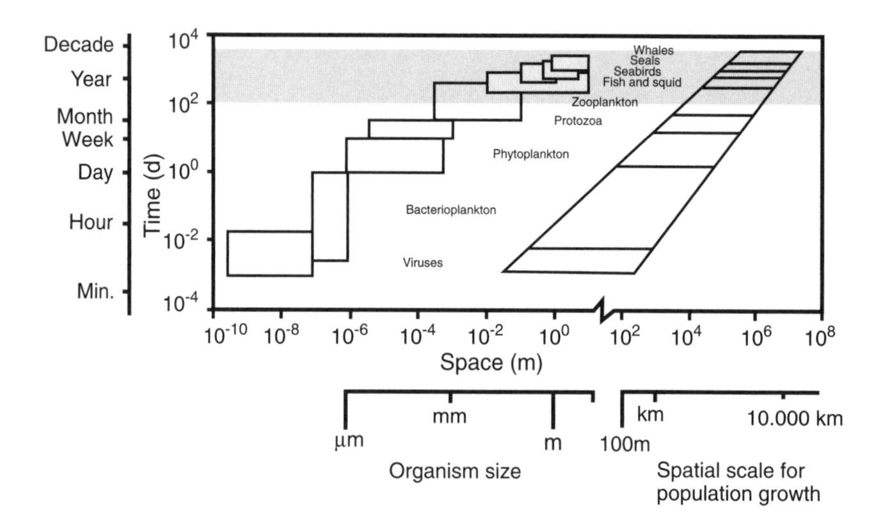

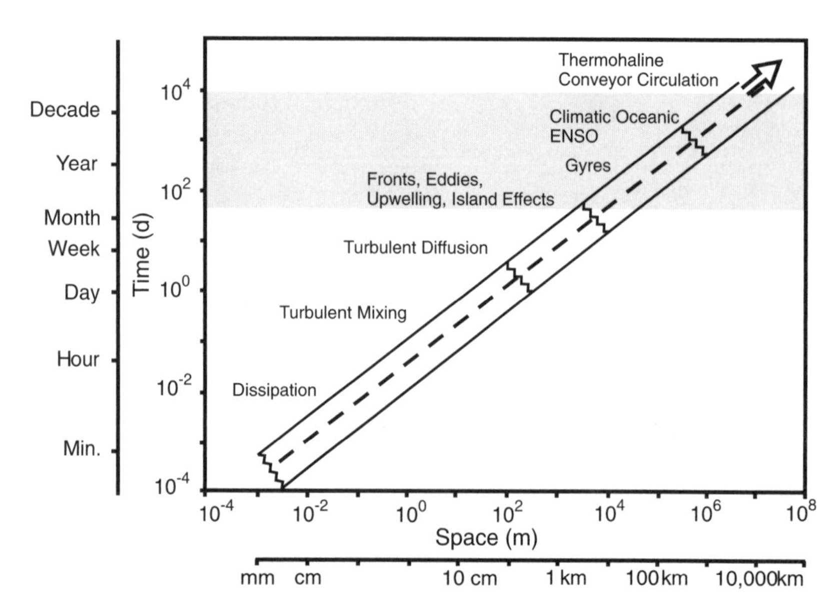

Fig. 9.1. The ecological scale for marine mammals (shaded areas) extends from months to decades, with the spatial scale defined by species' natural history. Oceanographic features such as fronts, eddies, gyres, and ENSO events may affect marine mammal prey (e.g., zooplankton and fishes) as well as the animals themselves.

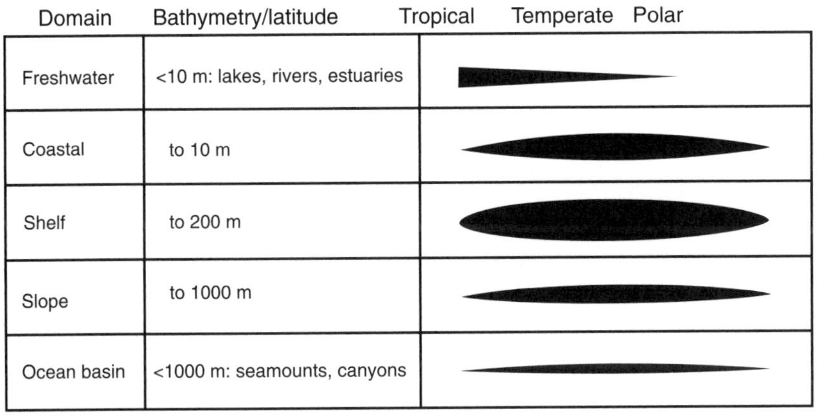

Domain	Bathymetry/latitude	Tropical Temperate Polar
Freshwater	<10 m: lakes, rivers, estuaries	
Coastal	to 10 m	
Shelf	to 200 m	
Slope	to 1000 m	
Ocean basin	<1000 m: seamounts, canyons	

Fig. 9.2. Marine mammal biogeography extends from freshwater to ocean basin habitats, but species distribution is nonuniform, a characteristic that complicates assessment of the effects of long-term environmental change.

slope, and basin) or proximity to shore (estuarine, neritic, and pelagic). Latitudinal domains are based upon the effect of solar radiation on the planet. Solar energy strikes the equator more directly than it strikes the poles, and the resultant gradient in warming sets concentric cells of atmospheric circulation in motion, which in turn drive the major ocean currents. Cooling is so pronounced at the poles that the ocean freezes, with sea ice expanding and receding to some greater or lesser extent each year. The moving ocean interacts with bathymetric features and, via mechanisms such as upwelling or eddies, nutrients are brought to the surface, making boundary zones (coast-shelf-slope) far more productive than deep basins. Estuaries, too, are productive because of the availability of terrestrial nutrients, with the result that estuarine and neritic environments in general are far more productive than pelagic ones. When large to mesoscale areas of enhanced productivity in the ocean, such as subpolar gyres (Springer et al. 1999), waters overlying seamounts (e.g., Haury et al. 1978), or tropical thermocline domes (Fiedler 2002), are added to the picture, it becomes clear that the marine environment is far from uniform.

Marine mammals are not uniformly distributed among the oceanographic domains; neither are they confined to only one (Fig. 9.2; Reeves et al. 2002a,b). Further, two species of seals, three species of river dolphins, and three manatee species are largely or completely confined to lakes, rivers, and estuaries. Of the oceanic species, nineteen are endemic to polar waters, although some of these also range to subpolar environs (Ainley et al. 2004). By far, most species occupy tropical to temperate waters, although many migrate to subpolar and polar waters to feed during seasonal periods of peak productivity. Coastal, continental shelf, and slope waters are disproportionately occupied, owing to their enhanced productivity, compared to the pelagic waters overlying ocean basins. Marine mammals aggregate over mesoscale features associated with productivity, such as blue whales at the Costa Rica Dome (Reilly and Thayer 1990), in the Southern California Bight (Fiedler et al. 1998), and near the Emperor Seamounts (Moore et al. 2002b). Although this crude attempt to "box" marine mammals into oceanographic domains fails to reflect the dynamic movements documented for some species, it does give a general sense of the number of species likely to be affected as environmental changes become manifest in the various regions.

ANTHROPOGENIC CONTRIBUTIONS TO LONG-TERM ENVIRONMENTAL CHANGE

Human activities can precipitate long-term environmental change on a global scale (Vitousek et al. 1997, Jackson et al. 2001). For example, there is little doubt that the wholesale harvest of many hundreds of thousands of whales, seals, sea turtles, and flightless seabirds during the commercial whaling era altered marine ecosystems in ways that are now impossible to measure. Although the interpretation of evidence that combustion of fossil fuels has contributed significantly to the warming of the planet remains contentious (e.g., McKitrick 2003), it appears that the rapid buildup of greenhouse gases over the past two centuries has amplified "natural" increases in temperature that may be part of a long-term planetary cycle (e.g., Keeling 1989, Knutson et al. 1999, Markgraf and Diaz 2000, Intergovernmental Panel on Climate Change 2001). At the poles, this warming has resulted in the progressive melting of glaciers and ice sheets and the dramatic thinning and unprecedented retreats of sea ice (e.g., Rothrock et al. 1999, Comiso 2002, Comiso and Parkinson 2004).

The story concerning pollution is equally dramatic. Chemicals used as fertilizers and pesticides foul aquifers, lakes, rivers, estuaries, and coastal environments. Some are borne atmospherically around the globe in a matter of days or weeks, sometimes precipitating out over especially cold or warm regions many thousands of miles from their source. Human-generated underwater noise is another pollutant on the rise (Andrew et al. 2002, International Whaling Commission 2004, Hildebrand this volume) and is perhaps one to which marine mammals are especially vulnerable as sound, not vision, is the primary sensory modality in the ocean. Finally, the combined removal of targeted and bycaught marine species around the world by commercial fisheries totals in the millions of tons each year (Food and Agriculture Organization of the United Nations 2000).

Marine mammal species are a common component of the bycatch, numbering hundreds of thousands of animals each year (e.g., International Whaling Commission 1994, Kuiken et al. 1994, Lopez et al. 2003, Marine Mammal Commission 2003, Read this volume). In addition to this direct mortality, commercial fisheries may compete for marine mammal prey (e.g., Merrick et al. 1995, Plagányi and Butterworth this volume), thus significantly altering the marine ecosystems upon which these animals depend.

Human activities that may contribute to long-term environmental change (e.g., chemical and noise pollution, bycatch, and commercial fishing) are not specifically treated in this chapter. Other chapters in this volume address habitat transformation (Ragen), contaminants (O'Hara and O'Shea), harmful algal blooms (Van Dolah), and disease (Gulland and Hall). The focus of this chapter is the potential effect of natural environmental variability, including long-term change, on marine mammals.

NATURAL VARIABILITY IN OCEAN ENVIRONMENTS

Most marine mammal species have evolved in an aquatic environment that exhibits a broad range of variability each year (seasonal cycles) and over multiyear to decadal to millennial scales (climatic cycles). Responses to this variability help to shape a species' life history strategy. For example, the long-range seasonal migrations of baleen whales represent a strategy that connects feeding grounds in productive subpolar and polar waters to breeding areas in less productive but comparatively stable temperate and tropical zones (e.g., Clapham 2001). How marine mammals respond to basin-scale, multiyear phenomena, such as El Niño and La Niña, or to multidecade regime shifts is less clear. In part, this is because these oscillatory environmental patterns have been described only recently (e.g., Thompson and Wallace 1998, Thompson et al. 2000, Wallace 2000, Overland and Adams 2001, Schwing et al. 2002) and also because any treatment of ecological response does not usually extend beyond trophic levels occupied by forage fish (e.g., Hunt and Stabeno 2002, Hunt et al. 2002, Chavez et al. 2003). Finally, although climatic processes occurring over longer periods (e.g., Little Ice Age, ca. 1300–1850; Younger Dryas cold period, ca. 11,500 years ago; formation of the Antarctic Circumpolar Current, ca. 25 Mya, in the late Oligocene) have surely influenced the biogeography and natural history of extant marine mammal species (e.g., Fordyce and de Muizon 2001), the temporal scale of these changes (10^3–10^6 years) is generally beyond the reach of conservation management and therefore outside the scope of this chapter.

Seasonal Cycles

Seasonal cycles are most pronounced at the poles, where the annual advance and retreat of ice, over scales of 1,000–10,000 km, dramatically changes the marine environment. The life histories of marine mammals that live near the poles are linked to these cycles, as are those of species that migrate there to feed. Cetaceans are largely excluded from the polar regions during winter, whereas most ice-adapted pinnipeds (and polar bears) remain and adjust their behavior to accommodate the extreme environmental shift. There are exceptions to this generalization, as recently shown in reports of overwintering Antarctic minke whales (*Balaenoptera bonaerensis;* Thiele and Gill 1999) and killer whales (*Orcinus orca;* Gill and Thiele 1997) in the Antarctic and acoustic detection of blue whales (*Balaenoptera musculus*) near the Antarctic Peninsula during the austral winter (Širovic et al. 2004). Conversely, cetaceans converge on polar seas in the summer to take advantage of the abundance of prey associated with these productive waters (Murphy 1995, Clarke and Harris 2003) although even then whales are often associated with frontal boundaries related to marginal ice zones (e.g., Kasamatsu et al. 1996, Thiele et al. 2000, 2004).

In subpolar and temperate domains, marine mammals respond to seasonal changes in prey availability, which are linked, via primary production, to seasonal cycles of mixing and upwelling (e.g., Croll et al. 1998, Forney and Barlow 1998, Springer et al. 1999) and zones of prey advection and concentration (e.g., Fiedler et al. 1998, Kenney et al. 2002, Moore et al. 2002a). Many of these productive zones are stable over time, such that species of large whales arrive predictably at specific locations to feed during certain months of the year (e.g., Kenney et al. 1995). Indeed, the migration patterns of baleen and sperm whales (*Physeter macrocephalus*) were well described, and exploited, by commercial whalers for nearly four centuries (e.g., Townsend 1935, Mackintosh 1946). Pinnipeds too may swim thousands of kilometers between breeding rookeries and prime foraging locations; northern elephant seals (*Mirounga angustirostris;* Stewart and DeLong 1995) and northern fur seals (*Callorhinus ursinus*) demonstrate this in the extreme (Loughlin et al. 1999).

Marine mammal responses to seasonal changes in the tropics are largely unknown, except in the eastern tropical Pacific (ETP). The tropical ocean domain is highly stratified. In the ETP, the dominant pattern of seasonal change is the shoaling of a thermocline ridge along 10° N latitude, associated with the north-south movement of a convergence zone between the trade winds (Reilly 1990, Fiedler 2002). Interannual variation exceeds seasonal variation in much of the ETP (Fiedler 1992) and is dominated by the quasi-periodic El Niño–Southern Oscillation (ENSO) cycle (Philander 1989). Reports of marine mammal responses to this variation come from intensive studies on monk seals (*Monachus schauinslandi*) and on four dolphin species associated with the commercial tuna fishery: short-beaked common (*Delphinus delphis*), pantropical spotted (*Stenella attenuata*), striped (*S. coeruleoalba*), and two forms ("whitebelly" and "eastern") of spinner (*S. longirostris*) dolphins (Perrin 1990). Monk seal pups exhibited improved body condition during El Niño years at Laysan Island and French Frigate Shoals, with enhanced survival noted at the latter site

(Antonelis et al. 2003). More subtle shifts in habitats for the four dolphin species were investigated using ordination statistics: common dolphins were found to be associated with cool, upwelling habitat, whereas both spinner forms were associated with warm tropical habitat, with the eastern form being more abundant toward the coast where the thermocline was shallowest (Fiedler and Reilly 1994, Reilly and Fiedler 1994).

In addition, the recent description of blue whale occurrence in the ETP, based on call detections, showed that whales arrive in tropical waters in winter and depart (or stop calling) in summer (Stafford et al. 2001, 2004). Notably, call types associated with blue whale populations in the Antarctic and both the northeastern and northwestern Pacific were recorded in the ETP. Although these examples coincide with cyclic climate change in the Tropics, they can show only a glimpse of how marine mammals might respond to long-term environmental change.

Decadal Cycles

Basin-scale climatic cycles with multiyear and multidecadal periods have been recognized for only about the past 20 years. In most cases, description of biological shifts preceded investigation of the associated physical forcing mechanisms (e.g., Mantua et al. 1997, Hare and Mantua 2000, Minobe 2002, Schwing et al. 2002). Now, climatologists and many marine biologists routinely refer to "regime shifts," whereby ocean ecosystems suddenly change location or structure. Examples include the multidecadal shift between a (cold) "anchovy regime" and a (warm) "sardine regime" in the northeastern Pacific (Chavez et al. 2003) and the shift from a "shrimp regime" to a "pollock regime" in the Gulf of Alaska (Andersen and Piatt 1999, Overland et al. 2004). Although the biological mechanisms responsible for such environmental change remain obscure, some argue that they must be relatively simple and direct to result in such wholesale shifts in dominant midtrophic species (Bakun 2001).

Hypotheses explaining marine mammal responses to the multidecadal regime shifts remain largely speculative (e.g., Frances et al. 1998), generally owing to the lack of data that extend over the period of environmental fluctuation. Studies of seabird population variability in response to climate-related changes in sea-ice extent in the sub-Arctic (e.g., Springer 1998) and the Antarctic Peninsula region (e.g., Croxall et al. 2002) are, by comparison, more empirically grounded.

Responses of marine mammals to the multiyear ENSO phenomena, also called El Niño (warm phase) and La Niña (cold phase), are better documented. In the North Pacific, El Niño conditions are initiated by an oscillation in atmospheric pressure cells that leads to a relaxation of trade winds. This allows warm water to cross the basin from west to east, effectively deepening the thermocline and "capping" the upwelling required for primary production (Philander 1989, Chang et al. 2004). The result is a marked downturn in productivity and survival at all trophic levels. These multiyear events, predicted and documented most recently in 1997–1998 (Schwing et al. 2002, Chavez et al. 2003), lead to dramatic declines in pinniped reproduction and survival and seem associated with declines in blue whale occurrence in feeding areas off California, as measured acoustically (Burtenshaw et al. 2004). Further, El Niño activity in the equatorial Pacific has been implicated in the initiation of the Antarctic circumpolar wave (ACW), a coherent pattern of sea level pressure, wind stress, sea-surface temperature (SST), and sea-ice extent that propagates eastward (anticyclonically) taking 8 to 10 years to make a circuit around Antarctica (White and Peterson 1996, Gloersen and White 2001, Liu et al. 2002a,b). The associated cyclic changes in sea-ice extent and associated productivity, in turn, can affect Antarctic seabird populations (e.g., Fraser and Hofmann 2003) as noted previously.

GLOBAL WARMING

Evidence of global warming now comes from a variety of disciplines. Although the proportion of warming caused by human activities remains an area of active debate (e.g., Alley et al. 2003), the warming itself is rapidly changing the marine environment, especially at the poles (e.g., Comiso 2002, National Research Council 2002, Clarke and Harris 2003, Johannessen et al. 2004). While extreme ice retreats and ice thinning make the news (e.g., Comiso and Parkinson 2004, Yu et al. 2004), most articles dealing with these observations relate the findings to underlying atmospheric forcing and paleoclimate records, with scant focus on ecological ramifications. To approach the topic of long-term effects of global warming on marine mammals, several complexities must be considered. First, warming is not uniform, even at the poles. For example, the western Arctic is experiencing extreme seasonal retreats (to 300 km offshore, Fig. 9.3) and reduction (–9% per decade, 1978–2000) in sea ice (Comiso and Parkinson 2004, Rigor and Wallace 2004), whereas the eastern Arctic has experienced ice advances (Stern and Heide-Jorgensen 2003) and lower-than-average temperatures over the past 10 years. In the Antarctic, warming is most acute along the peninsula and in the Weddell and Ross seas, where huge ice sheets are breaking up (Rind et al. 2001), even in the face of an overall increase in the length of the sea-ice seasons over the past 20 years (Parkinson 2001). These regional responses to global warming have effects at all trophic levels in the polar ecosystems.

Various scenarios have been advanced as to how global warming will affect the planet (Intergovernmental Panel on Climate Change 2001, National Research Council 2002, Arctic Climate Impact Assessment 2004, Walsh et al. 2004). A rising sea level accompanied by an increase in the severity and number of storms will no doubt take a toll on human coastal communities. But what about marine mammals? One possibility is an improvement in prey availability as a result of enhanced productivity supported by a decrease in sea ice and/or storm-induced ocean mixing followed by stratification. Early departure of sea ice has been posited to control zooplankton availability in the southeastern Bering Sea (Hunt et al. 2002) and in the Cape Bathurst polynya in the southeastern Canadian Beaufort Sea (Arrigo and van Dijken 2004), with maximum secondary production associated with late phytoplankton blooms in insolation-stratified open water.

Climate-change scenarios that include increasing storm events suggest a possible increase in nutrient availability in the upper ocean resulting from a deepened mixed layer in temperate and subpolar domains. If the mixed ocean then stratifies sufficiently to support a production cycle, this long-term change in the environment could enhance primary and secondary production, that is, generate more food for many marine mammals. Conversely, a warming ocean *without* storms can become more stratified, with a "capped" thermocline resulting in diminished production, as occurs during El Niño conditions along the eastern North Pacific. As in warming at the poles, these effects likely will not be uniform, with some areas of the ocean experiencing enhanced, and other areas diminished, productivity.

The effects of global warming will be felt most immediately by marine mammals in polar regions (Tynan and DeMaster 1997, Clarke and Harris 2003). For example, the extreme retreats of sea ice documented since the late 1990s in the western Arctic (Fig. 9.3), pose a challenge to ice-obligate marine mammals. Polar bears (*Ursus maritimus*), walruses (*Odobenus rosmarus*), and ice seals are the species most immediately at risk from these changes, as they require sea-ice habitat to hunt, haul out, rest, and give birth. Negative effects on polar bears have been documented in the Hudson Bay area, where sea ice has rapidly diminished, thereby disrupting the bears' annual hunting cycle (Stirling 2002, Derocher et al. 2004). Similarly, extreme ice retreats in the Chukchi Sea negatively affect walruses, as they must swim from sea-ice haul-outs to the productive shelf waters to feed. The Arctic ice seals, including ringed (*Pusa hispida*), ribbon (*Histriophoca fasciata*), bearded (*Erignathus barbatus*), and spotted (*Phoca largha*) seals and the Antarctic Weddell (*Leptochynotes weddellii*), Ross (*Ommatophoca rossii*), crabeater (*Lobodon carcinophaga*), and leopard (*Hydrurga leptonyx*) seals will similarly be affected by a loss or change in the quality of sea ice, with intensity of effect scaled to their dependence on ice for resting and reproduction.

Potential effects of global warming on cetaceans at the poles are harder to gauge although examination of Antarctic commercial whaling data suggests clear responses to variable ice conditions (de la Mare 1997). Extreme ice retreats in the western Arctic may open new foraging habitat for gray whales (*Eschrichtius robustus;* Moore et al. 2003) and potentially for the highly ice-adapted bowhead whale (*Balaena*

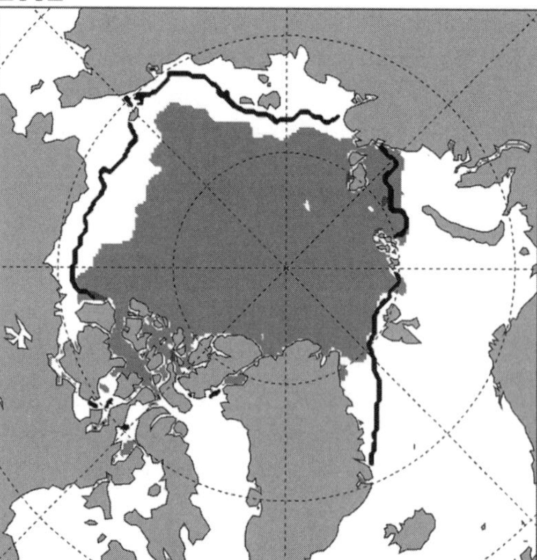

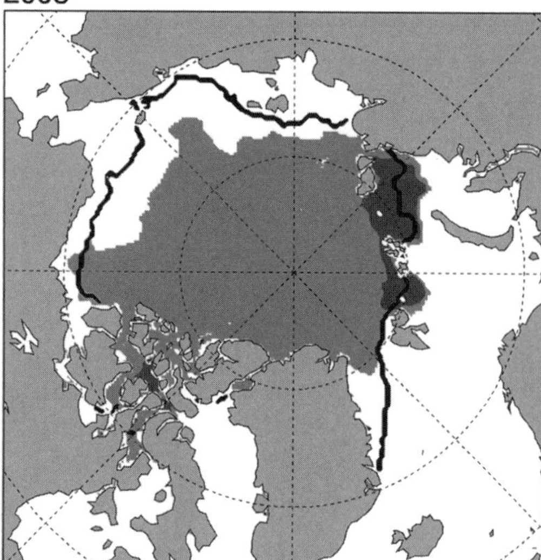

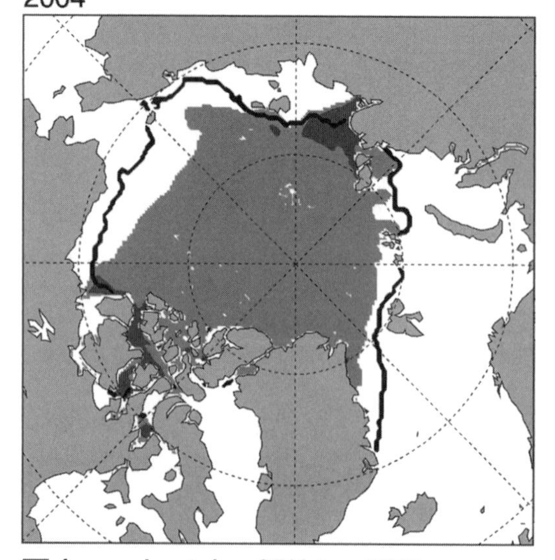

Fig. 9.3. Example of extreme Arctic sea-ice retreat in September 2002, 2003, and 2004 (see also Serreze et al. 2003). White areas are open water; shading of sea-ice changes is approximate. See also Sereeze et al. (2003) and Comiso and Parkinson (2004). (Image modified, with permission, from a color version provided by the National Snow and Ice Data Center.)

mysticetus; Moore 2000, Moore unpubl. data). Perryman et al. (2002) reported that fluctuations in gray whale calf production from 1994 to 2000 were positively correlated with the length of time that the primary feeding habitat was free of seasonal ice during the year prior to parturition. Thus, extreme ice retreats may extend feeding seasons for gray whales and thereby enhance recruitment in that species. Beluga whales (*Delphinapterus leucas*) and narwhals (*Monodon monoceros*) feed primarily on fish and crustaceans, and the degree to which these prey rely on ice-borne communities will help determine the nature and extent of the effects on the whales. At least one development that will have a negative impact on all Arctic cetaceans is the increase in underwater noise and the potential for ship strikes that will accompany increased commercial vessel traffic, anticipated with the opening of the Northwest Passage.

In the Antarctic, the dynamic relationship between sea ice and krill (*Euphausia superba*) will influence the effects of warming on prey availability for whales and seals. Krill density is positively correlated with extensive sea ice (Loeb et al. 1997), and years with low sea ice can be linked to downward fluctuations in predator populations (Croxall et al. 2002). For example, offshore the Antarctic Peninsula, years of extreme ice extent have been associated with high krill biomass and "good" years for penguins and seals (e.g., Trivelpiece and Trivelpiece 1998, Fraser and Hofmann 2003). As sea-ice habitat appears important to minke and blue whales (Kasamatsu et al. 1996, Thiele et al. 2000, 2004, Murase et al. 2002), their foraging opportunities may decrease with a loss of Antarctic sea ice. The association of ice, krill, and whales is complex (e.g., Nicol et al. 2000, Brierly et al. 2002) and was a central research theme in the recent large-scale multidisciplinary Southern Ocean GLOBEC studies (Širovic et al. 2004, Thiele et al. 2004).

MEASURING RESPONSES TO LONG-TERM ENVIRONMENTAL CHANGE: RESEARCH FOCI AND TOOLS

As noted at the outset, it is often difficult to judge the response of marine mammals to long-term environmental change because of the mismatch in scale between scientific research (months to decades, in localized areas) and marine mammal natural history (decades to centuries, often over large areas). In addition to being narrow in focus, research is often directed solely at marine mammal responses to anthropogenic sources of environmental change, with little or no regard to their "baseline" response to natural cycles of variability. Here I provide a brief overview of current research foci and tools, followed by two case-study summaries to elucidate attempts to measure marine mammal responses to environmental variability.

In the United States, research on marine mammals is largely framed by two statutes: (1) the Marine Mammal Protection Act (MMPA), passed into law in 1972 and revised in 1986 and 1994; and (2) the Endangered Species Act (ESA), which became law in 1973 (see Ragen this volume). Among other provisions, these two laws mandate reduced mortality and harassment of marine mammals incidental to human activities and the protection and conservation of marine mammal habitat. The mandate for habitat conservation provides the key link to research on marine mammal response to long-term environmental change. However, there has been little emphasis on determining what constitutes important or "critical habitat" for marine mammals because managers must devote the majority of available resources to surveys that support abundance estimates to provide a basis for determining allowable removals and incidental harassment.

Current research tools include visual censuses and surveys, acoustic censuses and tracking, photoidentification, biopsy-tissue sampling for genetic, pollutant, and (to some extent) diet analyses, radio and satellite tagging, and remote sensing. Various statistical analyses are applied to these data; analyses related to habitat evaluation are arguably the most poorly developed. In contrast, methods to estimate abundance of cetaceans using distance-sampling techniques applied to visual sighting-survey data and to model trends in abundance and population dynamics from these estimates have received considerable attention. Unfortunately, such analyses do not incorporate environmental parameters and, therefore, measurement or prediction of marine mammal responses to environmental change is not currently part of routine population assessment. Initial efforts to incorporate environmental parameters with visual sighting and whaling data using general additive models (e.g., Forney 2000, Gregr and Trites 2001) have met with limited success, perhaps in part because initial sampling designs did not take predictive statistical techniques into account. Plans are now under way within the National Marine Fisheries Service to integrate environmental variables to improve stock assessment methods in U.S. waters.

CASE STUDIES

The Intergovernmental Panel on Climate Change (2001) identified three large-scale impacts of global warming on the oceans: (1) increases in sea level and sea-surface temperature; (2) decreases in sea-ice cover; and (3) changes in salinity, alkalinity, wave climate, and ocean circulation. Unfortunately, as noted previously, sampling designs for most studies of marine mammals do not include an investigation of responses to environmental change or variability. Most studies are either too short or are too limited spatially and so do not encompass ecological scale as given by a species' natural history. Where responses have been explored, analyses have largely been post hoc, with results generally limited to correlations that do not establish cause and effect. Cases where habitat reduction can be measured provide clearer examples than those where a change in ecosystem structure can only be inferred. Two case studies are provided here representative of habitat reduction (polar bear) and habitat alteration (gray whales), both likely coincident with change in overall ecosystem structure.

Habitat Reduction: Polar Bears

The reduction and loss of habitat required by marine mammals are difficult to measure. For example, although a rise in sea level may contribute to a loss of haul-out habitat for the endangered Hawaiian monk seals on French Frigate Shoals in the northwestern Hawaiian Islands, changes are also influenced by variation in storm frequency, coral growth rate, and prevailing ocean currents (Antonelis et al. in press). The case of the habitat reduction associated with the loss of Arctic sea ice provides a somewhat clearer example, as summarized in what follows.

Rapid and dramatic reduction in Arctic sea-ice extent (Fig. 9.3) and thickness have been reported over the past two decades, with predictions of roughly a 20% reduction in annual ice extent by 2050 (Vinnikov et al. 1999, Parkinson and Cavalieri 2002, Comiso 2002, Serreze et al. 2003). Polar bears range across the circumpolar Arctic and rely on sea ice as a substrate for hunting seals, their primary prey. Possible effects of temporal and spatial reduction of sea ice on polar bears were first presented by Stirling and Derocher (1993), followed by a more thorough treatment by Derocher et al. (2004). In brief, the life history of polar bears depends largely on acquiring and storing body fat when hunting conditions are good and then using these stores during periods of low prey availability. Long-term studies of polar bears in the Canadian Arctic have documented substantial variation in body size and reproductive capacity that can be linked to interannual variation in ice conditions (Stirling 2002). Specifically, the breakup of annual sea ice in the western Hudson Bay is now occurring roughly 2.5 weeks earlier than it did 30 years ago. This shortens the amount of time that bears can hunt seals on the ice so they are forced to come ashore earlier with less body fat and to fast until sea ice reforms.

Declines in reproductive rate, subadult survival, and body mass are associated with periods of early ice breakup and late freeze-up. In addition to limiting hunting opportunities, a reduction in sea-ice habitat may limit female bears' access to dens and the overall movements of bears in their home ranges (Derocher et al. 2004). Each of these responses to habitat reduction are detrimental to the species. However, because changes in Arctic sea ice are regional and highly variable and the estimated size for sixteen of the twenty populations is poorly known, effects of sea-ice-habitat loss on circumpolar polar bear demographics are difficult to track.

Habitat Alteration: Eastern North Pacific Gray Whales

Migratory species may be particularly vulnerable to climate change as they must contend with ecosystem impacts on their winter and summer grounds and along the migration route (Ahola et al. 2004). The eastern North Pacific (ENP) gray whale population now numbers roughly 18,000 (Rugh et al. 2004), having recovered from a low of 3,000–5,000 individuals at the time that commercial whaling came to an end. This recovery, documented by a series of counts conducted over three decades, resulted in the removal of the population from the U.S. list of endangered species in 1994, the only such action taken to date for a marine mammal (Rugh et al. 1999). In 1999 and 2000, ENP gray whales stranded dead, or moribund, in unprecedented numbers (Gulland et al. 2004). Even with considerable effort devoted to sampling the carcasses, the causes of most of the deaths could not be determined. A downturn in prey availability in the northern Bering Sea that was due to ocean warming associated with the 1997–1998 El Niño was suggested as a cause (LeBoeuf et al. 2000), but because gray whales feed on a broad suite of prey over an extensive geographic range (Nerini 1984), this explanation seemed simplistic (Moore et al. 2001).

Retrospective analyses of abundance estimates suggest that the ENP gray whale population was approaching carrying capacity by the late 1980s (P. Wade pers. comm.). If so, and if the population remained near carrying capacity through the late 1990s, a sudden decline in marine ecosystem productivity caused by the 1997–1998 El Niño could have contributed to the whale mortality. A drop in ENP gray whale abundance estimates from a high of 27,958 (CV = 0.1) for the 1997–1998 migration, to 18,246 (CV = 0.09) for the 2000–2001 season, and to 16,848 (CV = 0.09) for the 2001–2002 season (Rugh et al. 2004) supports this view. Further, an assessment of gray whale body condition from photographs showed that the whales were thinner in 1999 and 2000, and calf counts were a third of those in the five years immediately prior to 1999 (Perryman et al. 2002). Only 21 gray whale strandings were reported in 2001 (Gulland et al. 2004), and by 2002 and 2003 calf counts had rebounded and northbound adult whales were fatter (W. Perryman pers. comm.). In combination, the evidence seems compelling that the 2-year gray whale "mortality event" was a response to the 1997–1998 El Niño warming.

Nevertheless, some questions posed by Moore et al. (2001) about the high ENP gray whale mortality remain unanswered. For example, what can be inferred from the mix of seemingly fat and emaciated whales that stranded in 1999–2000? Did disease contribute to the die off? And finally, if the die-off was a response to a decline in prey in the Chirikov Basin (LeBoeuf et al. 2000), why did the population not respond this way in earlier years (i.e., to the 30% loss of prey biomass reported in the late 1980s) (Highsmith and Coyle 1992)? Gray whales feed on a variety of prey and likely respond to local prey depletions by moving to alternate areas. Although alternate feeding areas are no doubt finite, provisional surveys of waters southeast of Kodiak Island, in the Gulf of Alaska, and north of the Bering Strait, in the south-central Chukchi Sea, suggest that whales are now feeding in these areas in larger aggregations than in the Chirikov Basin (Moore et al. 2003). Although the capability of gray whales to find and exploit concentrations of prey is supported by these observations and by their remarkable recovery, the key elements needed to establish causal relationships between climate change and gray whale population response are long-term measures of prey response to environmental variability and habitat alteration.

CUMULATIVE EFFECTS OF ENVIRONMENTAL CHANGE

In addition to climate warming, marine mammals must contend with multiple anthropogenic sources of long-term environmental change, including pollution (chemical and noise) and overfishing. The cumulative effect of all these factors on marine mammals is unknown, but each examined separately suggests that these animals must contend with an increasingly stressed ecosystem. For example, humans continue to dump chemicals into the atmosphere and ocean at an astonishing rate (Arctic Monitoring and Assessment Programme 1998, Reijnders et al. 1999), many of which are retained in the ecosystem for decades to centuries (see O'Hara and O'Shea this volume). Pollutant runoff to rivensheds results in especially high levels of chemicals in many estuarine and coastal habitats, sometimes reaching crisis proportions, as during harmful algal blooms (HABs; see Van Dolah this volume). Airborne pollutants can precipitate out thousands of miles from the source, as shown in the transport of polychlorinated biphenyls (PCBs) from Asia to the Arctic (Arctic Monitoring and Assessment Programme 1998). Such global-scale transport of these lipophilic chemicals is demonstrated by their entrainment in the blubber of beluga whales feeding in the remote Arctic basin, which have been shown to carry heavier organochlorine (OC) loads than whales that feed in the comparatively industrial area of Cook Inlet, Alaska (Krahn et al. 1999, Becker et al. 2000).

One means of measuring the cumulative effects of environmental change (natural and anthropogenic) on marine ecosystems is called "systemics," which compares fishery removals to calculated requirements by other top trophic consumers in the ecosystem (Fowler and Hobbs 2002, Fowler and Crawford 2004). Overall, fishery removals fall outside a Poisson distribution of natural variation by two orders of magnitude, which suggests that commercial fishing pressure may have a large and long-term effect on marine systems. Although this does not elucidate the complexities or trophic cascades that may be precipitated by fishery removals, it does suggest that human activities have altered, and may now be altering, the ecosystem in ways that will affect marine mammals negatively.

FUTURE RESEARCH

Conservation science requires simple "ecological rules of thumb" to prioritize and focus funding and action (Cote and Reynolds 2002). Addressing marine mammal responses to long-term environmental change requires research focused on predictive modeling of ecosystems, based on data from a suite of species selected for their life history characteristics. Decisions of priority have to be made regarding sources of environmental change and the threats they are perceived to pose. Goals have to be established regarding mitigative measures, and means to follow up, assess, and adjust those goals are needed (i.e., adaptive management). Two important points of departure for this discussion are (1) the importance of ecological scale in planning research and (2) the "false boundary" that currently exists in our thinking between natural and anthropogenic environmental change. Research and management should enhance our predictive capability in regard to marine mammal habitat: what do the animals require in terms of habitat and how do they respond to environmental change? Humans are part of the natural environment, and they induce environmental change that is part of the integrated whole. The ultimate goal should be to reduce the human footprint on marine mammal habitat to the greatest extent possible.

Ecological versus Management Scale

Funding for marine mammal science is often linked to management concerns, and therefore objectives are defined by regulations that are politically driven. Given this situation, science often serves, but does not inform, policy decisions. If concerns regarding the impact of long-term environmental change on marine mammals are ever going to be addressed in a serious or conclusive manner, it will require research across a range of temporal and spatial scales, most of them exceeding the 1-year fiscal horizon of federal budgets. In other words, marine mammal science has to be conducted at appropriate ecological scales, not shaped to fit regulatory and budget cycles. To achieve that, clear objectives have to be established at the outset of multiyear research programs, and there have to be mechanisms for assessing progress toward those objectives and for making appropriate adjustments to the research plan when necessary.

Natural versus Anthropogenic Variability

Current research focuses almost exclusively on responses of marine mammals to anthropogenic environmental change. In part, this is because human actions can be regulated and ecosystem variability cannot, and, in part, it is because (as the term "anthropogenic" implies) the changes are at the human scale. What is missed by this focus is that the anthropogenic scale of impact on the ecosystem is integrated with broader-scale natural variability. Natural and anthropogenic variability are not independent. They can act synergistically or in opposition, but they represent "one state" to the individual animal and affect (or not) fitness at that level. If we are to measure responses to anthropogenic influences such as global warming, high-intensity sonars, or commercial fishing, we must understand how marine mammals respond to "natural" long-term environmental change, such as regime shifts and ENSO. Multiresolution sampling designs, which nest fine-scale experimental field studies within information about broad-scale processes (e.g., Hewitt et al. 2002, Wang et al. 2002), should be considered to address linkages between marine mammal responses to anthropogenic and natural variability.

RESEARCH RECOMMENDATIONS

To be proactive with regard to the potential impact of environmental change on marine mammals, proposed research must approach questions more broadly (e.g., Cote and Reynolds 2002). Research resources (funding and access) will always limit what can be investigated, but if a broad plan is set forth at the outset, much can be achieved over time. Overarching themes to improve research include definition of focus (threats and plan of study), improvement of sampling and analytical tools, and integration of marine mammal research into broad-scale, multidisciplinary research programs with nested fine-scale sampling.

Define Focus

Research should have an ecological focus, with a clear definition of anticipated types and causes of environmental change. For example, as it is widely agreed that the Earth is warming and that the effects of warming are amplified at the poles, expanded long-term studies of marine mammals in the Arctic and Antarctic are clearly needed. Because all species cannot be studied, effort should focus on those that can be sampled most reliably and, if possible, where a time series of data already exists. In the Arctic, candidate species include beluga and bowhead whales, walruses, ringed seals, and polar bears. Each of these species samples, and reflects, the Arctic ecosystem at a different trophic level. Therefore, if all are studied, data will be available on the effects of warming on marine mammals that reflect different ecological linkages (i.e., trophic cascades) in the system.

In the Arctic, particular care should be taken to include traditional ecological knowledge (TEK) in research planning. Native peoples rely on marine mammals for subsistence and cultural fulfillment, making many of them keen observers. One way to link the two seemingly divergent modes of observation—science on one hand and the direct experience of locals on the other—might be to nest fine-scale sampling (based on traditional knowledge) within broad-scale sampling (based on oceanographic research programs) to address agreed-upon questions of how seals and whales, for example, are responding to environmental change. Similar strategies could be worked out for other threatened marine mammals. There are already examples of successful combinations of TEK and traditional science (e.g., for beluga and bowhead whales in Alaska; beluga whales and narwhals in Canada; dugongs (*Dugong dugon*) in Australia and Thailand, and others), as there is now a growing awareness of the role that indigenous peoples play in conserving wildlife resources (Riedlinger and Berkes 2001, Berkes 2004, www.nativescience.org). Here, I suggest formal adoption of this approach, with the goal being a model of marine mammal responses to environmental variability at nested ecological scales.

Improve Tools

The technology (hardware and software) that supports marine mammal research is bound to improve. In other words, better satellite tags, acoustic recorders, GIS mapping, and other computer-based instrumentation will become available, as will more powerful analytical techniques for genetic investigations and the complex chemistry required for pollutant and fatty acid assays. Perhaps even more pressing than these types of advances, however, is the need for improved planning tools, most importantly conceptual, explanatory, and predictive models of marine mammal ecology (Burnham and Anderson 1998). To date, research has focused almost exclusively on "explanatory" models. For example, numerous papers from many parts of the world (e.g., Weir et al. 2001) show that habitat for marine mammals is related to features that enhance productivity or entrain prey, such as bathymetry, topographic complexity, sea-surface temperature (SST), primary production (chlorophyll-a), and eddies. Such insights should make it possible to: (1) integrate marine mammals into conceptual models of the marine ecosystem, and (2) apply predictive modeling techniques to assess patterns of marine mammal occurrence and response to environmental variability. Predictive modeling tools, such as general linear models (GLMs) and general additive models (GAMs) (Guisan et al. 2002, Lehmann et al. 2002), offer great promise, but they have yet to be fully developed for marine systems. A recent study using a GAM to investigate linkages between yellowfin tuna distribution and oceanographic variables in the Atlantic, based on catch and bycatch data, demonstrated the effectiveness of this analytical tool (Maury et al. 2001). Although application of GAMs (and GLMs) is in its infancy with regard to marine mammals (e.g., Forney 2000), these tools are well suited to investigating and comparing the roles of density and climatic variation in the population dynamics of large, long-lived mammals (e.g., Coulson et al. 2000).

Promote Integration

To understand the roles of marine mammals in marine ecosystems, sampling must be integrated within broad-scale, multidisciplinary oceanographic studies. For example, the National Science Foundation (NSF) has supported long-term environmental research (LTER) at specific sites for the past 20 to 30 years. One such LTER site is located at Palmer Station offshore the Antarctic Peninsula. Although marine mammal sightings in this area have been recorded during surveys on an ad hoc basis, they have never been fully integrated into the study. It would be simple to formally add this analysis to the ongoing study, and it should be included in the planning of other such long-term observations sites (e.g., the Bering Strait long-term observatory). Further, marine mammal sampling can be easily integrated into large-scale oceanographic programs. A current example is the inclusion of acoustic and visual sampling of cetaceans in the GLOBEC program (Širovic et al. 2004, Thiele et al. 2004).

A similar approach should be explored to formally integrate TEK, where it is available, into marine mammal sampling designs. For example, interviews with subsistence hunters can provide insights about marine mammal behavioral ecology that are not obtainable through reliance on standard survey sampling designs (e.g., Huntington 2000, Riedlinger and Berkes 2001). Regular dialogue with Native subsistence hunters should be sought, for insights into both changing physical habitats and animal condition and health. Sampling at integrative scales can be achieved (e.g., Hewitt et al. 2002) and will strengthen our understanding of marine mammal behavioral ecology and responses to environmental variability.

SUMMARY

In summary, research on marine mammal responses to long-term environmental change must extend across a range of spatial and temporal scales. Interdisciplinary research, including the use of TEK, is required and should be guided by a predictive modeling framework. This type of research and associated adaptive management can be accomplished only with reliable multiyear financial support. Most urgently, long-term studies should be initiated that capture existing conditions in polar regions, where climate warming is ongoing and is predicted to be amplified in the near future. For example, in the Arctic we should: (1) establish baselines for marine mammal population abundance and trends, (2) investigate the effects of ice retreat and warming along the trophic pathway leading to marine mammal prey, and (3) measure chemical contaminant and underwater noise levels in advance of environmental changes associated with anticipated increases in commercial vessel traffic and fishing. In the Antarctic, where there are much longer marine mammal time series, but where commercial fishing and tourism are already ongoing, we should focus on development of predictive models to investigate the ecological role of marine mammals in that highly productive ecosystem. Where feasible, these polar-focused studies should integrate historical data and conduct retrospective analyses to extend temporal assessments as far as possible; assessment of ambient noise in the Arctic via analyses of military and industry-held datasets is one example of this approach. Lakes, rivers, estuaries, and coastal regions are also undergoing rapid change because of warming and should be an immediate focus in order to capture marine mammal responses to habitat alterations, especially as populations in these areas are often very isolated.

Further development of tools (e.g., passive acoustics, satellite tagging, and remote sensing) that extend focal-animal and habitat sampling capability over space and time is essential. Finally, and most important, the planning of long-term marine mammal research should begin with integration of purpose among scientific disciplines, including TEK. The goal of such planning should be development of testable hypotheses that can guide adaptive management and support species conservation. Predictive models can provide a framework for sampling design and hypothesis testing, at nested scales, from the combined multidisciplinary data. Ultimately, the outcome of any research effort directed at determining marine mammal responses to long-term environmental change will be successful only if clear goals can be agreed upon, the complexities of multiple factors can be integrated, and multiyear support is made available.

ACKNOWLEGMENTS

Lloyd Lowry provided constructive comments and discussion that contributed to the final version of this chapter, while John Hildebrand provided guidance on an earlier version of the manuscript. Karna McKinney provided excellent graphics support, and the editing skills of Marcia Muto, Bill Perrin, and Randy Reeves further improved the manuscript.

BARBARA L. TAYLOR

10

Identifying Units to Conserve

THREE HUMAN GENERATIONS AGO, my great-grandparents paused in their westward move while herds of migrating buffalo crossed the Missouri River in front of their steamboat. Long before the birth of my father, the migratory herds of the North American Great Plains were a distant memory, and the plains ecosystems had been forever changed. Similar hunting zeal resulted in the near-extermination of the great migrating "herds" of whales in my lifetime. We humans now choose what we will conserve into the future. We choose whether the blue whale species will be represented by multiple healthy populations throughout its former range or by a single remnant population. The question of how to define "units to conserve" is not a question of how to define a species, but rather how to define a vision of the future.

In the United States our vision for natural resource management is implemented through various treaties, laws, and regulations in which we assume that discrete conservation units can be delineated from animal and plant populations that are by their nature continuous. Where do we draw the line? By analogy, where do we draw the line in saying that someone is "family"? It is easy to say that your sibling or parent is family, but what about a second cousin or someone who shares the same last name but comes from a different country? Drawing the categorical boundaries is often difficult even though we recognize that we are members of a hierarchy of relatedness ranging from immediate family to our racial makeup to being a human and even to being a primate and a mammal. Generally, as distance within the hierarchy increases, it is increasingly easy to place the categorical boundary unambiguously. For example, telling a vertebrate from an invertebrate is simple, whereas telling human races apart genetically is not. From an evolutionary perspective, the strength of the categorical boundary increases with the depth of the evolutionary branch in the tree of life.

Management of animals and plants usually occurs at or below the species level. Scientists still have some difficulty defining and identifying species, more difficulty at the subspecies level, and much more at yet lower levels. Despite the difficulties of defining units at lower levels, properly categorizing units is necessary if we want to be able to implement our vision of the future and reach our objectives. Success depends on three steps: (1) defining "units" that match ideals, which is largely a policy/societal choice; (2) identifying the units in nature, which is a scientific undertaking; and (3) making necessary management decisions and taking action as needed. Although these steps seem discrete, there is a good deal of necessary iteration among them to get the full process operating effectively. Nonetheless, scientific identification of appropriate conservation units is essential for achieving society's long-term objectives and vision. For example, and as described later in this chapter, failure to recognize the "southern resident" population of killer whales off the U.S. West Coast as a distinct population segment likely would have reduced management efforts to bring about its recovery and increased its chance of extinction. Although I am concerned primarily with the science of identifying such units, I begin by reviewing the existing hierarchy of units to conserve both in the United States and at international levels for marine mammals.

Both U.S. and international laws and treaties have been influenced by past excessive human exploitation of marine mammals. Eight of the species originally listed under the U.S. Endangered Species Act (ESA) were large whales. Despite decades of protection, many parts of the ocean remain devoid of certain whale species that once were found there. Thus, even though the oceans may appear to us to have no barriers to movement, the animals that live there do perceive barriers. In fact, recent research using photographic identification, branding, radio and satellite-tagging, and genetic methods have indicated that most marine mammals have a very distinct population structure, which may be most evident in their patterns of movement. Some, like humpback whales, appear not dissimilar to migratory herds of caribou with distinct summering and wintering grounds to which they return faithfully. Others, like harbor seals, roam when they are young but usually settle down as adults near their birth site. Should our vision be to maintain healthy numbers of marine mammals throughout their range, scientists and managers have to understand their population structure in order to manage human impacts such as hunting or incidental kills in fisheries. Similarly, if we want to understand the magnitude of the effects of pollutants from a particular point source on marine mammals in the vicinity, that "vicinity" must be described for different species, and it will clearly differ for resident species compared to migratory species.

Conservation science provides the tools and knowledge needed to implement society's vision of the future. A vision of maintaining a healthy ecosystem complete with top predators, such as Steller sea lions, killer whales, and sea otters, requires using science to understand how these populations are structured. I discuss three hierarchical levels of structural units: (1) species, (2) subspecies and distinct population segments (DPSs), and (3) stocks (Fig. 10.1). The scientific literature uses many terms to refer to different levels of structure, and I briefly review these and how they fit into the three levels discussed here. I refer to all the terms that embody structure at different levels as "units to conserve" (UTCs). All these levels have been deemed worthy of conservation in different national and international laws although some laws, such as the ESA, focus on conservation primarily at the species and subspecies levels. All these terms are

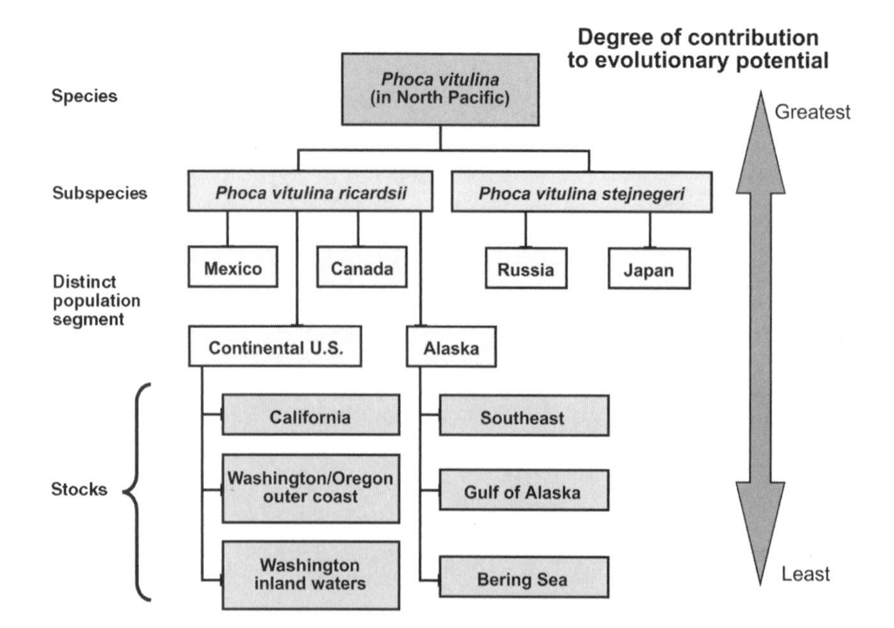

Fig. 10.1. Hypothetical schematic of the hierarchical relationships of different units to conserve, using harbor seals as an example. As DPSs have not been defined for harbor seals, this example is hypothetical and draws DPS boundaries solely according to the criterion of national boundaries. The stocks shown are those in 2003 stock assessment reports. For simplicity, only the North Pacific is shown.

explained in detail later in the chapter, but, briefly, units at the species and subspecies levels are on different evolutionary trajectories and therefore represent important evolutionary potential. Stocks are units whose population dynamics are essentially independent of neighboring stocks (called demographically independent) but *may or may not* represent important evolutionary potential. Thus, units shown in the hierarchy in Figure 10.1 are ordered from greatest (species) to least (stocks) in terms of the contribution they make to evolutionary potential.

The purpose of this chapter is to review conservation unit concepts, consider how these concepts pertain to marine mammal conservation, and suggest where research focus is most needed. I illustrate concepts and problems associated with identifying UTCs with case studies on killer whales and harbor seals. These cases indicate the most important research and management needs regarding UTCs:

- Adequate recognition and treatment of the scientific uncertainties involved.
- Corresponding development and testing of analytical tools to define stocks in a probabilistic manner.
- Development of working UTC definitions that incorporate scientific uncertainty in a precautionary manner.

UNITS TO CONSERVE: SPECIES, SUBSPECIES, AND DISTINCT POPULATION SEGMENTS

Laws and treaties make an implicit assumption that scientists have already properly defined species and subspecies. For example, the ESA defines species [Section 3(15)] as "any subspecies of fish or wildlife or plants, and any distinct population segment of any species of vertebrate fish or wildlife which interbreeds when mature." Similarly, the Convention on the International Trade of Endangered Species of Wild Fauna and Flora (CITES) and the Red List, developed by IUCN–TheWorld Conservation Union, assume accurate definitions. Clearly, mistakes in implementing these laws and treaties are likely if species and subspecies remain unidentified. Nevertheless, scientists are still identifying new species and subspecies of cetaceans, including large whales. For example, northern right whales, listed in the original ESA, have recently been split into two species, North Atlantic right whales and North Pacific right whales (Rosenbaum et al. 2000), based on genetic data. This departs from previous general practice in taxonomy (the naming of species and subspecies), which has used primarily morphology, especially skull measurements, to designate species.

Designation of a species without corroborating morphological data, as for the right whale case, remains controversial among taxonomists. This instance demonstrates why traditional taxonomy lags behind and may not be appropriate for species designations involving marine mammals: adequate skull collections are not, and often will not become, available. To promote more timely progress in cetacean taxonomy, a workshop was held in April 2004 to develop and agree on new definitions and criteria for species and subspecies (Reeves et al. 2004). Workshop participants agreed that (1) multiple species concepts should be acknowledged; (2) two congruent lines of evidence should be required to define a species, which would ideally incorporate both morphological and genetic data but could be based on two independent genetic markers; (3) only a single line of appropriate evidence should be required to define a subspecies and this may be a genetic marker; and (4) subspecies could be either a geographic form or an incipient species. Because they validate the use of genetic data in taxonomic designations, these guidelines pave the way for more timely resolution of taxonomic uncertainty in many marine species for which substantial skull morphometric datasets are unlikely to be obtained in the foreseeable future.

Species (excerpted from Reeves et al. 2004)

Definition: The workshop acknowledged that both major species concepts—the Biological Species Concept (BSC) and the Phylogenetic Species Concept (PSC), as well as their various subapproaches—have merit and should be considered relevant and useful in cetacean taxonomy. It was agreed that the different approaches to species delimitation should be employed in a flexible and pragmatic way, with the basic aim of using proxies to identify irreversible divergence.* Multiple lines of evidence are essential, and what ultimately matters is that a convincing argument is provided demonstrating irreversible divergence.

Criteria: Both morphological data and genetic data can be taken as proxies for reproductive isolation and irreversible divergence. It is possible, however, for individual morphological characters to be convergent and for the data from one genetic locus not to reflect phylogenetic history because of homoplasy or natural selection. Therefore, a finding of congruent divergence for each of multiple distinct kinds of data should be taken as strong support for species designations. Such distinct kinds of data could include morphological data together with genetic data or data from multiple independent genetic loci. In the case of morphological data, any phenotypic character is acceptable. Ideal datasets, including both morphological data and data from multiple loci, can provide not only a large amount of information for decisions regarding species, but also the information necessary to assess the uncertainty associated with that decision.

Data on geographical ranges and on behavior (e.g., feeding behavior and vocalizations) can complement morphological and genetic data and serve as useful lines of evidence in species delimitation. Given the difficulties of knowing the degree to which geographical distribution and behavior actually reflect genetic divergence, however, these kinds of data should not be the primary basis of such delimitation.

---

* Although hybridization does occur in cetaceans, these rare events do not lead to separate species merging to become one and hence are consistent with the term irreversible divergence.

Subspecies (excerpted from Reeves et al. 2004)

Definition: A critical distinction between species and subspecies involves the question of reticulation,* or reversibility. In the case of a subspecies, it may not be possible to demonstrate that the population is on an independent evolutionary trajectory without reticulation, while such demonstration is a requirement for species status. Because subspecies (and ESUs†) are on a continuum, it should be no surprise that distinctions are often problematic and require judgments by the investigator as to the strength of a given factor or suite of factors. Thus, the subspecies concept may be construed as broad enough to contain two types of entities: (1) populations that are not quite far enough along the continuum to be judged as species, and (2) populations that should be species but for which not quite enough evidence is yet available to justify their designation as such.

Thus far, cetacean subspecies have been geographical forms that are noticeably different. Therefore, designations have been based on a combination of morphology and distribution. In the context of this workshop, attention was drawn to the potential for bringing genetic evidence, including neutral markers, into the subspecies definition. It was suggested that for many cetacean species, the difficulty of bringing together, over a reasonable timescale, the large, representative series of osteological specimens needed for definitive morphological comparisons is effectively insurmountable. This is true, for example, for many of the elusive, offshore beaked whale species, the wide-ranging killer whales, and all of the large whales. Thus, for taxonomy at the subspecies level to be relevant for conservation, the range of evidence that can be used needs to be broadened to include genetic markers.

Criteria: In addition to the use of morphology to define subspecies, the subspecies concept should be understood to embrace groups of organisms that appear to have been on independent evolutionary trajectories (with minor continuing gene flow), as demonstrated by morphological evidence or at least one line of appropriate genetic evidence. Geographical or behavioral differences can complement morphological and genetic evidence for establishing subspecies. As such, subspecies could be geographical forms or incipient species.

Participants in the cetacean taxonomy workshop also produced a prioritized list of species for which we most need further taxonomic research (Table 10.1), based on both taxonomic uncertainty and conservation status (Table 10.2; Reeves et al. 2004). Most of the seventeen species given highest priority ranking are found in non-U.S. waters, especially coastal dolphins in and near Asia. Proper identification of species and subspecies is critical to international conservation activities such as those undertaken by the International Whaling Commission (IWC), required by the CITES, and in accordance with the Red List maintained by the IUCN. In some instances critically endangered subspecies or even species remain unrecognized and therefore are more vulnerable to human activities. In other instances human activities may be restricted unnecessarily because of inadequate understanding of population structure.

There are actually two types of units that are below the species level but are still likely to be on a different evolutionary path: subspecies and distinct population segments (DPSs). Both units are used to conserve the essential genetic variability for future evolutionary potential. That is, conservation of both units is intended to maintain sufficient evolutionary potential that the species can respond to future environmental challenges through adaptation. The DPSs (defined later in the chapter) are recognized as "species" in the ESA and are equivalent to the evolutionarily significant unit (ESU), an earlier designation that was used primarily for salmon (Dizon et al. 1991, Waples 1991, Moritz 1994). I use only the term DPS, which for the purposes of this discussion is the same as an ESU. The division between subspecies and DPSs is unclear given the new criteria for subspecies for cetaceans, as described previously, but it seems likely that they differ by a matter of time elapsed since separation. For subspecies, the most closely related units have been separated long enough that morphological differences may have accrued. The DPSs, in contrast, are experiencing sufficiently low gene flow that local adaptation *may* occur. A rule of thumb often used to distinguish DPSs, which is also used by the IUCN to define "regional populations" (Gärdenfors et al. 2001) of global species, is that the level of gene flow between or among them is less than one disperser per generation. The number of different terms used (subspecies, ESU, regional population, and DPS) can lead to confusion, but they all indicate units that are important to the evolutionary potential of the species and receive the same level of protection under the various national laws and international treaties as a full species.

Criteria to be used to designate a DPS were established in the joint agreement of the agencies charged with implementing the ESA—the U.S. Fish and Wildlife Service (FWS) and the National Marine Fisheries Service (NMFS). In the "Policy Regarding the Recognition of Distinct Vertebrate Population Segments" (U.S. Fish and Wildlife Service 1996), the FWS and NMFS concluded that DPSs should be determined based on three sequential considerations: (1) the discreteness of the population relative to the rest of the species, (2) the significance of the population segment to the species, and (3) the population segment's conservation status in relation to the ESA's standards for listing (i.e., is the population segment endangered or threatened when treated as if it were a species?).

A population segment of a vertebrate species may be considered discrete if it satisfies one of the following criteria. First, it is markedly separated from other populations of the same taxon as a consequence of physical, physiological, ecological, or behavioral factors. Genetic distinctness has been used as a proxy for the aforementioned factors. Or second, the group is delimited by international governmental

* In this context reticulation pertains to gene flow between two evolutionary lineages.

† "ESUs" are evolutionarily significant units (see Dizon et al. 1991, Waples 1991, and Moritz 1994).

Table 10.1 Cetacean species and subspecies, with priority rankings for taxonomic research

Species (common name)	Species or subspecies (Latin name)	Risk/taxonomic uncertainty	Subspecies (common name)
Bowhead whale—3	*Balaena mysticetus*	(H,L)	
North Atlantic right whale—3	*Eubalaena glacialis*	(H,L)	
North Pacific right whale -3	*Eubalaena japonica*	(H,L)	
Southern right whale—3	*Eubalaena australis*	(M,L)	
Pygmy right whale—3	*Caperea marginata*	(L,M)	
Gray whale—3	*Eschrichtius robustus*	(H,L)	
Blue whale—1	*Balaenoptera musculus*	(M,H)	
	2	(M,M)	Common blue
	2	(M,M)	Pygmy blue whale
	1	(M,H)	Northern Indian Ocean blue whale
	2	(M,M)	Antarctic blue whale
Fin whale	*Balaenoptera physalus*		
	3	(L,M)	Northern Hemisphere fin whale
	2	(M,M)	Southern Hemisphere fin whale
Sei whale	*Balaenoptera borealis*		
	2	(M,M)	Northern Hemisphere sei whale
	2	(M,M)	Southern Hemisphere sei whale
Common Bryde's whale—1	*Balaenoptera brydei*	(M,H)	
Pygmy Bryde's whale—1	*Balaenoptera edeni/omurai?*	(H,H)	
Common minke whale—1 or 2	*Balaenoptera acutorostrata*		
	3	(M,L)	North Atlantic minke whale
	1	(M,H)	North Pacific minke whale
	2	(M,M)	Dwarf-form minke whale
Antarctic minke whale—3	*Balaenoptera bonaerensis*	(L,L)	
Humpback whale—2	*Megaptera novaeangliae*	(M,M)	
Sperm whale—3	*Physeter macrocephalus*	(L,L)	
Pygmy sperm whale—3	*Kogia breviceps*	(L,M)	
Dwarf sperm whale—3	*Kogia sima*	(L,H)	
Amazon River dolphin—1 or 2	*Inia geoffrensis*		
	2	(M,M)	Amazon dolphin
	1	(H,M)	Orinoco dolphin
	1	(H,M)	Bolivian dolphin
Gangetic dolphin—1	*Platanista gangetica*		
	[1]	(H,H)	Ganges dolphin
	[1]	(H,H)	Indus dolphin
Franciscana—2	*Pontoporia blainvillei*	(M,M)	
Yangtze River dolphin (Baiji)—3	*Lipotes vexillifer*	(L,H)	
Baird's beaked whale—2	*Berardius bairdii*	(M,M)	
Arnoux's beaked whale—3	*Berardius arnuxii*	(L,L)	
North Atlantic bottlenose whale—2	*Hyperoodon ampullatus*	(M,M)	
Southern bottlenose whale—3	*Hyperoodon planifrons*	(L,M)	
Indo-Pacific beaked whale—3	*Indopacetus pacificus*	(L,L)	
Shepherd's beaked whale—3	*Tasmacetus shepherdi*	(L,L)	
Cuvier's beaked whale—1	*Ziphius cavirostris*	(H,M)	
Hector's beaked whale—3	*Mesoplodon hectori*	(L,L)	
True's beaked whale—3	*Mesoplodon mirus*	(L,M)	
Gervais' beaked whale—3	*Mesoplodon europaeus*	(L,M)	
Sowerby's beaked whale—3	*Mesoplodon bidens*	(L,L)	
Gray's beaked whale—3	*Mesoplodon grayi*	(L,L)	
Pygmy beaked whale—3	*Mesoplodon peruvianus*	(M,L)	
Andrews' beaked whale—3	*Mesoplodon bowdoini*	(L,L)	
Spade-toothed whale—3	*Mesoplodon traversii*	(L,L)	
Hubbs' beaked whale—2	*Mesoplodon carlhubbsi*	(M,M)	
Ginkgo-toothed beaked whale—3	*Mesoplodon ginkgodens*	(L,M)	
Stejneger's beaked whale—1	*Mesoplodon stejnegeri*	(H,M)	
Layard's beaked (Strap-toothed) whale—3	*Mesoplodon layardii*	(L,L)	
Perrin's beaked whale—3	*Mesoplodon perrini*	(L,L)	
Blainville's beaked whale—1	*Mesoplodon densirostris*	(H,M)	
Narwhal—2	*Monodon monoceros*	(M,M)	
Beluga or white whale—3	*Delphinapterus leucas*	(M,L)	
Finless porpoise—1	*Neophocaena phocaenoides*	(H,H)	
	[1]	(H,H)	Indian Ocean finless porpoise
	[1]	(H,H)	Western Pacific finless porpoise
	[1]	(H,H)	Yangtze River finless porpoise

continued

Table 10.1 continued

Species (common name)	Species or subspecies (Latin name)	Risk/taxonomic uncertainty	Subspecies (common name)
Harbor porpoise	*Phocoena phocoena*		
	2	(M,M)	Black Sea harbor porpoise
	2	(M,M)	North Atlantic harbor porpoise
	2	(M,M)	Eastern North Pacific harbor porpoise
	2	(M,M)	Western North Pacific harbor porpoise
Spectacled porpoise—2	*Phocoena dioptrica*	(M,M)	
Vaquita—3	*Phocoena sinus*	(H,L)	
Burmeister's porpoise—2	*Phocoena spinipinnis*	(M,M)	
Dall's porpoise—2	*Phocoenoides dalli*	(M,M)	
Commerson's dolphin—2	*Cephalorhynchus commersonii*		
	2	(M,M)	South American Commerson's dolphin
	3	(L,M)	Kerguelen Commerson's dolphin
Chilean dolphin—3	*Cephalorhynchus eutropia*	(M,L)	
Heaviside's dolphin—3	*Cephalorhynchus heavisidii*	(M,L)	
Hector's dolphin—3	*Cephalorhynchus hectori*		
	[3]	(M,L)	South Island Hector's dolphin
	[3]	(H,L)	North Island Hector's (Maui's) dolphin
Short-beaked common dolphin—2	*Delphinus delphis*	(M,M)	
Long-beaked common dolphin—2	*Delphinus capensis*		
	[2]	(M,M)	Indo-Pacific common dolphin
	[2]	(M,M)	Long-beaked common dolphin
Pygmy killer whale—3	*Feresa attenuata*	(L,M)	
Short-finned pilot whale—2	*Globicephala macrorhynchus*	(M,M)	
Long-finned pilot whale—3	*Globicephala melas*	(L,M)	
	[3]	(L,L)	North Atlantic long-finned pilot whale
	[3]	(L,M)	Southern Hemisphere long-finned pilot whale
Risso's dolphin—2	*Grampus griseus*	(M,M)	
Fraser's dolphin—2	*Lagenodelphis hosei*	(M,M)	
Atlantic white-sided dolphin—3	*Lagenorhynchus acutus*	(L,L)	
White-beaked dolphin—3	*Lagenorhynchus albirostris*	(L,L)	
Peale's dolphin—3	*Lagenorhynchus australis*	(L,L)	
Hourglass dolphin—3	*Lagenorhynchus cruciger*	(L,M)	
Pacific white-sided dolphin—2	*Lagenorhynchus obliquidens*	(M,M)	
Dusky dolphin—2 or 3	*Lagenorhynchus obscurus*		
	[2]	(M,M)	South American dusky dolphin
	[3]	(L,M)	South African dusky dolphin
	[3]	(L,M)	New Zealand dusky dolphin
Northern right whale dolphin—3	*Lissodelphis borealis*	(L,L)	
Southern right whale dolphin—3	*Lissodelphis peronii*	(L,M)	
Irrawaddy dolphin—1	*Orcaella brevirostris*	(H,H)	
Killer whale—1	*Orcinus orca*	(M,H)	
Melon-headed whale—2	*Peponocephala electra*	(M,M)	
False killer whale—2	*Pseudorca crassidens*	(M,M)	
Tucuxi—1	*Sotalia fluviatilis*		
	[1]	(M,H)	Marine tucuxi
	[1]	(M,H)	Freshwater tucuxi
Atlantic humpbacked dolphin—1	*Sousa teuszii*	(H,M)	
Pacific humpbacked dolphin—1	*Sousa chinensis*	(H,H)	
Pantropical spotted dolphin—2	*Stenella attenuata*	(M,M)	
	3	(M,L)	Eastern Pacific offshore spotted dolphin
	2	(M,M)	Hawaiian spotted dolphin
	3	(M,L)	Eastern Pacific coastal spotted dolphin
Clymene dolphin—3	*Stenella clymene*	(L,M)	
Striped dolphin—2	*Stenella coeruleoalba*	(M,M)	
Atlantic spotted dolphin—2	*Stenella frontalis*	(M,M)	
Spinner dolphin—2	*Stenella longirostris*	(M,M)	
	[2]	(M,M)	Gray's spinner dolphin
	[3]	(M,L)	Eastern spinner dolphin
	[3]	(M,L)	Central American spinner dolphin
	[3]	(M,L)	Dwarf spinner dolphin
Rough-toothed dolphin—2	*Steno bredanensis*	(M,M)	
Common bottlenose dolphin—1	*Tursiops truncatus*	(M,H)	
Indo-Pacific bottlenose dolphin—1	*Tursiops aduncus*	(M,H)	

Note: Priority ranking is as follows: 1, high; 2, medium; 3, low priority (see text and Table 10.2). Species with at least one high-priority subspecies are indicated in bold. Rankings in brackets were inadvertently omitted from Reeves et al. (2004) but are added here using Table 10.2. Subspecies mainly from Rice (1998).

Table 10.2 Basis for integrating taxonomic uncertainty and conservation risk to rank species by importance of taxonomic research in relation to conservation

		Taxonomic uncertainty		
		High	Medium	Low
Conservation risk	High	1	1	3
	Medium	1	2	3
	Low	3	3	3

Note: Priority ranking is as follows: 1, high; 2, medium; 3, low priority (see Table 10.1). Taxonomic uncertainty takes into account distributional discontinuities, especially between ocean basins.

boundaries within which there are differences in control of exploitation, management of habitat, conservation status, or regulatory mechanisms that are significant in light of Section 4(1)(1)(D) of the Act.

A population segment that satisfies at least one of the above criteria for discreteness is looked at in terms of its biological and ecological significance for the species. This consideration may include, but is not limited to, the following: (1) persistence of the discrete population segment in an ecological setting that is unusual or unique for the taxon; (2) evidence that the loss of that segment would result in a significant gap in the range of the taxon; (3) evidence that the segment represents the only surviving natural occurrence of a taxon that may be more abundant as an introduced population outside its historic range; and (4) evidence that the segment differs markedly from other populations of the species in its genetic characteristics. As circumstances are likely to vary considerably from case to case, it is not possible to describe prospectively all the classes of information that might bear on the biological and ecological importance of a discrete population segment.

Finally, if a population segment is discrete and significant (i.e., it is a distinct population segment), its evaluation for endangered or threatened status will be based on the ESA's definition of those terms and a review of the factors enumerated in Section 4(1) of the Act. It may be appropriate to assign different classifications to different distinct population segments of the same vertebrate taxon.

ILLUSTRATIVE CASE OF SPECIES/SUBSPECIES/DPS DEFINITION FOR MARINE MAMMALS

In 2001 NMFS was petitioned to list southern resident killer whales (*Orcinus orca*) as threatened or endangered under the ESA. This group of animals, now numbering around eighty individuals, summers in inland waters near Seattle and Vancouver. Photographic records collected over the past three decades indicate that these individuals interact exclusively with one another. Genetically and behaviorally, they group with a fish-eating killer whale "form" called "resident" killer whales that range from Washington to at least the Kamchatka Peninsula in Russia. After NMFS concluded that an ESA listing might be warranted for the group called southern residents, the agency conducted a status review using a biological review team composed of government scientists with diverse backgrounds.

In the first round of deliberations, the team determined that southern residents met the criterion for "discreteness" under the joint policy on DPSs. However, the determination of "significance" was far more difficult, largely because of issues surrounding killer whale taxonomy (Krahn et al. 2002). Correctly identifying the killer whale taxon proved critical because the criteria used to evaluate "significance" of a DPS are defined relative to other populations within that taxon. The inability of the team to achieve a consensus on "significance" exposed a number of critical issues involving the treatment of uncertainty in defining the unit to conserve.

In the face of uncertainty, the field of taxonomy is evidentiary as opposed to precautionary, which means that, traditionally, a new species is not named without very strong supportive evidence. In cases where data are poor, the burden of proving taxonomic status falls on the scientist, who must gather the data required to make a strong case for naming a species. In the case of killer whales, the typical requirement of examining a large number of adult skulls could result in taxonomic inaction for decades because skulls can only be obtained from rare strandings of these animals. For some forms of killer whales, skulls may never be obtained. For example, a group of killer whales was photographed killing several sperm whales 50 miles off the central California coast (Pitman et al. 2001). These whales not only did not match any photographic identification catalogs in the North Pacific (which primarily catalog whales found in coastal waters) but also had unique scarring from cookie-cutter sharks. These whales have never been seen again. These animals may represent a pelagic form of killer whales, but skulls may never be found because dead animals will likely sink long before washing up on a beach where they can be collected.

A case is being made for multiple species of killer whales in the Antarctic (Pitman and Ensor 2003). These Antarctic whales look quite different and exhibit different dietary specializations. In the North Pacific, the fish-eating resident form differs genetically from the mammal-eating transient form, despite having overlapping ranges where interbreeding is possible. Although lacking the quantity of data normally required to name species, the biological review team did find that the current designation of one global species for killer whales is probably inaccurate. A similar conclusion was later drawn by a group of specialists convened to address killer whale taxonomy (Reeves et al. 2004): "A straw poll within the working group indicated little support for the premise that one or more new species could be described on present evidence. Nevertheless, a majority of participants expressed the opinion that more than one species of killer whale exists and will eventually be described and named."

The types of data that are likely to be available in a timely manner are genetic data and morphological data that do not require dead animals (such as photographs of coloration patterns and simple metrics like the shape of the dorsal fin and perhaps lengths when aerial photogrammetry is possible). Under the new definitions noted previously, these data are acceptable for use in describing new species and subspecies and, consequently, faster progress in killer whale taxonomy is expected.

If a unit of animals is recognized as discrete, determining its biological and ecological significance can be confounded by uncertainty regarding the "geographical range" of the species to which it belongs. What is the range of the unperturbed "resident" or "fish-eating form" of killer whale? Uncertainties concerning "range" fall into two main categories: the knowable and the unknowable. The knowable uncertainties pertain to the current range of resident killer whales, which now extends from Puget Sound to the Kamchatka Peninsula, at least in the summer months. The resident form also may be found farther to the west, perhaps ranging to other parts of Russia and Japan, where salmon, thought to be their primary prey, occur. The winter range of residents, including southern residents, remains unknown. These gaps in knowledge can be filled by more extensive sampling (photographic and genetic) and through satellite-tagging of known residents.

It is more difficult, if not impossible, to fill knowledge gaps concerning historical distribution. Most marine mammal populations have been greatly reduced in the recent past from various human impacts (overhunting, bycatch, habitat destruction, and prey reduction). Our understanding of distributions of relatively unperturbed populations is poor, and our vision of what is "normal" is very much influenced by distributions observed in only the past few decades. In conservation biology, the redefining of "normal conditions" by each new generation of scientists according to what they observed early in their careers is called the problem of shifting baselines (Pauly 1995, Tegner and Dayton 1997). Southern residents and their prey have both declined in the past few decades, and the possibility that range contraction has already occurred is a plausible one. At present, the primary range of southern residents is Puget Sound–San Juan Islands, but they could have occupied areas that formerly sustained much larger salmon runs, such as waters off Washington, Oregon, and California. Although it may be possible to reconstruct historical range through genetic examination of teeth from museum collections, it is possible that no data are available, and historical range becomes "unknowable." In this case, a "significant portion of the range" can be based on either current range or range inferred from suitable habitat. The problem of "shifting baselines" is pronounced for marine populations where historical distributional data are scarce. It is particularly problematic for species like killer whales, where presence/absence types of data cannot be used because different potential taxa (such as the fish-eating "residents" and the mammal-eating "transients") cannot be readily identified by nonexperts and hence are often referred to as cryptic species.

The final DPS "significance criterion" relates to evidence that the southern residents differ markedly from other populations in genetic characteristics. Although a considerable number of genetic samples have been taken from resident and transient killer whales, interpretation of those samples and the genetic differences between the groups varied widely among scientific experts (Reeves et al. 2004). Thus, killer whales are a particularly good illustrative example of the problem of reducing biological complexity to simple categories for use in a legal framework.

Uncertainty in interpreting genetic data arises because the amount of genetic differentiation that develops between or among demographically isolated groups of animals depends on the numbers of animals in the groups, their social structure, and the length of time these groups tend to remain in existence. Far from providing a tool that allows determination of species status like a litmus test, genetic markers reflect the complex biology of the animals and require scientists to account for that biology when they interpret the genetic data.

Killer whales illustrate the interesting interplay between a species' natural history and its genetic patterns. Apparently, there is a selective advantage for killer whales to become dietary specialists and adapt their social organization accordingly. In the North Pacific, there are at least three types of genetically distinct killer whales. Residents are fish-eaters that specialize in salmon and form relatively large, strictly matrilineal pods (ten to twenty individuals). Both males and females nearly always remain within the pod of their birth, but most mating occurs with individuals from other pods (Barrett-Lennard 2000). Transients are mammal-eaters that are found in much smaller groups. It is unlikely that these groups are strictly matrilineal because it is common to see one or two males alone for long periods. Little is known of the third form, offshores, except that the group size is large and some individuals have been observed eating fish.

Nuclear and mitochondrial DNA data are consistent with little to no current gene flow among these forms, and the magnitude of the differences is quite large (larger than for many acknowledged species). However, interpretation of significance remains difficult. Small populations genetically differentiate more rapidly than large populations through a process called genetic drift, where gene frequencies drift to different levels through the random process of inheritance. Killer whales have low genetic diversity and small populations so the rate of genetic drift is likely to be high.

Further, it is possible that metapopulation dynamics could make founder events likely. Within metapopulations, local populations may go extinct and be recolonized on a relatively frequent basis. When small populations are founded from a larger population, a phenomenon called "lineage sorting" can occur. Imagine that there is a large population that has lived in an ocean basin for many thousands of generations. This population would contain many genetic haplotypes, which are represented by strings of letters for different nucleotide base pairs and are inherited like family names. Some haplotypes are old and differ by many letters from newer haplotypes that differ from one another by only a single letter. If new populations are created from the large old population, there is the possibility that the new ones will contain different frequencies of haplotypes. If these new populations are small and drift such that they end up with only a single name (lineage), then interpretation of the relationship among these new populations can be incorrect without an understanding of the history. For example, if one population ended up with an "old" haplotype and its neighbor ended up with a "new" one, then it might be incorrectly inferred that they had been separated for a time long enough to develop all the letter changes (mutations) between the haplotypes. The correct relationship (that they were recently founded from a large population) can only be reached by considering lineage sorting. However, because we do not know the history, it is also possible that two neighboring populations with very different haplotypes actually have been separated for a very long time. A possible precautionary approach would be to assume that the populations are in fact very different as long as that hypothesis remains plausible.

The southern resident killer whale biological review team was reconvened following the cetacean systematics workshop. After considering the findings of the workshop, most of the scientists agreed not only that more than one species or subspecies was likely under the new definitions, but that North Pacific residents and transients probably belong to separate subspecies (Krahn et al. 2004). Although taxonomic uncertainty remains, the team used a system of likelihood points to vote for different plausible scenarios and ultimately concluded that southern residents did meet the criteria to be defined as a DPS. The National Marine Fisheries Service then announced its intention to list this DPS as threatened under the ESA.

To return to the definition of DPS, the uncertainty associated with interpreting genetic data precludes a litmus test for determining when those data indicate a "marked difference." Instead, a checklist of the type of genetic differences that contribute to evolutionary significance together with guidelines on how to treat uncertainty would prove more beneficial. For example, if fewer than one disperser per generation is consistent with the ability of populations to maintain local adaptations, then a guideline for "marked difference" such as "a 10% chance of fewer than ten dispersers per generation" would allow the incorporation of uncertainty, including factors such as lineage sorting. Of course, putting genetic results in a probabilistic context requires a case-specific modeling approach.

UNITS TO CONSERVE: STOCKS

Species, subspecies, and DPSs may contain another level of population structure involving groups of animals that are essentially independent demographically. In other words, the internal dynamics within each group are far more important to the group's maintenance than immigration from neighboring groups. In fact, the degree of exchange or connectivity among such groups is an important determinant of overall population structure and the vulnerability of both the individual groups and the overall metapopulation to human effects. Consider, for example, a case in which subsistence hunters from a particular village want to maintain hunting of harbor seals in perpetuity within a few hours of the village by boat. The immediate concern of the hunters probably would not be preserving the evolutionary legacy of harbor seals, but rather would more likely be making sure that the number of seals they kill is not greater than can be replaced by the combination of local births and seals swimming in from nearby locations.

The connectivity among groups within a larger metapopulation can be symbolized with water bottles linked by unknown levels of flow among the bottles (Fig. 10.2). When the bottles are full, each group has the maximum number of animals that the habitat can support. The water taps indicate drains from groups resulting from human-caused mortality. When the drain rate from each bottle is the same (i.e., where human-caused mortality is proportional to density across the range), it is not necessary to know the bottle structure or the level of flow between the bottles (Fig. 10.2a). This depicts the case where human-caused mortality is proportional to density across the range. In contrast, Figure 10.2b has a heavy drain from only a single bottle in the bottle "range," illustrating cases where incidental kills are restricted to areas near a particular location or human activity. If the objective is to keep all the bottles at a level of at least 50%, then scientists and managers need to know both the "structure" (how many bottles and how much water is in each bottle) and the "connectivity" (i.e., the level of flow among bottles).

Whether management efforts are focused on the overall metapopulation or the demographically independent groups within that population depends on the management objective, which also determines the level of population maintained (10, 50, or 75%). Hence, these demographically independent groups are sometimes called management units. In marine mammal management, they are usually called stocks. Under the U.S. Marine Mammal Protection Act (MMPA), a population stock is a part of a management system that aims to maintain healthy marine ecosystems.

The definitional boundaries between stocks and the next level up at DPSs remain blurred. For example, the "regional populations" of the IUCN (Gärdenfors et al. 2001) are defined both in genetic terms at levels consistent with the DPS concept and at a metapopulation level in terms of "probability of recolonization," which is more consistent with a level

Fig. 10.2. Population structure depicted as a connected system of water bottles with removals from the system shown as drains.

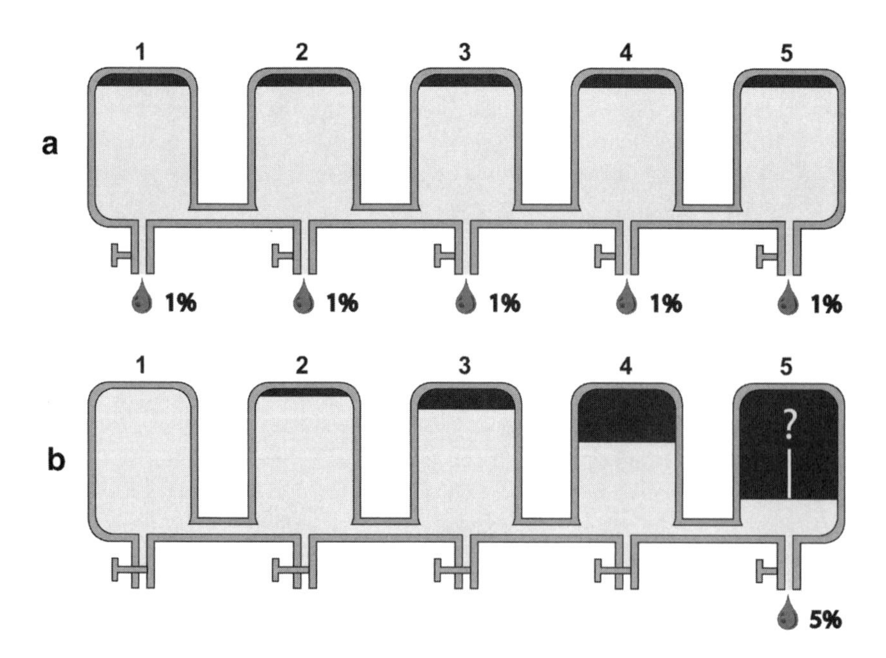

of dispersal typical of stocks. To avoid such confusion, I use stock to mean a unit with a level of demographic independence needed to meet a specific management objective.

The MMPA aims to preserve marine mammal stock structure. The Act specifies that endangered or depleted species "and populations stocks should not be permitted to diminish beyond the point at which they cease to be a significant functioning element in the ecosystem of which they are a part, and consistent with this major objective, they should not be permitted to diminish below their optimum sustainable population." The Act further states that "the primary objective of their management should be to maintain the health and stability of the marine ecosystem. Whenever consistent with this primary objective, it should be the goal to obtain an optimum and sustainable population keeping in mind the carrying capacity of the habitat." As guidance to interpreting these management objectives, the Act defines "optimum sustainable population" (OSP) with respect to any population stock as "the number of animals which will result in the maximum productivity of the population or the species, keeping in mind the carrying capacity of the habitat and the health of the ecosystem of which they form a constituent element." By regulation, NMFS defined populations to be at OSP when they were between carrying capacity (K) and the maximum net productivity level (MNPL) (Gerrodette and DeMaster 1990). Furthermore, the Act defines "population stock" as "a group of marine mammals of the same species or smaller taxa in a common spatial arrangement that interbreed when mature."

Unfortunately, managers have found it problematic to use the criterion "interbreed when mature" for most species. If we interpret the phrase to represent the degree of genetic interchange, then nature presents us with a continuum. Some geographically separate groups of animals may exchange members at the rate of one per generation and others at the rate of 1% per year. If we restrict our definition of a stock to those groups exchanging individuals at the rate of only a few individuals per generation, we will likely have stocks that (1) are distributed over large geographic ranges with disparate habitats, and (2) include multiple groups exhibiting a high degree of demographic independence. Such groups may also exhibit an important degree of ecological independence.

As illustrated in Figure 10.3, improper definition of stocks may undermine management objectives and goals. In the figure, *a* is the pristine distribution where width represents abundance and length geographic distance. Constrictions in this schematic represent limited movement such that this distribution could be described as a series of population

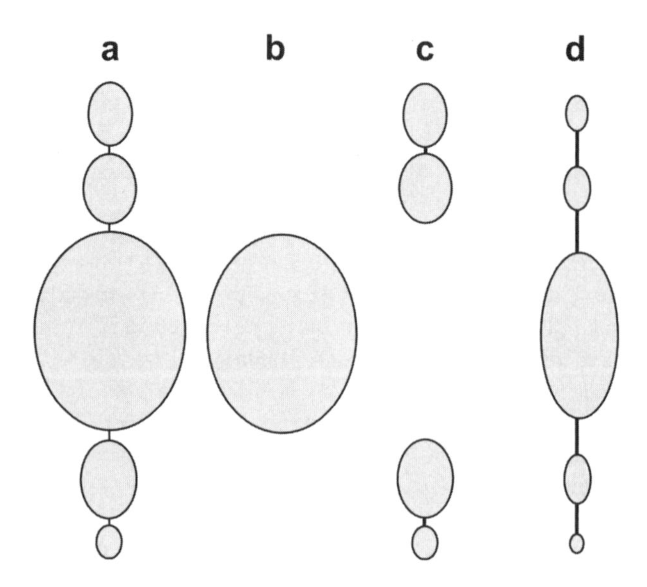

Fig. 10.3. Distribution of pristine populations (a) versus potential distributions after 50% of the total abundance is removed (b–d). Width represents abundance; length represents distance.

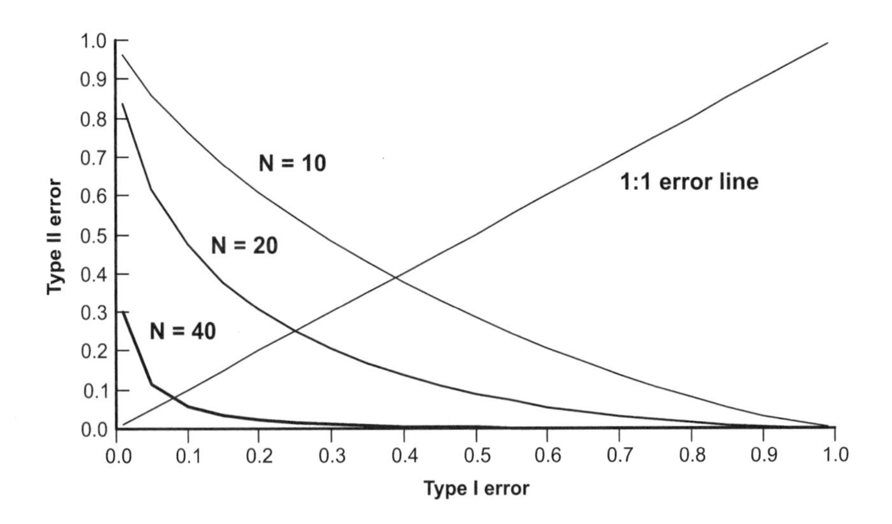

Fig. 10.4. Error trade-off curves for a null hypothesis of panmixia between two populations and an alternate hypothesis of a dispersal rate of 0.75% a year between two populations. The alternate hypothesis is true. Using the typical significance criterion of a Type I error (α) of 0.05, the Type II error would be 0.85, 0.60, and 0.10 for sample sizes (n) of 10, 20, and 40, respectively. Using this criterion means that the scientist is promoting Type II/Type I error ratios of 17 (0.85/0.05), 12 (0.85/0.05), and 2 (0.10/0.05) for the different sample sizes. In other words, when $n = 10$ the scientist, by using $\alpha = 0.05$, is seventeen times more willing to incorrectly lump populations than to incorrectly split them. Another alternate decision framework is deciding to equalize the Type I and Type II errors, which is shown in the 1:1 error line. Using this decision process would result in using $\alpha = 0.40$ (rather then $\alpha = 0.05$) when $n = 10$.

stocks connected by dispersal (the aggregate is the metapopulation). Reduction of abundance by 50% could result in any of the other distributions: *b,* range contraction; *c,* range fragmentation; or *d,* range maintenance. Although all may meet the goal of maintaining population stocks within OSP (i.e., about 0.5*K*), *b* and *c* probably do not meet the ecosystem goal. The 1994 amendments to the MMPA allow regulation of human-caused mortality through the calculation of potential biological removals (PBRs) for each affected stock. One element of the equation used to calculate PBR is an estimate of the stock's abundance. However, because there are no "rules" for defining population stocks, any of the above alternatives (i.e., range contraction, fragmentation, or maintenance) could occur depending on the distribution of human-caused mortality. Taylor (1997) developed quantitative methods to determine the dispersal rate (or flow) needed to maintain desired population levels assuming that population stock structure and abundance as well as human-caused mortality are known.

To meet MMPA objectives, the implementing agencies must draw lines on a map to represent stock population boundaries for all forty-eight marine mammal species that occur in U.S. waters. Available data for making such population boundary decisions range from very crude distributional data to very detailed data on movement, morphology, genetics, and distribution. Most of the time, however, the agencies must make their decisions in the face of considerable uncertainty.

A Type I error results from the incorrect splitting of stocks. It is an "overprotection" error and may result in unnecessary restriction of human activities. A Type II error results from the incorrect lumping of stocks. It is an "underprotection" error and may result in a failure to detect a stock falling below MNPL or, worse, its extinction. To calculate the probabilities of making these errors, management objectives must be defined quantitatively.

Methods have been suggested that allow full presentation of the uncertainty in the data without a specific definition of UTC. For example, Taylor and Dizon (1996) suggested the use of error trade-off curves, which have the important advantage of not forcing the researcher into making a decision as to what error ratio is appropriate for management (Fig. 10.4). Taylor and Dizon (1999) argue that in matters of population structure, policy must precede science because the data can only be properly interpreted once the policy decision concerning the level of population structure that is being sought has been made. Although this is true, use of error trade-off curves at least allows scientists to present results without having to choose an error ratio. However, as seen in Figure 10.4, either a choice has to be made regarding an appropriate dispersal rate or scientists must present a range of possibly relevant dispersal rates, which becomes computationally burdensome.

Participants in a workshop to provide guidelines for implementing the MMPA concluded that stocks must be identified carefully to ensure that the Act's ecosystem goals are met (Wade and Angliss 1997). The stock definition section of this report states:

> Many types of information can be used to identify stocks of a species: distribution and movements, population trends, morphological differences, genetic differences, contaminants and natural isotope loads, parasite differences, and oceanographic habitat differences. Evidence of morphological or genetic differences in animals from different geographic regions indicates that these populations are reproductively isolated. Reproductive isolation is proof of demographic isolation, and thus separate management is appropriate when such differences are found. Failure to detect differences experimentally, however, does not mean the opposite. Dispersal rates, though sufficiently high to homogenize morphological or genetic differences detectable experimentally between putative populations may still be insufficient to deliver enough recruits from an unexploited population (source) to an adjacent exploited population (sink) so that the latter remains a functioning element of its ecosystem. Insufficient dispersal between populations where one bears the brunt of exploitation coupled with their inappropriate pooling for management could easily result

in failure to meet MMPA objectives. For example, it is common to have human-caused mortality restricted to a portion of a species' range. Such concentrated mortality (if of a large magnitude) could lead to population fragmentation, a reduction in range, or even the loss of undetected populations, and would only be mitigated by high immigration rates from adjacent areas.

Therefore, careful consideration needs to be given to how stocks are defined. In particular, where mortality is greater than a PBR calculated from the abundance just within the oceanographic region where the human-caused mortality occurs, serious consideration should be given to defining an appropriate management unit in this region. In the absence of adequate information on stock structure and fisheries mortality, a species' range within an ocean should be divided into stocks that represent defensible management units. Examples of such management units include distinct oceanographic regions, semi-isolated habitat areas, and areas of higher density of the species that are separated by relatively lower density areas. Such areas have often been found to represent true biological stocks where sufficient information is available. There is no intent to define stocks that are clearly too small to represent demographically isolated biological populations, but it is noted that for some species genetic and other biological information has confirmed the likely existence of stocks of relatively small spatial scale, such as within Puget Sound, Washington, the Gulf of Maine, or Cook Inlet, Alaska.

In the decade since the PBR scheme was incorporated into the MMPA, relatively few changes have been made to the stock assessment reports regarding stock definition. Some changes have been made using genetic and distributional data as evidence, but accumulation of genetic data has been slow. Even when a respectable sample size has been analyzed, analytical methods remain nonoptimal for defining stocks (see the harbor seal case later in the chapter). As a result, most species within the five regions covered by stock assessment reports are considered to be one single stock. For those single-stock species, underprotection errors are possible, whereas overprotection errors are not.

ILLUSTRATIVE CASE OF STOCK DEFINITION FOR MARINE MAMMALS

Harbor seals can be used to illustrate several points regarding stock definition: (1) analytical methods for defining stocks remain untested and are more likely to underestimate the number of stocks than the converse, and (2) the importance of accurate stock definition is connected to levels of risk of depletion because errors have a graver consequence when risk is high.

The state of analytical methods for detecting and delimiting stock structure is illustrated with an ongoing study in Alaska. The decline of harbor seals in some regions of Alaska is a conservation concern because the causes are unknown. In addition, harbor seals are an important subsistence resource for many Alaska Native coastal communities, with an estimated annual take of 2,200–2,800 seals (Wolf 2001). Management objectives are thus concerned not only with maintaining harbor seals as functioning elements of their ecosystem but also meeting agreed co-management goals to ensure that this species remains a sustainable resource (MMPA as amended 1994, Alaska Native Harbor Seal Commission and National Marine Fisheries Service 2000).

NMFS currently recognizes three separate stocks of harbor seals in Alaska, identified primarily on the basis of regional differences in trends in abundance using data collected prior to 1994 (Small and DeMaster 1995, Hill et al. 1997, Angliss and Lodge 2002). At the time of their designation, however, it was recognized that large gaps existed in our knowledge of dispersal and movement patterns and stock structure, and it was recommended that more information be collected to define more meaningful management units (Small and DeMaster 1995). Over the past decade, a large body of research has been conducted that greatly improves our understanding of harbor seal stock structure, including further trend studies and directed studies of patterns of movement and population genetic structure (summarized in O'Corry-Crowe et al. 2003).

Like many marine mammals, harbor seal distribution is continuous, and samples are obtained opportunistically. Although more than 800 samples have been collected, large sampling gaps remain, and some areas of high density are represented with relatively few samples. This leaves the analyst with two problems: defining hypothetical units to initiate analysis and treating sample gaps.

The analysis of genetic data to identify stocks often employs hypothesis testing, which requires the analyst to stratify the data into initial units (the hypotheses that are being tested). This is problematic for species with continuous distributions. Martien and Taylor (2003) showed that hypothesis testing to assess stock structure is biased and results in too few stocks when used for species that are continuously distributed with animals at the ends of the range isolated from each other by distance. They showed that the strongest statistical evidence for stock structure was obtained by dividing the range in half, even if the true structure contained many stocks. This resulted from two factors: (1) testing for only two strata directly compared the individuals that differed the most from opposite ends of the range (statistically speaking this comparison has the largest effect size or, in other words, the difference between the hypotheses has the greatest possible magnitude), and (2) each stratum has the largest possible sample size with further subdivision resulting in fewer samples in each stratum. (Statistical comparisons with the greater sample sizes have greater precision.) In addition, even when statistical hypothesis testing indicates the presence of stock structure, it does not provide evidence for the location of the boundary between the stocks.

O'Corry-Crowe et al. (2003) avoided this pitfall by starting with many initial units and using clustering methods to group units that could not be distinguished as demographically independent. These analyses revealed a minimum of twelve demographically isolated units (estimated to have less than 0.5% dispersal per year among neighboring units). These units were consistent with tagging data that revealed low levels of movement and with trend data that revealed, for example, three different trends even within southeastern Alaska (where three units were proposed based on genetic data that were consistent with the trend data).

In spite of this evidence for more stock structure than previously recognized, data are still insufficient for determining stock boundaries because significant gaps remain in the sampling distribution. For example, it is not clear where the boundary lies between Glacier Bay and the Copper River Delta, which includes Yakutat and Icy Bay. Although Glacier Bay has apparently experienced a strong decline in the past decade and there are local concerns about the status of seals in Yakutat, there are still insufficient samples upon which to base scientific advice concerning stock identity of seals in Yakutat. Thus, despite improvement, management still faces the problem of how to define stocks in the face of uncertainty.

The degree of attention paid to stock definition in different regions has often been commensurate with risks. The large declines in abundance in some parts of Alaska have resulted in the extensive studies just discussed. In contrast, trends in harbor seal abundance are increasing or stable on both coasts of the contiguous United States. There are three stocks defined from Puget Sound south to the California-Mexico border and only one stock on the U.S. East Coast. Given the scale of harbor seal movements, it is very likely that currently recognized stock structure is too coarse in these areas. Because abundances are currently stable or increasing throughout the range, there is no negative consequence of stock structure errors. There is, however, still some level of risk in the sense that should a sudden decline begin, the agency would have to act with little understanding of stock structure.

CONCLUSIONS

Research needs regarding UTCs are as follows: (1) adequate treatment of uncertainty in taxonomy by application of the precautionary principle and shifting the burden of proof regarding taxon identification, (2) adequate funding to do the science needed to advance marine mammal taxonomy, and (3) development and testing of tools to define UTCs in a probabilistic manner. Managers have to work together with scientists to develop working definitions of UTCs that incorporate scientific uncertainty in a precautionary manner.

Taxonomy

Recent agreements on definitions and criteria for species and subspecies and prioritization for taxonomic research (Tables 10.1, 10.2) should facilitate more rapid advancements in marine mammal taxonomy. Nevertheless, the burden of gathering data still falls on scientists with little to no funding. Thus, although these definitions are improvements, they still do not explicitly incorporate the treatment of uncertainty and therefore remain biased toward underprotection errors. Progress in taxonomy in the next few years will indicate whether more precautionary definitions and criteria are warranted.

To reduce uncertainty, greater long-term support is needed for research on existing samples and for collection of and research on additional samples from the world's oceans. The needs for current collections are detailed in Reeves et al. (2004) and include integrated database management on a global scale, development of better methods of preservation, and development of better nuclear markers.

Analytical Tool Testing and Development

The scientific challenges with respect to defining UTCs are to gather data that allow UTC definition and to develop analytical methods that minimize errors in UTC definition. Geneticists use a number of methods to analyze their data, but none of these has been tested to see how well they actually detect demographically independent units (stocks). The same can be said for DPSs, although errors are expected to be more minor because the larger degree of genetic differentiation results in higher statistical power to identify DPSs. Methods to analyze genetic data were developed to address evolutionary questions where the levels of genetic differentiation are expected to be strong. An effort is now underway to test how these methods perform using simulated data where the true population structure is known (International Whaling Commission 2004). Scientists also have to develop better methods of integrating disparate types of data in a rigorous fashion. For example, data on distribution, trends in abundance, contaminant levels, morphology, timing of migration or reproductive events, acoustics, telemetry, and genetics all contribute to understanding population structure but relate to that structure on different scales (both spatial and temporal). Perhaps some of these data can contribute to initial hypotheses (or prior distributions in a Bayesian framework) that can be used in model choice exercises.

Ideally, research results should be used to guide conservation decisions. We need more research methods designed specifically for such applied problems. We also have to develop analytical tools that allow researchers to design studies of population structure better; for example, they should be able to say how many samples and how many genetic markers will be needed to provide a given level of certainty about dispersal rates. Scientists interested in detecting trends in abundance have such tools, which allow them to make a preliminary estimate of their ability to detect a given trend with a certain power after a number of surveys. However, in part because the definition of UTC remains nebulous, researchers are presently unable to say how many samples/markers will be needed to identify UTCs; moreover, after a study, they cannot yet estimate the probability that, given their sample size and distribution, they would have been able to detect multiple stocks that existed in the range of their study.

Incorporating Scientific Uncertainty into Management Definitions of Stock

The management/policy challenge is to phrase policy in a fashion that allows the best use of our knowledge while acknowledging the impact of our ignorance (see Goodman this volume). To better define stocks, our definitions have to incorporate uncertainty both when there are sufficient data and when data are either poor or entirely lacking. A sample rule-of-thumb definition would be "a stock is a population for which the best estimate for the dispersal rate with a neighbor is less than *x*%/year, where *x*%/year is considered to be demographically trivial."

A probabilistic definition provides more flexibility. These definitions all have a statement of probability linked to a desired state in a given time period. Such definitions were used in the PBR scheme, which incorporated uncertainty to proscribe precautionary management (Taylor et al. 2000, Taylor and Wade 2000). Incorporation of uncertainty was accomplished through use of quantitative criteria such as "a 90% chance of a population being greater than 50% of historical numbers in 100 years." The statement of probability results in situations with less certainty receiving more conservative management, which is needed to ensure reaching the desired management state.

ACKNOWLEDGMENTS

This work was funded by the Southwest Fisheries Science Center. Ideas were developed in fruitful discussions within the Protected Resource Division, with special thanks to Jay Barlow, Susan Chivers, Andy Dizon, Rick LeDuc, Karen Martien, Greg O'Corry-Crowe, Bill Perrin, and Paul Wade. The manuscript was improved through reviews by Karen Martien, Bill Perrin, Tim Ragen, and an anonymous reviewer.

DANIEL GOODMAN

11

Adapting Regulatory Protection to Cope with Future Change

THE WORLD'S HUMAN POPULATION is growing rapidly, and over a long period the growth rate itself has been increasing. Since the early 1970s, when the bulk of the present legal framework for U.S. environmental regulation was enacted, the U.S. population has grown by about one-third. The world's population, growing considerably faster, now exceeds a staggering 6 billion people.

The footprint of the world's human population on world resources is growing faster than the population because of advances in technology, the spread of technology, and the rising expectations of the less-developed, faster-growing areas. Worldwide, the cumulative effects of human activities include wholesale conversion of relatively natural landscapes to agricultural and settled landscapes, massive redirection of water supplies, extreme overexploitation of theoretically renewable resources, alteration of the composition of the atmosphere, and, probably, alteration of the climate.

I see these effects posing six categories of serious threat to marine mammals:

1. Erosion of carrying capacity because of intensified exploitation of ocean fishery resources affecting prey availability to marine mammals (see Plagányi and Butterworth this volume).
2. Loss of specific essential habitat types because of climate change—most notably contraction of sea ice (see Moore this volume).

3. Loss of specific essential habitat types because of direct human encroachment, such as coastal development (see Ragen this volume).
4. Rising incidental mortality owing to interactions with fisheries and vessel traffic (and, on a smaller scale, potential for increased disturbance owing to tourism) (see Read this volume).
5. Increased risk of epidemics owing to greater worldwide transport of vector organisms, compounded by the emergence of more virulent and resistant disease agents in the world population of domestic animals and in the wake of aquaculture operations (see Gulland and Hall this volume).
6. Increased risk of mortality, reproductive failure, and immune system compromise owing to direct global pollution (e.g., PCBs), local pollution events (e.g., oil spills), and outbreaks of toxic algae that may increase in frequency or severity because of ecological disturbance (see O'Hara and O'Shea this volume, Van Dolah this volume).

This list, as a set of cause-and-effect relationships and predicted trends that are candidates for scientific investigation, suggests a research agenda that is fairly overwhelming in magnitude. This list, also, is no secret. These issues have been recognized before, the causal factors are important for reasons apart from their implications for marine mammals, and research is underway, at some level, on all of them. So I am not sure that I have much more to say from this perspective, except to reiterate that more research (and more research funding) would be better.

From the standpoint of a desire to reduce threats by directly reducing the magnitudes of the causal factors, this list suggests a conservation agenda that is pretty discouraging. The prospects for dramatically reducing human population growth, global pollution, climate change, and activities associated with economic development, on a broad scale, are not very bright. I am not going to say we should not try, but I think it would be a mistake to pin all our conservation hopes on efforts to influence global, or even U.S., policy to stem the human tide.

Prospects are better for achieving some conservation successes through the design and implementation of conservation measures that anticipate these negative trends and put protection and mitigation measures in place that will sustain marine mammal populations in the face of these trends. The key here is the word "anticipate."

Current approaches to making regulatory decisions such as determination of optimum sustainable population (OSP), listings under the U.S. Endangered Species Act (ESA), designation of critical habitat, and evaluation of reasonable and prudent alternatives to avoid jeopardy often have not taken sufficient account of worsening trends on a time scale of decades. As a consequence, we may make decisions that seem adequate for the present, but are almost certainly inadequate for the long run. At the same time, many kinds of protective measures, such as establishing reserves or buffer populations, are easier and cheaper to implement if they are undertaken sooner rather than later. All this argues for the importance of finding ways to factor deteriorating trends into our conservation decision making in order to respond effectively and efficiently to a predicted future that, in fact, will be less conducive than the present for survival of many marine mammals.

POLICY DEVELOPMENT AS A RESEARCH ENTERPRISE

The Marine Mammal Commission's August 2003 consultation, which was the origin of this volume, was about research, not policy. However, a focus on policy here is justified because the policy needed has a large technical content and its development requires a substantial commitment of research talent.

Effective management guidance cannot be guaranteed by a policy that consists merely of exhortation to pursue conservation goals because conservation goals may conflict with other visions and with economic goals having to do with differing uses of resources. These other visions and economic interests are vigorously defended and promoted by their respective constituencies. Therefore, effective policy requires explicit standards to ensure fairly predictable implementation despite the conflicts. Ideally the standards refer to measurable and objective criteria to minimize the discretionary element in each decision.

This ideal is only partially realized in the flagship federal legislation that concerns us. The four most important laws from the perspective of marine mammal management are the National Environmental Policy Act (NEPA), the Fishery Conservation and Management Act (FCMA, which, after reauthorization and amendment, is now referred to as the Magnuson-Stevens Act, MSA), the Endangered Species Act (ESA), and the Marine Mammal Protection Act (MMPA), all of which date from the early to mid-1970s.

Of the four, NEPA is disappointingly the farthest from the ideal I propose. The NEPA requires that environmental consequences of contemplated government action be evaluated rather thoroughly, but it does not require that decisions about such actions be driven directly by the outcome of the evaluation. The language of the Act directs federal agencies to ensure that "environmental amenities and values may be given appropriate consideration in decision making" (42 U.S.C. 4321). That is to say, compliance requires documentation that the environmental consequences were "considered," but it does not stipulate what weight they must be given. In this sense, NEPA is essentially about process rather than substance and thus provides fertile ground for very contentious litigation about whether the process of evaluation has been thorough enough. For those, like myself, concerned about the substance of environmental standards, it is an ongoing frustration that the primary effect of this

NEPA litigation is delay rather than predictable and rational guidance of decisions. All other things being equal, it is still a good thing that government actions with environmental implications be subjected to such NEPA scrutiny. But the atmosphere of decision by litigation is not good for it raises the costs of decision making and is corrosive to the kind of trust in process that is necessary for long-range environmental planning to be accepted as a credible norm.

The MMPA, by contrast, articulates standards that are intended to strongly influence substantive aspects of decisions. The two most important classes of actions covered by these standards are broad prohibitions on intentional take (with important exceptions) and the requirement for specific authorization for inadvertent ("incidental") take. Such authorization requires determinations that the stock will not be "disadvantaged" by the incidental take and that it is not below its "optimum sustainable population" level (OSP). Early amendments allowed for authorizing some incidental take on stocks below OSP provided consequences for the affected stock are "negligible," in the sense that the level of incidental take allows the population to recover to OSP without untoward delay.

As originally enacted, the MMPA included only circular definitions of OSP (which at that time was not a recognized scientific term), and it provided inadequate guidance on quantitative standards for determining whether a proposed incidental take would "disadvantage" the stock. The early amendments did not quantify what consequences should be considered "negligible." These insufficiencies became apparent during early attempts at implementation of the Act.

For a large fraction of the species covered by the MMPA, the responsibility for implementation falls on the National Marine Fisheries Service (NMFS). The first major systematic efforts to clarify implementation of the MMPA were a series of meetings involving the Marine Mammal Commission and NMFS to arrive at an operational definition for OSP. The definition that was adopted and written into regulations was that OSP represents a range between the population's maximum net productivity level (MNPL) and its carrying capacity (K), both of which terms have definite scientific meaning. Subsequently, this crucial definition was written into amendments to the law itself.

The second major systematic efforts at clarification were directed at the determination of negligible levels of incidental take in connection with commercial fisheries. By the time this effort was initiated, experience with attempts at determination of OSP in practice had revealed difficulties connected with data scarcity, complicated modeling, and the need to make decisions under conditions of scientific uncertainty. The "potential biological removal" (PBR) system attempted to address these difficulties, for the authorization of incidental take in connection with commercial fisheries, through a decision rule that was embodied in a simple data-driven formula that could function even with minimal data and responded to uncertainty through use of a confidence limit in one of its key inputs.

The formula for PBR is so simple and abstract that it is not intuitively obvious that its application would in fact achieve the goals expressed verbally, albeit vaguely, in the MMPA. The process that culminated in the adoption of the PBR system involved extensive and detailed modeling exercises to evaluate the performance of this decision rule under various circumstances. These evaluations demonstrated that the average performance was consistent with a concrete interpretation of the intent in the MMPA and was acceptable to the Marine Mammal Commission and NMFS. The two agencies participated in a series of meetings to develop specific definitions and statements of goals, propose operational approaches, and consider the results of the performance evaluations. The PBR system was subsequently submitted as part of the NMFS legislative proposal package to Congress when the MMPA was considered for reauthorization. After intensive briefings before congressional staff and committees of Congress itself, the PBR system was written into the amended law.

The point of this historical review is that the clarification, rectification, and operationalization of policy, as carried out in the definition of OSP and the development of the PBR system, required a large investment in research. Many of the scientists who conducted the labor-intensive performance evaluations of proposed management systems also were central participants in the many meetings to develop a scientific consensus on proposed statements of policy. In its ongoing responsibilities under MMPA and ESA, NMFS relies upon these same scientists for various scientific determinations, such as stock assessments.

A much greater commitment of scientific personnel is needed to refine policy and develop decision rules that will achieve the goals of that policy in the face of the challenges of the future. Notwithstanding improvements in the MMPA through the definition of OSP and development of the PBR system, some significant loose ends remain in the MMPA—most notably in connection with protection of habitat and ecosystem productivity and the relation of depleted status to the possible circumstance of declining K. The ESA and the MSA embody even more unfinished business.

Smooth implementation of the ESA awaits, most notably adoption of consistent quantitative standards, clarification of population units that are objects of protection, and reconciliation of scientific uncertainty regarding the adequacy of proposed management measures with the demand that "reasonable and prudent alternatives" assuredly alleviate "jeopardy." The MSA has received considerable attention from NMFS in promulgating National Standards Guidelines to operationalize "optimum yield" (OY) with respect to the Act's goal of extracting the maximum sustainable yield (MSY) "on a continuing basis . . . from each fishery," but similar attention has not been given to operationalizing the Act's goal that OY should be reduced from MSY for "ecosystem protection" and to ensure that "irreversible or long-term adverse effects on fisheries and the marine environment are avoided."

CHARACTERISTICS OF A GOOD DECISION RULE

Ideally, a decision rule operates entirely from objective data inputs so that there is no discretionary latitude. But we also want assurance that a decision rule will achieve the "right" result. If there were no uncertainty about the inputs and no uncertainty about the consequences of actions, a good decision rule would essentially function as a lookup table for choosing the correct action to achieve the desired result for the observed circumstances.

When there is uncertainty about the inputs and/or the consequences of actions, it may not be possible to specify an action that guarantees a desired outcome. Or, if such a sure-thing action does exist, it may represent an extreme kind of "overkill" in that the action must contain a large enough "margin of safety" to compensate for the uncertainty. Often, a spectrum of actions is available to represent a spectrum of magnitudes for the margin of safety, so that the range of choice of actions is, in effect, a range of choices for the probability of the desired outcome.

Acting to create a margin of safety generally imposes costs (the costs of implementing the more intense action and the costs of frequently overreacting). The cost, in theory, may or may not be necessary in the given case (although this cannot be known at the time) because uncertainties give rise to "false alarms" with some frequency. Thus, those bearing the costs may be reluctant to accept imposition of the action with the largest available margin of safety. By the same token, the parties concerned with the conservation outcome are inclined toward large margins of safety because their priority is assuring the conservation outcome, regardless of implementation costs.

A rational choice of the margin of safety takes account both of the costs of implementation when the action selected is the correct one for the true (but imperfectly known) state of affairs and the costs of implementation and damage control when the action selected proves to have been mistaken. Very roughly, there are two main classes of mistakes—false alarms and failure to respond to a genuine need for action—and two main classes of ultimately correct choices—responding when action is genuinely needed and recognizing when action genuinely is not needed. More generally, there may be gradations of appropriate action.

Statistical decision theory provides a rigorous mathematical approach for selecting the best action under conditions of uncertainty (Berger 1985). This body of theory has its least problematic applications in economic decision making, where the same party bears the costs and reaps the benefits of an action and costs and benefits are readily measured in the same units—usually money. Nevertheless, the theory has powerful applications to environmental decision making as well (Goodman and Blacker 1998). The salient result from this theory is that the configuration of the costs of actions and the costs of mistakes entirely determines the decision rule that maximizes the average net benefit (benefit minus cost). This decision rule is a formula that selects the course of action based on the data. In this formula, the level of uncertainty in the available information plays a role as an intermediate quantity calculated from the data, but the critical levels of uncertainty for tripping the decision are determined by the cost structure.

For a binary decision (choosing between two courses of action), the optimal form of the decision rule is very simple: it is to select the action that would be appropriate for perfect knowledge of a particular status whenever the probability of that status, based on available data, exceeds a critical threshold that is determined by the cost structure. In other words, the margin of safety is quantified operationally as a critical probability of the true status, and the technically correct value for this critical probability does not vary with the case-specific availability of information but is determined entirely by the costs of actions and the costs of mistakes. When the decision is among more than two actions, the decision rule uses the respective probabilities of all the relevant possible status assignments, and the operating formula is whether some weighted sum of the respective probabilities falls between some critical thresholds. The critical thresholds (and the weightings) are still determined entirely by the cost structure and do not vary with the case-specific availability of information.

When the theory is applied to public policy, the costs and benefits do not always have obvious dollar equivalents but instead represent formalization of "values," which function as weights for balancing desired outcomes versus failure to achieve those outcomes. That is to say, the quantification of costs and benefits must refer to some measure of how much society cares, collectively, about the pertinent outcomes. Ideally, the expression of how much society cares should be encapsulated in the legislation that establishes a policy, which would automatically determine the appropriate margins of safety for the decisions that derive from this policy. In each particular case, the available data would be evaluated statistically, the statistical certainty of the assignments of the case to various categories would be compared to the critical probability (or probabilities), and the results of the comparison would determine the decision.

Although this may seem technically complicated, it still makes intuitive sense. The margin of safety embodied in the critical threshold probability defines an "error tolerance" that should correctly reflect society's values. The comparison of the case-specific certainty to the critical threshold(s) ensures that, in the long run, the frequency of errors complies with that tolerance. The calculation of the case-specific certainty is a wholly technical undertaking, which should be a matter of technical competence rather than esoteric legal or philosophical interpretation. The use of the calculated case-specific certainty automatically adjusts the decision process, so the decision rule functions in accordance with societal values regardless of how much or how little information is available for the particular case.

When there is enough information to make the predicted outcome highly certain, the decision rule operates according to that prediction. When the case-specific information is extremely sparse, so that the predicted outcome is very uncertain, the rule decides in favor of whichever option is given "the benefit of the doubt" by societal values. When the case-specific information is intermediate in its predictive power, the decision rule weighs the relative probabilities of the respective outcomes and the respective seriousness of being right and being wrong, so as to achieve a balance between overreacting and underreacting that is consistent with societal values. A side benefit of this kind of decision rule is that it provides an automatic incentive for collecting more data whenever the implementation cost for the margin of safety is higher than the cost of collecting data to reduce uncertainty.

THE PROPER FORM FOR A REGULATORY STANDARD

The decision rule in the PBR system has these desired properties and satisfies this type of rationale. The allowable incidental take is a product of three terms: a lower 20% confidence limit on the estimate of the population size N_{min}; half the estimate of the population's maximum productivity R_{max}; and a "recovery factor" F_r. In the absence of compelling population-specific information to the contrary, the R_{max} value is 0.04 and 0.12 for cetaceans and for pinnipeds, respectively. The value of the recovery factor for endangered populations is 0.1. The performance evaluations that justified this formula concluded that the formula satisfied the following three standards: with 95% probability a population at MNPL will not be below MNPL 20 years later; with 95% probability a population at 30% of *K* will not be below MNPL 100 years later; with 95% probability the time to recovery of a depleted population is not delayed by more than 10%. All three of these standards follow the same logical form: a statement of a desired state and an error tolerance for achieving it.

These three standards are the foundation of the PBR system. In principle then, any decision system that can be shown to satisfy these standards should be compatible with the goals of the MMPA for regulating incidental take and, logically, such a decision system should be an acceptable alternative to the actual PBR formula.

Given such latitude, the basis for selecting among alternative acceptable decision systems might reasonably be implementation cost. In a data-poor situation, it is unlikely that the performance of the PBR formula itself could be improved upon. Moreover, given the history of investment in the analysis to document the performance of the PBR formula, it is unlikely that the costs of analysis of case-specific alternatives for a data-poor case would be worthwhile. In a data-rich situation, on the other hand, it is quite plausible that an alternative analysis might allow for decisions that satisfy the required error tolerance at a lower implementation cost. This would be possible if the alternative analysis achieved a higher level of certainty by using available case-specific information that is not used by the PBR formula and capitalized on that higher level of certainty by reducing the margin of safety accordingly (Goodman 2002b). As a practical matter, then, the merits of pursuing an alternative in a given case are revealed by weighing the expected cost saving from the reduced margin of safety against the costs of conducting the alternative analysis, including the costs of the performance evaluations to verify that the case-specific analysis satisfies the standards.

We see, then, that the standard is more fundamental than the decision rule. The standard is the basic expression of policy defining the desired state, which is the expression of what we, collectively, want. The error tolerance is the expression of the priority that we put on the desired state, in the context of possibly competing goals and costs of implementation. The decision rule, by contrast, is merely instrumental: it is one way, among possibly many, of satisfying the standard. Because harvest plans or recovery plans and similar environmental plans of action are multidimensional, with time schedules, spatial components, and perhaps "if-then" contingencies or junctures for adjustment, the spectrum of possible decision rules is potentially infinite. For this reason, as a practical matter, is it a good idea to have a comparatively simple default decision rule ready to prevent initial paralysis of the decision process. But for the same reason, it is important to be open to an alternative case-specific decision rule, if one is found, that is demonstrably more efficient than the default (i.e., operates at lower cost) in the context of the data availability for that case while still demonstrably satisfying the underlying standard.

From this perspective, it might have been better if (1) the three standards that the PBR decision rule was designed to satisfy had been written into the legislation (rather than residing as they do in the legislative history), and (2) the current PBR decision rule had been written into regulation by NMFS as a default option for complying with the standard while allowing for alternative decision rules that can be shown to comply with the standard based on performance analysis. The present practice in applying PBR allows case-specific information about stock productivity (R_{max}) to be used instead of default values if the quality of that information warrants. The technicalities for determining the adequacy of the case-specific information for R_{max} relative to the default have yet to be formalized administratively.

Bayesian statistical methods are the natural technical solution for combining case-specific information with a default encapsulated in a "prior distribution." This use of a distribution, rather than a point estimate as in the PBR, for the default provides a mechanism for realistically representing uncertainty in existing information. In the development of the PBR formula, the real and admitted uncertainty in R_{max} was addressed by tuning other safety margins, such as the "recovery factor" and the confidence level for N_{min}. But the awkwardness that this causes for scientifically clean refinement of the decision rule beyond the default formula is a reminder of Berger's (1985) admonition that it is better to set up the solution correctly in the first place than to try to prop up a flawed procedure with a collection of individually reasonable but ad hoc fixes.

At the time the PBR system was developed, there was a recoil away from case-specific modeling analyses: Experience with dueling models in the courts had been frustrating; modeling exercises for many species were considered to be largely empty because of inadequacies of the data; and NMFS had a sense that it was necessary, as a practical matter, to mass produce assessments on an assembly-line basis that could be derailed by case-specific complications. In this atmosphere, the attraction to a simple recipe like the PBR formula is understandable. But the shortcoming of this system is its inflexibility when more information is available. As the scientific knowledge for crafting decision rules, analyzing their performance, and conducting statistical inference with sparse and poor-quality data has advanced since that era, perhaps now there would be merit to revisiting the choice of convenience that gave primacy to the simple (default) decision rule over the general standard.

Simpler decision rules also can be psychologically attractive. The regulated community, in particular, often just wants to pay the ticket and go home, and rules expressed as simple body counts may appear more concrete and predictable than rules phrased in the abstractions of probability. But these body count rules beg the questions of what standard they should satisfy and how one determines whether they satisfy it.

With the biological reality of big animals and small, fast-reproducing and slow, early maturing and late, long life span and short, numerous and rare, and so on, it is obvious that one body count rule will not perform reasonably for all species. So the apparent predictability to the regulated party of a regulatory condition premised on a body count does not confer a similarly predictable consequence for the affected stock, and it does not confer predictable consistency among stocks. If a body count rule is going to be adjusted on a case-specific basis, what is the underlying logic for the adjustment? Furthermore, if an underlying logic is identified, then that is the real standard, which might as well be made explicit.

A QUANTITATIVE STANDARD FOR ENDANGERED SPECIES ACT LISTING

Most actual environmental legislation and regulations fall short of the ideal of an unambiguous, data-driven decision rule expressed as a stated critical probability for an estimated quantity exceeding a stipulated threshold, justified in terms of a basic standard stating a quantitative error tolerance for attainment of a desired quantitative condition. This shortcoming holds even for the most important and far-reaching laws. Consider the Endangered Species Act. The language of the Act defining the central decision about "listing" a population as "endangered" states that a species should be so classified if it is "in danger of becoming extinct throughout all or a significant portion of its range."

The ESA classification is an important decision because a population listed as endangered automatically is given a broad range of protections. But the Act provides very little guidance for this decision. Although extinction is unambiguous, the criterion that the population is "in danger of becoming extinct" introduces an element of prediction over some time horizon as well as an element of probability since danger can refer to something that "may happen" as distinct from something that definitely will happen. Similarly, the Section 4 definitions of factors that may form a basis for listing as endangered include "the present or threatened destruction, modification or curtailment of its habitat or range" where the word "threatened" again indicates that a potential may be sufficient cause. So, the natural questions for operationalizing the listing criterion are: (1) how distant a time horizon for predicted possible extinction is still a matter for concern, and (2) how certain does the predicted extinction have to be to warrant a listing as endangered. These questions, as a matter of uniform policy, remain unresolved.

The legislation itself, supported by the legislative history and upheld by the courts, established the principle that a population listed as endangered under ESA should receive the benefit of the doubt in a "jeopardy" decision—the burden of proof is to show that a proposed action will not "jeopardize the continued existence" of the population or cause "significant adverse modification" of its "critical habitat." In such a decision for a listed population if there is a high level of uncertainty, the presumption is that there will be jeopardy, and so a "reasonable and prudent alternative" must be substituted for the proposed action. But this burden-of-proof standard is not specified for the equally important decision of whether a candidate species should be listed.

There is a growing general trend in conservation policy of shifting the burden of proof to assure a substantial probability that the conservation goal of natural resource legislation will be served under circumstances of uncertainty. An early example of assigning some burden of proof under MMPA was a court ruling that incidental take cannot be authorized in the absence of information to demonstrate that the population will not be disadvantaged. This precedent was set in *Committee for Humane Legislation* v. *Richardson*, 1976, in which Judge Richey struck down a "general permit" for incidental take of dolphin in the tuna fishery, issued by the Department of Commerce in lieu of an assessment of the stocks and the take on a claim that there were insufficient data to support such an assessment. The presumption in favor of a high burden of proof that a conservation goal will not be compromised is often called the "precautionary principle" and is often invoked in discussions of conservation policy, but it is not actually written into any U.S. environmental legislation.

The 1998 NMFS National Standards Guidelines for implementation of the MSA espouse a "precautionary approach" and set the "target reference point," which the fishery management plans are intended to achieve on average, conservatively enough so that the MSA goal of not overfishing, with respect to MSY, will be violated only rarely. This draws on a definition of precautionary approach developed in the 1995 United Nations Agreement on Straddling Fish Stocks and Highly Migratory Fish Stocks.

All this argues that there should be a low error tolerance for failing to list a species that may actually need the protection of ESA to avoid extinction. But that still does not give us an actual number for the critical probability. What about the appropriate time horizon? If the extinction process operated consistently over time so that, for a given population, the instantaneous extinction probability were constant over time (until extinction actually occurred), the choice of time horizon would merely be a matter of the choice of units for expressing the same essential rate. For example, a constant instantaneous extinction rate of 12% per year will confer a probability of 54.9% for surviving 5 years, 30.1% for surviving 10 years, 4.98% for 25 years, 2.48% for 50 years, 0.000614% for surviving 100 years, and so on. This equivalence, however, is a mathematical property that depends on constancy of the instantaneous extinction rate.

Biologically, there are two primary classes of mechanisms that cause the instantaneous probability of extinction to vary over time, most often in the direction of increasing probability of extinction. One class is internal to the dynamics of the population and is driven by age distribution effects. It is commonly the case that survival rates of the very young are the component of the life history that is most sensitive to adverse conditions. Thus for a long-lived species, a downturn in conditions may effectively depress recruitment to near zero so that the probability of extinction within several generations is virtually 100% if the poor conditions persist. But at the time of the onset of the adverse conditions, some considerable fraction of the population will be individuals that are in their prime and less affected by these conditions. Under these circumstances, for a time corresponding approximately to the adult life expectancy, the existing population will decline at a slow rate, roughly equivalent to the adult mortality rate, and so, for this period, the probability of extinction is virtually zero. The point is that even with total recruitment failure, the adult count will initially decline at the average adult mortality rate, and the probability of extinction in the very short term is near zero, although it is certain after the passage of one generation. Therefore, the negligible probability of extinction and the slow rate of population decline during the first generation following the onset of unfavorable conditions are poor guides to the very high extinction probability and high rate of population decline that will be manifested thereafter.

The second class of mechanisms is long-term trends of population decline that is due to deteriorating circumstances external to the population. Consider a population numbering 500 individuals that is declining steadily at 7% per year. This population will reach essentially certain deterministic extinction in 79 years and has almost no probability of extinction in a much shorter time. Or consider a population of a species with a short generation time that has a mean per capita birth rate of 5% annually, a natural mean per capita mortality rate of 2%, and a human-caused mortality rate that is currently 1% but that is growing linearly in time at a rate that is adding 0.2% in mortality each year. Initially, this population will be growing at a rate of 2% a year, but within 10 years the growth rate will have declined to zero and after 10 years the population will decline at an accelerating rate. If such a population starts with 1,000 individuals, it will top out at 1,115 individuals in 10 years and will reach deterministic extinction in 88 years.

We see then that the appropriate time horizon for stating an extinction rate standard should be long enough to wash out transient age distribution effects (several generations) and to encompass pertinent trends in the population and its environment. As a practical matter, the time horizon also should be long enough to allow for effective mitigating action to halt or reverse adverse trends before their consequences for the population raise the probability of extinction even higher. Another way of saying this is that the listing criterion should give an early enough warning to allow time for practical intervention. A recent NMFS workshop to consider delisting criteria for large whales concluded that a 100-year time horizon was appropriate for the "endangered" classification (Angliss et al. 2002).

Once a time horizon has been chosen, it defines the units in which the critical probability of extinction will be expressed. The test of whether the chosen probability really is consistent with our interpretation of the intent of the ESA is whether the long-term extinction rate it implies is consistent with our interpretation of the intent of the Act.

It certainly is easy to identify some rates that are not reasonable. Consider an unimaginative path of least resistance being to default to a 95% certainty (the standard significance test in experimental science) and to default to a 5-year time horizon (the typical period for regulatory review—indeed, ESA recovery plans are reviewed on that schedule and ESA biological opinions are revisited on that schedule). A constant 95% probability of persistence for 5 years is equivalent to an instantaneous extinction rate of 0.01026 per year, which may look like a small number but actually corresponds to a species half-life of 67.6 years, which is a disconcertingly high extinction rate. If, at the time of the American Constitutional Convention, the Founding Fathers had adopted this rate as a policy standard for managing a resource portfolio of ten stocks of whales and had managed accordingly, there would be a 31.5% probability that all of those stocks would be extinct by now. Surely, that is not consistent with the intent of the ESA to conserve species for future generations.

Arguably, if the intent of the ESA is to prevent human-caused extinctions, the appropriate critical extinction rate should be some value not much larger than the normal background extinction rate. The NMFS workshop on delisting criteria for large whales recommended a 1% probability of extinction within a 100-year time horizon. This corresponds to an instantaneous extinction rate of 0.000101 a year and a species half-life of 6,817 years. This low extinction rate is still considerably higher than the normal background rate. Nevertheless, if the Founding Fathers had adopted this standard, the per-species probability of surviving from the time of the Constitutional Convention to the present would be 97.9%. The probability that ten stocks managed accordingly would all have gone extinct in this time is 0.000000000000002%.

In actual practice, most recovery criteria are still of the body count type. If these all were justified by a case-specific population viability analysis showing that for the population of interest the criterion satisfied a common underlying standard of the form "not to exceed x% probability of extinction within y years," where x and y are common fixed values, this would ensure consistency of policy. Absent that consistency, it is hard to imagine how the different body count recovery criteria might be justified. Consider the cluster of evolutionarily significant units (ESUs) of ESA-listed salmon in the Pacific Northwest. There are, at the time of this writing, twelve such listed populations in the Columbia River Basin alone and ten more in other rivers in Washington, Oregon, and Idaho. If differing recovery criteria are adopted for each one and analysis shows that these confer different probabilities of persistence of the population for some reference period, say 100 years, then the constituencies bearing the costs of the recovery measures are being held to different standards in the different respective ESUs. Would not those being held to a more stringent standard be motivated to argue for a better deal? Inconsistency (or absence) of standards is a continuing invitation to litigation.

A PBR-LIKE APPROACH FOR POPULATION VIABILITY ANALYSIS

If the basic standard adopted for recovery becomes a specified probability of persistence for a substantial duration, such as 100 years, decisions will hinge on determining whether a given set of interventions offers assurance that this standard will be met. Population viability analysis (PVA) is the label given to methods of prospectively investigating probabilities of extinction or survival. But, in fact, the methods themselves have not yet been standardized. This again raises the specter of dueling models and hints that attempts at wholesale use of PVA for ESA decision making could run afoul of the same obstacles—data scarcity, methodological contention, and expensive case-specific modeling efforts—that motivated development of the PBR system as an alternative to case-specific modeling.

The best solution, I believe, would be to adopt a quantitative standard of the form recommended above and to develop a default PVA recipe, in the spirit of PBR, that is automatically the starting point for each analysis of whether a particular population's circumstances satisfy that standard. Learning from the PBR experience, we could facilitate a seamless transition from the default to a more complex analysis that capitalizes on case-specific data availability by framing the default PBR as a Bayesian model from the start and capturing the default parameter information in prior distributions.

The research effort to achieve consensus on the default PVA recipe and on the appropriate priors for parameters will occupy a fair number of practitioners for a considerable time.

CONTINGENCIES AND EXPERIMENTATION

The embarrassing consequence of an appropriately stringent criterion for listing as endangered, such as a 1% probability of extinction in 100 years, is that in some cases we are hard pressed to institutionalize fixed management measures in the present that will assuredly keep the probability of extinction below the critical value over that long a time horizon unless we resort to extraordinarily precautionary measures. Costs of such draconian protection can affect implementation decisions under ESA.

The legislative intent in ESA is clearly that the decision whether to list should not be influenced by economic considerations. However, the law allows that economic considerations may be weighed in the designation of critical habitat once a population is listed as endangered and in the selection of reasonable and prudent alternatives (RPAs) when an action is found to pose jeopardy to a listed population, provided the selected alternative does alleviate jeopardy. In principle, the escape valve of the ESA Endangered Species Committee (popularly called the "god committee") can, under Section 7(h), allow an action posing jeopardy to proceed if the committee determines that the social and economic costs of alleviating the jeopardy, or prohibiting the action, are too great.

In fact, the Endangered Species Committee mechanism has never yet been called into play to allow an action to proceed without removing jeopardy (although Congress has directly exempted some particular cases). This attests to the political currency of the protective role of ESA. But at the same time, it must be said that discussions of possibly invoking the Endangered Species Committee do arise in connection with the perceived need to keep the implementation costs of RPAs "within reason." The bottom line is that the costs of ESA protection do matter.

The factors that can cause adequate fixed management measures to require an extraordinary margin of safety derive from two sources of uncertainty. One is uncertainty about environmental changes and the other is uncertainty about the effectiveness of actions intended to raise the persistence probability for a population.

As scientists we are becoming increasingly aware of the large influence of shifts of oceanographic conditions on the biology of many species, on community species composition, and on the productivity of entire ecosystems. These changes, such as the Pacific Decadal Oscillation, tend to occur erratically, so we cannot predict the shifts and can only assign them probabilities. The shifts are infrequent enough that the attendant regimes often persist for many years. Because the different regimes may have very different implications for the demography of an endangered population, a set of management measures appropriate for the current state may be ill-suited to the conditions when a regime changes. Fixed management measures are a poor strategy for erratically shifting conditions.

Similar considerations arise in connection with episodic hazards such as rare storms, epidemics, red tides and the like, and even some human-caused hazards such as oil spills, whose long-run frequency of occurrence may be known to some extent but whose actual timing is unpredictable. For these unpredictable, infrequent hazards, there may be some preemptive measures that can be implemented in advance, such as establishing multiple spatially distributed populations to spread the risk from spatially limited episodic hazards and establishing protected reserves at sites that, for one reason or another, are less subject to the adverse effects of the episodic events or that actually are likely to be favored locations when a regime shift occurs. But if opportunities for preemptive measures are too restricted, the efficient strategy is to prepare for contingencies: develop plans for responding to the various episodic hazards, prepare the logistics and necessary resources in a state of readiness, and obtain in advance institutional authorization so that as soon as conditions indicate the need the response can be launched.

The second kind of uncertainty that works against efficiently assuring a high enough probability of survival for a long time horizon is uncertainty about the effectiveness of our management actions. There are populations that have experienced long periods of decline where the ultimate causes of the decline are unresolved. In those cases, reversing the decline by intervening directly in its cause is not a ready option. There may be multiple hypotheses about the cause of the decline. The costs of attempting to mitigate for all of them could be very high and some of the efforts might work at cross-purposes.

When uncertainty about causes and effects is so high, the strategy of an experimental approach is attractive. Experiments can usually be carried out on a scale that is smaller than full deployment of a recovery action, which can help control costs and limit risk. Well-designed experiments could reveal which management measures, among those tried, hold more promise for halting or reversing the population decline. After the results of experimentation have reduced the uncertainties, it would be possible to implement more selective management actions with a much higher probability of successful outcomes.

Consider a population that numbers 30,000 and is declining at a steady and unexplained rate of 5% per year. The steady decline, especially if it has been going on for an extended time, is legitimate cause for concern; and in some ways, the uncertainty about the cause is even more threatening than if the cause were definitely known. But with these numbers, the decline could continue for another 35 years and still leave the population above 5,000. If experiments were allowed, and resources for such experimentation and attendant monitoring were available, 35 years ought to be enough time to get a lot of answers about the causes of the decline and effective measures to reverse it. A reduction from 30,000 to 5,000 animals is not something that we would endorse if it were avoidable (at tolerable cost). But, the increased risk of extinction from allowing a decline to 5,000 followed by sustained population growth managed through the improved information gained in the interim may not be very great, provided we were certain enough that, one way or the other, the 35 years of experimentation would resolve the management questions and certain enough that the management solutions would be adopted and effectively implemented.

Scope for Experimentation under the ESA

The ESA has some provision for experimentation. Section 10 allows for scientific research permits that may cause "take" but not "jeopardy" if it is determined that the research will be beneficial to conservation of the population. Under this provision, some degree of short-term harm to the population is allowed if the harm is more than offset by a prospect for its benefit. The jeopardy restriction serves as a minimal standard for how much short-term harm is allowed and how the trade-off against longer-term benefit is weighed. The experiment is not permitted if it poses jeopardy as defined in the Act (a more stringent standard may apply to scientific research permits).

Section 10(j) allows for establishing experimental populations of a species listed as endangered. The purpose of the experimental population must be to further the conservation of the listed species, but the special experimental status reduces the level of protection. The management of experimental populations is exempted (except in national parks or national wildlife refuges) from Section 7 prohibitions on "jeopardy" and from the requirement to designate "critical habitat." If it is determined that the experimental population is "not essential to the continued existence of the species," its management is further exempted (except in national parks or national wildlife refuges) from the automatic Section 9 prohibitions on "take" although in practice experimental populations are usually protected against most take under discretionary provisions applying to threatened populations. The status of "experimental population" applies only to a population that is deliberately established outside of the current species range and continues to be geographically separate from the rest of the species. If such a population subsequently becomes demographically connected to the rest of the listed population, it loses its experimental status and receives the same level of protection as the rest of the listed population.

If we try to abstract a common logic for the ESA provisions for scientific research and for experimental populations, I think it is that experimentation is not allowed if it worsens the situation of the population by an amount that would trigger a "jeopardy" determination. Similarly, once an experiment that may benefit the listed population has been initiated, it may not be abandoned if that action would trigger a "jeopardy" determination by eliminating the potential for benefit.

Experiments as Reasonable and Prudent Alternatives

We note that the explicitly sanctioned experimentation under the ESA does not mention experimentation as a "reasonable and prudent alternative" (RPA) to alleviate jeopardy. The courts, in fact, have ruled that for a proposed RPA to be a valid remedy for jeopardy, the proposed actions must be "reasonably certain to occur." In actual cases, the "reasonably certain to occur" standard has been applied to the question of whether there is a high enough certainty that the proposed actions will be implemented. To my knowledge, this standard has not yet been applied to the question of whether the proposed actions will have the hoped-for consequences for the population.

There has been great reluctance on the part of the responsible agencies to propose experimental RPAs. I am aware of only one major biological opinion that has incorporated frank experimentation—the 2000 biological opinion for operation of the Columbia River hydrosystem in connection with jeopardy to listed populations of salmon. The costs of at least one of the original candidate mitigation actions in this case were high enough to raise thoughts of convening the Endangered Species Committee. The RPA adopted by NMFS was less costly. In the biological opinion, NMFS acknowledged uncertainty about the effectiveness of its preferred mitigation actions and committed to a "check-in" schedule for reviewing the results of effectiveness monitoring. The RPA does state a clear "measurable and objective" quantitative criterion, based on a measured response of the population, according to which effectiveness will be judged. The RPA does not commit very specifically to the mitigation actions that will be undertaken in the event that the monitoring reveals that the mitigation actions have not been effective, stating rather that if this occurs, all options will again be opened for consideration at that time.

As it turns out, the court ruled this biological opinion invalid on grounds that it fails the "reasonably certain to occur" standard (*Committee for Humane Legislation* v. *Richardson,* 414 F. Supp. 297, 305 [(D.D.C. 1976]). My reading of the judge's opinion is that the ruling had to do with insufficient assurance that the proposed actions would be implemented (several of the proposed actions were not under the agency's control) and was not a question of uncertainty about their effectiveness.

Mechanisms are needed to deal rationally with the uncertainties about causes of population declines and uncertainties about eventual effectiveness of planned conservation measures. Otherwise, in cases that combine high economic stakes with high levels of genuine scientific uncertainty about the effectiveness of proposed mitigation actions, the demand for fixed RPAs that will assuredly alleviate jeopardy simply asks for more certainty than can be had with the available information. Operating under these premises can encourage pressures that are not healthy for the credibility of the science.

The technical challenge is to devise a system that will allow experimentation when necessary without increasing the risk to the population. The policy challenge is to embed this in a legal framework that confers predictable implementation.

The *theory* of how to solve the technical challenge—the formal mathematical theory of adaptive management (Walters and Hilborn 1976)—has been known for over a quarter of a century. It offers a framework, combining the statistical and operations research disciplines of decision theory and dynamic programming, to guide the experimentation according to a continuously updated risk/benefit analysis. The analysis takes account of current uncertainties and weighs each experimental action according to three simultaneous characteristics: (1) its prospects for directly benefiting the population, (2) its prospects for directly harming the population, and (3) its prospects for providing new information that can then be employed to benefit the population. The formal mathematical analysis in effect considers alternative decision trees where branches are characterized both by probabilities and by "if-then" statements governing the actions that will be taken depending on events. In the formal mathematical theory, each stage of the process employs analyses with explicit decision rules for all possible contingencies and, with quantification of consequences weighted for their probabilities, for all possible future paths through the decision tree. In theory, this works very well.

Case histories involving public sector resource management plans that were called "adaptive management" are a rather different story. The practical problem has been that nonrigorous, nonmathematical versions of the *language* of management by improvisational experimentation are excessively attractive for a variety of political and bureaucratic reasons; but without the rigorous analytical guidance, the implementations are prone to spin out of control, and the results are disappointing. Review of the track record has led to the judgment that the many failures were primarily from institutional causes (Ludwig et al. 1993). Successful adaptive management requires a highly disciplined setting.

Carefully worked-out cases are needed to demonstrate technically how a rigorous adaptive approach to experimental RPAs would perform and to encapsulate the lessons learned in a rule system that, if adopted as policy, would set standards that would guarantee acceptable performance. This is a large undertaking.

Recovery Criteria under an Adaptive Recovery Plan

The option for crafting adaptive recovery plans may generate enthusiasm for premature declaration of victory in a recovery effort. To guard against this, we first need moral clarity about the purpose of the ESA. The general purpose is to prevent an unnaturally high rate of extinctions; for listed populations specifically, the purpose is to hold both the extinction probability and the extinction frequency to an acceptably low level. By this measure, I think the ESA has been a grand success. "The purposes of this Act are to provide a means whereby the ecosystems upon which endangered species and threatened species depend *may be conserved,* to provide a program for *the conservation of such endangered species and threatened species*" (emphasis added). Preventing extinctions does amount to "conserving" the species.

This appraisal runs counter to the critique that the ESA is a failure because so few delistings have been achieved. That is the wrong measure. Moreover, if I am correct that the future will bring ecological circumstance less favorable than the present for the survival of many species, then delisting scorecards will become more and more misleading as an indicator of the success of our policy. *Survival* of listed species should be held as the measure of success.

Now, an acceptable adaptive recovery plan will, by definition, outline a decision tree of protective measures, monitoring, and contingencies that confers a probability of extinction low enough to meet the stated quantitative standard. Should adoption of such a plan, with assurances that it will be adhered to, constitute a basis for delisting? I think not. One of the four causes described in the legislation as a basis for ESA listing is the inadequacy of existing regulations. Thus, one of the necessary tests for delisting should be an examination of the population's persistence probability if ESA protection were lifted. If delisting removes the regulatory guarantee that the adaptive recovery plan will be implemented, delisting would effectively cause the extinction probabilities to revert to the higher extinction probabilities that apply in the absence of the recovery plan.

BEST SCIENCE

The call for "best science" is now a common feature of environmental legislation. The ESA requires that decisions make use of the best available commercial and scientific data. The MMPA requires that decisions be based on the best scientific evidence available. How should we recognize "best science," what is "best data," and what are the implications for selection of appropriate error tolerances in a quantitative standard?

This language might seem like an invitation to the courts to enforce a confidence requirement on scientific inferences used in regulatory decision making or to set their own standards for data quality. Indeed, the long history of dolphin stock assessments for stocks involved in the eastern tropical Pacific tuna purse seine fishery has been strongly influenced by a court decision that the tuna vessel observer data must be used in the analysis, notwithstanding the view of NMFS scientists at that time that the data were unreliable (*American Tunaboat Association* v. *Baldrige,* 738 F.2d 1013 [9th Cir. 1984]). Another instance where a federal judge ruled on a statistical technicality arose in litigation over the Environmental Protection Agency's (EPA) classification of cigarette smoke in indoor air as a carcinogen based on its analysis of epidemiological data on passive exposure to tobacco smoke (*Flue-Cured Tobacco Cooperative Stabilization Corporation* v. *Environmental Protection Agency,* 4 F. Supp. 2d 435 [M.D.N.C. 1998]). There the judge ruled against EPA's departure from the usual practice in experimental science of requiring a 95% confidence limit to achieve "statistical significance." (EPA used a lower confidence level.) The decision was subsequently vacated on appeal based on other grounds (*Flue-Cured Tobacco Cooperative Stabilization Corporation* v. *Environmental Protection Agency,* 313 F.3d 852 [4th Cir. 2002]). In some other venues, the best scientific evidence standard has been interpreted by the court as being met by a "preponderance of evidence" standard of proof, which might mean any probability greater than 50%.

These interventions by the courts are worrisome since they seem disconnected from the logic of setting error tolerances as an expression of policy to reflect societal priorities, and the preponderance of evidence standard is contrary to the decision theoretic approach of using the confidence level in a decision rule to adjust the margin of safety according to the certainty afforded by the data. Note that a preponderance of evidence standard corresponds to use of a point estimate as the best estimate, essentially discarding the information quantifying the uncertainty of that estimate. The potential for confusion of this sort would be greatly reduced if legislation stated the appropriate error tolerances or explicitly stated the basis for the agencies to set error tolerances in regulation.

Best science might best be interpreted as using the best statistical techniques to analyze all the available data, weighting each data source according to its quality. This may sound like a daunting, and perhaps megalomaniacal ambition, but in fact recent Bayesian approaches demonstrate that it is practical. I say this at a time when the uses of Bayesian inference are unquestionably increasing in conservation biology and fisheries management, but professional consensus on best statistical practices, or even on the best way to conduct a Bayesian analysis, has not yet been reached. I think the right answers are "out there" in the existing literature, but it will take some concerted effort—again, in the form of extensive worked-out examples, research to demonstrate performance characteristics, and communication among scientists in a forum such as symposia and workshops devoted to the topic—to accelerate formation of a working professional consensus.

THEORETICAL OVERVIEW

Growth of the world's human population and an increase in that population's industrial, agricultural, and fishing activities will create a future that poses even greater conservation challenges than the present. One strategy for coping with that future is to take it into account in present conservation planning.

Much conservation activity is the result of government regulation. Because regulation is coercive, it is often contested. Uncertainties in a regulatory situation provide grounds for contention.

Knowledge of trends notwithstanding, our particular knowledge of the future is uncertain. We face a quandary, then, in that future environmental deterioration warrants anticipatory responses in regulation, but regulation can be difficult to implement effectively in the face of uncertainty. The solution is to frame regulation in terms of standards that incorporate an explicit, quantitative burden-of-proof provision.

The canonical form for such a standard is a statement of a desired state of reality and a stipulated error tolerance for attainment. This canonical form gets to the heart of the matter, capturing an expression of what we want along with an expression of the priority that we accord it. Theoretically, the canonical form encourages consistency and efficiency since it allows flexibility in the search for ways to comply with the essential standard. Psychologically, however, because the canonical form is stated in very abstract terms, it is not readily embraced by those without technical training. Decision rules bridge this gap by stating how a decision process is determined by concrete operations on measurable and objective data. The technical challenge in crafting a decision rule is to ensure that it does perform in a way that meets the underlying standard and to provide a mechanism for effectively using as many data as are available in each case.

The present reality is that much of our most important conservation legislation provides only vague standards and few decision rules. An effective regulatory framework for coping with a hostile future requires development of explicit quantitative standards and an array of decision rules for efficiently satisfying those standards. In order for this to happen, a large research commitment must be made to craft efficient decision rules, to document their functioning, and to illustrate and communicate the results in terms that are concrete and vivid enough to engender a policy consensus. The experience in developing the PBR system provides some indication of the magnitude of this research undertaking.

SPECIFIC RESEARCH AGENDA

The specific topics that I think should receive high priority for research efforts and for workshops to facilitate consensus are as follows:

1. A consistent quantitative standard for ESA listing.
2. A consistent quantitative standard for ESA jeopardy.
3. A consistent quantitative standard for MMPA negligible impact.
4. A consistent quantitative standard for MMPA OSP in conditions of declining carrying capacity.
5. A consistent quantitative standard for the ecosystem protection provisions of the MSA.
6. An adaptive framework for ESA RPAs.
7. An option for experimentation in ESA RPAs.
8. An adaptive framework for ESA recovery plans.
9. A standard default Bayesian PVA.
10. Minimal standards for conducting a Bayesian analysis.
11. A catalog of relevant Bayesian prior distributions.

Such work can provide the regulatory decision structure with improved analytical tools for making conservation decisions in ways that anticipate changing, and often intensifying, challenges to population persistence and that, in an orderly and consistent way, will select conservation actions to meet the challenge.

TIMOTHY J. RAGEN,
RANDALL R. REEVES,
JOHN E. REYNOLDS III,
AND WILLIAM F. PERRIN

Future Directions in Marine Mammal Research

THE MARINE MAMMAL PROTECTION ACT of 1972 (MMPA) is one of a series of laws passed in the late 1960s and 1970s designed to moderate the effects of human activities on the natural environment. The MMPA grew out of controversies surrounding the annual killing of hundreds of thousands of dolphins in the eastern tropical Pacific tuna fishery, the annual killing and skinning of hundreds of thousands of harp seal pups in Canada, and the failure of the international community to address the decimation of large whales from commercial hunting. In passing the Act, the U.S. Congress found that, as a matter of policy, "marine mammals have proven themselves to be resources of great international significance, esthetic and recreational as well as economic, and it is the sense of the Congress that they should be protected and encouraged to develop to the greatest extent feasible commensurate with sound policies of resource management and that *the primary objective of their management should be to maintain the health and stability of the marine ecosystem*" (16 U.S.C. 1362, Section 2; emphasis added).

In the three decades since the MMPA was passed, considerable progress has been made in the science and management of marine mammals. A moratorium, albeit controversial and not complete, has been implemented on the commercial harvest of large whales. Reported mortality of dolphins in the eastern tropical Pacific has been reduced to fewer than 2,000 animals annually. The number of marine mammals killed incidentally in U.S. fisheries has been reduced significantly under Sections 117 and 118 of the Act. These sections require assessment of marine mammal stocks, monitoring and reporting of serious injury and mortality in fisheries, categorization of fisheries according to the level of inflicted serious injury and mortality, determination of the effect on each stock based on the potential biological removal (PBR) concept, and implementation of take-reduction efforts where PBR levels are exceeded. Other marine mammal species that had been driven to low numbers by commercial hunting or other human activities have exhibited partial or full recovery (e.g., the California gray whale, the northern elephant seal, the California sea lion, the western Arctic stock of the bowhead whale, and some populations of blue whales and humpback whales). Traditional knowledge of Alaska Natives is being introduced into scientific and management efforts for marine mammals in Alaska through co-management agreements that foster Native involvement. A host of significant advances have been made in scientific research methods (e.g., surveys, diet analysis), technology (e.g., satellite-linked transmitters, physiological and environmental monitors), and disciplines (e.g., genetics, taxonomy, health). Moreover, both scientists and managers are expanding the nature of their investigations and activities to address the role and significance of marine mammals within marine ecosystems.

In spite of such progress, a number of marine mammals are now at the center of controversies that pit conservation against other human values and goals (e.g., economic, security). Some of these controversies involve species listed as endangered or threatened under the Endangered Species Act (ESA):

- *North Atlantic right whale.* Numbering about 350 animals and potentially declining, North Atlantic right whales are being injured and killed by ships strikes and entanglement in fishing gear. Shipping and fishing are major industries in the western North Atlantic.
- *Florida manatee.* Numbering at least 3,300 with evidence of population growth over the past three to four decades, manatees are at the center of conflicts arising from coastal development and boating activities in Florida's nearshore waters. Over the past decade, the human population in Florida also has grown at the rate of about 1,000 persons each day and abundance is projected to double by around 2030.
- *Hawaiian monk seal.* Numbering about 1,300 animals, the Hawaiian monk seal is poised for further decline in the Northwestern Hawaiian Islands. It is, however, starting to populate the main Hawaiian Islands, where its presence requires effective management and education of Hawaiians and visitors to avoid conflicts on beaches and in nearshore waters.
- *Steller sea lion.* Numbering about 35,000 animals, the western stock of Steller sea lions has declined by 80% or more over the past three decades. The role of commercial fisheries in this decline remains an unresolved source of controversy.
- *Southern sea otter.* Numbering about 2,500 animals, the southern sea otter population has failed to recover as expected and is in potential conflict with a number of nearshore activities including, in particular, fisheries.

Other controversies involving species and stocks designated as depleted under the MMPA include the following:

- *Northeastern offshore spotted dolphin, eastern spinner dolphin, coastal spotted dolphin.* Numbering about 640,000, 450,000, and 145,000, respectively, these stocks are estimated to be at about 19–28, 44, and 32%, respectively, of historical abundance. Nonetheless, they have failed to recover as expected in spite of large reductions in reported annual mortality in eastern tropical Pacific tuna fisheries. Evidence suggests that their failure to recover may be due, at least in part, to cryptic changes in survival and reproduction associated with fishery methods involving chase and capture of dolphins to catch associated tuna. Such effects are central to the debate as to whether the tuna fisheries should be labeled as "dolphin-safe."
- *North Atlantic coastal bottlenose dolphin.* Numbering about 37,000 in total, this stock was designated as depleted following an unusual mass-mortality event along the U.S. East Coast in 1987–1988. The stock has subsequently been the focus of take-reduction efforts to reduce fisheries-related injury and mortality. Based on available scientific evidence, the stock is being split into multiple stocks whose individual status is unclear. These new stocks require additional protection from incidental catch in fisheries, disease, contaminants, and harmful algal blooms.
- *Cook Inlet beluga whale.* Numbering about 350 animals, this stock was reduced to low levels by ill-managed subsistence harvests in the 1980s and 1990s. It was designated as depleted with the assumption that a moratorium on the harvest would lead to recovery. To date, the population shows no signs of recovery and appears to be in need of additional protections, particularly in light of expected coastal development in the Cook Inlet area.
- *Southern resident killer whales.* Numbering about 80–85 animals, southern resident killer whales occur in Puget Sound and the waters around the southern end of Vancouver Island off the U.S. and Canadian west coast. Their numbers were reduced in the 1960s and early 1970s by live capture for aquaria, began to recover from the mid-1970s to the mid-1990s, and then began a sharp decline that may be due to loss of prey (salmon), the effects of contaminants, disturbance from whale-watching and other vessel activity, or some combination of these risk factors.

Similar controversies exist in the waters of other countries:

- *Asian river dolphins.* Probably numbering only in the tens of individuals, the Yangtze river dolphin, or baiji, may well become extinct within our lifetimes. It has declined due to a suite of human-related activities in the Yangtze River including vessel traffic, fishing, pollution, habitat degradation, and waterway management. The Indus river dolphin is thought to number at least 1,000 individuals. The abundance of the Ganges river dolphin is unknown but thought to be greater than the Indus river dolphin. Both of these subspecies have been extirpated from large regions of their habitat and are faced with a wide range of risk factors including waterway management, pollution, habitat fragmentation and degradation, fisheries, and hunting.

- *Western North Pacific gray whales.* Numbering about 100 animals, the western gray whale population is on the verge of extinction. Its primary foraging area is near the northeastern shore of Sakhalin Island in the Okhotsk Sea, where oil and gas deposits have been discovered and exploitation has begun. These operations pose a number of risks to the gray whale, all of which could potentially hasten the population toward extinction.
- *Vaquita.* Numbering in the low hundreds, the vaquita, a species of small porpoise, is also threatened with extinction as a result of incidental mortality in fishing gear in the northern Gulf of California.
- *Mediterranean monk seal.* Numbering about 450 to 525 animals and continuing its decline, the Mediterranean monk seal is extremely fragmented in distribution and vulnerable to a range of human activities, including fisheries interactions, shooting, disturbance, and habitat degradation. Recovery actions are confounded by the lack of international cooperation with regard to research and management.
- *Okinawan dugong.* Probably numbering fewer than 100 animals, and possibly fewer than 50, the habitat of the Okinawan dugong population is just offshore of the proposed building site for a new U.S. Marine Corps Air Station. The site will be determined by the government of Japan, which intends to replace the existing air station at Futenma.

These controversies were anticipated by the framers of the MMPA, ESA, and NEPA. Such controversies raise the questions of whether science and management are currently adequate to achieve our conservation goals and whether they will be adequate when future human population growth and consumption increase demands and impacts on marine ecosystems even further. These questions motivated the chapters of this volume and the consultation they were intended to inform and guide.

THE MARINE MAMMAL COMMISSION CONSULTATION: EMERGENCE OF STRATEGIES AND COMMON THEMES

In 2002 the U.S. Congress provided funding to the Marine Mammal Commission to address these questions. In 2003 the Commission consulted with leading scientists from six countries to identify and explore future directions for marine mammal research. The consultation focused on the threats described in the various chapters in this volume. In 2004 the Commission completed an overview report to Congress and a detailed summary of meeting discussions and recommendations for each threat. Both of these documents are available from the Marine Mammal Commission.

At the Commission's consultation, each of the issue-oriented presentations stimulated vigorous discussion and debate that led to important research recommendations, as described in the detailed meeting summary. In addition, however, common themes or strategies emerged from these presentations and the subsequent discussions, and we believe those strategies are central to the question of whether scientific and management efforts will be sufficient for guiding marine mammal and marine ecosystem conservation in the future. For that reason, we devote the rest of this chapter to those strategies.

1. *Develop long-term, multidisciplinary programs suitably scaled to ecosystem complexity.* Our understanding of natural ecosystems has evolved from a simple, static, "balance-of-nature" paradigm to one reflecting their dynamic, complex character. This new paradigm recognizes that ecological change is caused by multiple factors and occurs on a range of spatial and temporal scales. Under this new paradigm, maintaining the health and stability of marine ecosystems is more complicated conceptually and practically and the collection of adequate "baseline" information for assessing human impacts is a greater challenge. Still, collection of such information is an indispensable step for assessing human impacts on marine ecosystems, which may be additive to or synergistic with natural changes.

Understanding the complex dynamics of multifaceted, variable marine ecosystems and accounting for diverse human effects requires multidisciplinary research (e.g., oceanography, marine mammal and fishery biology, invertebrate biology, physiology, ecology, and various social sciences). Research must be tailored to match the temporal and spatial scale of complex ecosystem dynamics. Multidisciplinary research and assessment strategies require better communication and coordination among previously isolated disciplines, expansion of existing monitoring programs, and new programs where none currently exist. The California Cooperative Oceanic Fisheries Investigations (CalCOFI) program exemplifies a comprehensive, long-term research approach that should be continued and replicated to study other marine ecosystems. Well-managed marine protected areas (MPAs) are needed as controls for distinguishing between natural phenomena and anthropogenic effects. A comprehensive national strategy is needed to set priorities for research and assessment, measure progress, and secure support over the required long periods and large areas.

2. *Ensure that population and ecosystem assessment programs are sufficient to inform management decisions regarding current and future threats.* Existing assessment efforts are, in many

cases, not sufficient to describe the status, trends, and ecology of marine mammal populations, the effects of human activities on them, and the status of the ecosystems of which they are a part. Basic information on such things as abundance, demography, distribution, mortality, reproduction, and health is lacking for most populations, including some at great potential risk from human impacts. Declines of 50% or more could go undetected for some populations under the present inadequate monitoring systems. More rigorous assessment programs are needed for marine mammals. The programs must be appropriately scaled temporally and spatially, and they must involve multidisciplinary approaches that relate marine mammal status and trends to natural and human-altered ecosystem dynamics.

3. *Develop and validate specific, measurable, and robust management standards to achieve the conservation goals of the Marine Mammal Protection Act and related legislation.* The existing management standards set to achieve the goals of the MMPA and related legislation (e.g., the ESA) often lead to controversy because they (1) lack sufficient specificity, (2) cannot be reliably measured for natural populations, or (3) vary as a function of human activities and therefore do not provide stable or suitable references for assessing the effects of those activities. The "optimum sustainable population" (OSP), an important standard in the MMPA, has been estimated for only a few marine mammal populations. Scientists and managers generally assume the OSP to be a fraction of the environmental carrying capacity, but carrying capacity is often unknown for marine mammal populations and, in some cases, may have been artificially reduced by human activities. The ESA standards of "jeopardy" and "adverse modification" are similarly vague and controversial. More specific, measurable, and robust standards must be developed and validated to guide management and ensure that conservation goals are met.

4. *Identify marine mammal conservation units essential to ecosystem health and function.* Marine mammal species often exist as multiple population, stock, or demographic units with limited interaction among them. These units can vary in distribution, status, trends, vital rates (survival and reproduction), life history characteristics, and genetics. Although subtle, such variation implies different ecological and evolutionary roles for these units. Killer whales in the North Pacific, for example, comprise three "ecotypes" that, among other things, differ in foraging strategies. Their respective roles in marine ecosystems clearly differ, depending upon whether they prey on fish, marine mammals, or both. The identification and conservation of such units (whether they are called populations, stocks, subspecies, or ecotypes) is essential to maintaining the natural function of healthy ecosystems.

5. *Increase international cooperation in studying and addressing human-related threats.* Many threats to marine mammals and marine ecosystems result from the activities of more than one nation, and many marine mammal populations occur in the waters of two or more nations. Hence, marine mammals and threats to them are most effectively studied and managed through international cooperation. Cooperative arrangements are needed to address such multinational issues and may range from informal sharing of information on transboundary stocks to highly structured agreements to study and manage resources in international waters (e.g., the Convention for the Conservation of Antarctic Marine Living Resources). Cooperation increases both the knowledge base pertaining to threats to marine mammals and marine ecosystems and the resources available to study and manage those threats.

6. *Properly assess and communicate the strengths and limitations of the scientific process, including measures of uncertainty that are an essential element of high-quality science.* Management decisions made under the MMPA and related legislation are to be based on the "best available science," but because of the difficulties involved in studying marine ecosystems, that science may be associated with a high degree of uncertainty. Accounting for such uncertainty is an essential element of risk analysis and informed decision making. To reliably guide decision makers, the best available scientific information must include appropriate descriptions of uncertainty (i.e., how good is the information?). Such descriptions are needed to judge whether scientists are able to detect significant human-related effects when they occur and to assess the likelihood that decision makers will impose unnecessary regulation of human activities or allow activities with excessive environmental impacts.

7. *Address ultimate as well as proximate causes of environmental problems.* Research on the factors affecting marine mammals and marine ecosystems often focuses almost entirely on proximate rather than ultimate causes. Yet, virtually all of the threats discussed at the consultation are ultimately related to the size, growth rate, consumption patterns, and behavior of the Earth's human population. Some threats may be mitigated or resolved by technological advances, but others are less likely to be resolved technologically and will probably worsen over time. In the foreseeable future, the human species is projected to grow in number and further increase its consumption of resources. Maintaining the health and stability of marine ecosystems requires focused long-term research on our own expanding population, shifts in distribution, and patterns of consumption (i.e., our ecological footprint). Such research is needed to elucidate our impacts and provide guidance on how to limit and compensate for them. It should be linked to long-term assessments of marine biodiversity, ecosystem resilience, the loss of marine habitats through development and contamination, consumption of renewable and nonrenewable resources, and human activity in the marine environment.

RISKS OF INADEQUATE RESEARCH AND MANAGEMENT

The Commission asked participants to predict the consequences of not pursuing a more integrated, holistic, and anticipatory marine mammal research agenda. They identified the following consequences:

- The goals of the MMPA, the ESA, and other environmental legislation will likely not be met, and marine ecosystems will continue to deteriorate.
- Some marine mammal populations will persist, perhaps in large numbers, but many of those that are presently endangered will decline to extinction, as has already occurred for the Steller's sea cow, North Atlantic gray whale, Caribbean monk seal, and Japanese sea lion.
- Management and recovery efforts will remain reactive rather than proactive, and will be confounded by uncertainty and controversy.
- Controversies will be fueled by our inability to distinguish anthropogenic effects from natural phenomena.
- In the absence of clear, unambiguous evidence of human impacts, economic demands will force governments and management agencies to compromise conservation objectives, and this will lead to further losses of biodiversity and ecological integrity.
- Remedies will continue to focus on proximate rather than ultimate causes, and short-term conservation successes will be offset by long-term conservation failures.
- Long-term degradation of marine ecosystems will increasingly limit socioeconomic alternatives, as has already been witnessed in many overfished ecosystems.
- Alaska Natives and other indigenous people will have to drastically modify, if not abandon, subsistence aspects of their cultures.
- The natural character of marine ecosystems will remain unknown and eventually become unknowable.
- Ultimately, we will pass on to our children a world diminished in its diversity, its options, and its biological wonder.

IMPLEMENTATION OF THE RECOMMENDED STRATEGIES

The Marine Mammal Commission believes that implementation of the strategies recommended herein is essential if we are to resolve the threats to marine mammals and marine ecosystems, to avoid the adverse future consequences anticipated by the consultation participants, and to achieve the goal of maintaining healthy, stable marine ecosystems without imposing unnecessary constraints on human activities. Implementation requires an investment beyond current levels of support for research and assessment. Like preventive medicine, such investment will prove to be cost effective over time. Pending congressional approval and direction, the Marine Mammal Commission is prepared to assist in the implementation of these recommended strategies.

Who should bear the added cost of the essential research? This question is not a scientific issue, per se, but was discussed by consultation participants because it has a bearing on whether or not the necessary work can be done. At present, much of the burden for carrying out such work falls on offices and divisions within the National Marine Fisheries Service and Fish and Wildlife Service. Their budgets have been and are presently insufficient for implementing the recommended strategies. In general, consultation participants supported the view that the cost of implementing these recommendations should be borne by those who stand to benefit financially from activities that pose a threat to marine mammals and marine ecosystems.

Finally, science alone does not and cannot resolve the threats that have been described. Solutions must reflect societal values, whether cultural, economic, aesthetic, or conservation oriented. Science provides both knowledge that can shape those values and tools for estimating the costs and benefits of particular courses of action. Proactive science, in particular, can inform the public and decision makers of the effects of certain actions before social, economic, and environmental crises arise. The Marine Mammal Commission hopes that its efforts to foster thoughtful, carefully directed, proactive science will be useful in preventing such crises, shaping human values, and maintaining the health and stability of marine ecosystems.

Literature Cited

Aanes, R., B. E. Saether, F. M. Smith, E. J. Cooper, P. A. Wookey, and N. A. Oritsland. 2002. The Arctic Oscillation predicts effects of climate change in two trophic levels in a high-arctic ecosystem. *Ecology Letters* 5:445–453.

Abraham, W. M., A. J. Bordelais, J. R. Sabater A. Ahmed, T. A. Lee, I. Serebriakov, and D. G. Baden. 2005. Airway responses to aerosolized brevetoxins in an animal model of asthma. *American Journal of Respiratory and Critical Care* 171:26–34.

Acevedo-Gutierrez, A., D. A. Croll, and B. R. Tershy. 2002. High feeding costs limit dive time in the largest whales. *Journal of Experimental Biology* 205(12):1747–1753.

Acevedo-Whitehouse, K., F. Gulland, D. Greig, and W. Amos. 2003. Disease susceptibility in California sea lions: inbreeding influences the response of these animals to different pathogens in the wild. *Nature* 422:35.

Agreement for the Conservation of Small Cetaceans of the Baltic and North Seas. 2000. Proceedings of the Third Meeting of Parties to ASCOBANS, Bristol, U.K., 26–28 July 2000. ASCOBANS Secretariat, Bonn, Germany.

Aguilar, A. 2000. Population biology, conservation threats and status of Mediterranean striped dolphins (*Stenella coeruleoalba*). *Journal of Cetacean Research and Management* 2:17–26.

Aguilar, A., and A. Borrell. 1994. Abnormally high polychlorinated biphenyl levels in striped dolphins (*Stenella coeruleoalba*) affected by the 1990–1992 Mediterranean epizootic. *Science of the Total Environment* 154:237–247.

Aguirre, A. A., T. M. O'Hara, T. R. Spraker, and D. A. Jessup. 2001b. Programs monitoring the health and conservation of marine mammals and their ecosystems. In *Conservation Medicine: Ecological Health in Practice* (A. A. Aguirre, R. S. Ostfeld, C. A. House, G. M. Tabor, and M. C. Pearl, eds.). Oxford University Press, New York, NY.

Aguirre, A. A., R. S. Ostfeld, C. A. House, G. M. Tabor, and M. C. Pearl (eds.). 2001a. *Conservation Medicine: Ecological Health in Practice.* Oxford University Press, New York, NY.

Ahola, M., T. Laaksonen, K. Sippola, T. Eeva, K. Rainio, and E. Lehikoinen. 2004. Variation in climate warming along the migration route uncouples arrival and breeding dates. *Global Change Biology* 10(9):1610–1616.

Ainley, D. A., C. T. Tynan, and I. Stirling. 2004. Seabirds and marine mammals and ice. In *Sea Ice: An Introduction to Its Physics, Chemistry, Biology and Geology* (D. N. Thomas and G. S. Dieckmann, eds.). Blackwell Publishing. Oxford, U.K.

Alaska Bulletin. 1986. Cadmium in Walrus—St. Lawrence Island. *Alaska Bulletin* 22:2, 19 Dec. 1986.

Alaska Bulletin. 2002. Statewide Mercury Hair Biomonitoring Program. *Alaska Bulletin* 11. Available at www.epi.hss.state.ak.us.

Alaska Bulletin. 2003. PCB blood test results from St. Lawrence Island. *Alaska Bulletin* 7(1). Available at www.akepi.org.

Alaska Community Action on Toxics. www.akaction.net.

Alaska Eskimo Whaling Commission. www.uark.edu/misc/jcdixon/Historic_Whaling/AEWC/AEWC.htm.

Alaska Native Harbor Seal Commission. www.fakr.noaa.gov/protectedresources/seals/actionplan.htm.

Alaska Native Harbor Seal Commission and National Marine Fisheries Service. 2000. Harbor seal co-management agreement. 12 pp.

Alaska Native Tribal Health Consortium (ANTHC). 2002. *Diabetes among Alaska Natives.* Vol. 5 (5). ANTHC, Sept./Oct. 2002.

Albers, P. H., and T. R. Loughlin. 2003. Effects of PAHs on marine birds, mammals and reptiles. In *PAHS: An Ecotoxicological Perspective* (E. T. Douben, ed.). John Wiley & Sons Ltd., New York, NY.

Alexander, K. A., and M. J. Appel. 1994. African wild dogs (*Lycaon pictus*) endangered by a canine distemper epizootic among domestic dogs near the Masai Mara National Reserve. *Journal of Wildlife Diseases* 30:481–485.

Allen, M. R., P. A. Stott, J. F. B. Mitchell, R. Schnur, and T. L. Delworth. 2000. Quantifying the uncertainty in forecasts of anthropogenic climate change. *Nature* 407:617–620.

Alley, R. B., J. Marotzke, W. D. Nordhaus, J. T. Overpeck, D. M. Peteet, R. A. Pielke Jr., R. T. Pierrehumbert, P. B. Rhines, T. F. Stocker, L. D. Talley, and J. M. Wallace. 2003. Abrupt climate change. *Science* 299:2005–2010.

Alverson, D. L., M. H. Freeberg, J. G. Pope, and S. A. Murawski. 1994. A global assessment of fisheries bycatch and discards. FAO Fisheries Technical Paper No. 339, U.N. Food and Agriculture Organization, Rome, Italy. 233 pp.

Anderson, D. M. 1994. Red tides. *Scientific American* 271:52–58.

Anderson, D. M., P. M. Glibert, and J. M. Burkholder. 2002. Harmful algal blooms and eutrophication: nutrient sources, composition, and consequences. *Estuaries* 25:704–726.

Anderson, D. M., and A. W. White (eds.). 1989. *Toxic Dinoflagellates and Marine Mammal Mortalities.* Proceedings of an expert consultation held at the Woods Hole Oceanographic Institution, Woods Hole, MA. Technical Report WHOI-89-36.

Anderson, D. R., K. P. Burnham, and W. L. Thompson. 2000. Null hypothesis testing: problems, prevalence, and an alternative. *Journal of Wildlife Management* 64:912–923.

Anderson, P. J., and J. F. Piatt. 1999. Community reorganization in the Gulf of Alaska following ocean climate regime shift. *Marine Ecology Progress Series* 189:117–123.

Anderson, R. M. 1979. Parasite pathogenicity and the depression of host population equilibria. *Nature* 279:150–152.

Andrew, R. K., B. M. Howe, J. A. Mercer, and M. A. Dzieciuch. 2002. Ocean ambient sound: Comparing the 1960's with the 1990's for a receiver off the California coast. *Acoustic Research Letters Online* 3(2):65–70.

Angliss, R. P., and K. L. Lodge. 2002. Alaska marine mammal stock assessments, 2002. U.S. Department of Commerce, NOAA Technical Memorandum NMFS-AFSC-133. 244 pp.

Angliss, R. P., G. D. Silber, and R. Merrick. 2002. Report of a workshop on developing recovery criteria for large whale species. NOAA Technical Memorandum NMFS-OPR-21.

Anon. 1999. Sea siren. *Nature Conservancy* 49(3):38.

Antonelis, G. A., J. D. Baker, T. C. Johanos, R. C. Braun, and A. L. Harting. In press. Hawaiian monk seal (*Monachus schauinslandi*): status and conservation issues. Third Northwestern Hawaiian Island Symposium, Honolulu, HI.

Antonelis, G. A., J. D. Baker, and J. J. Polovina. 2003. Improved body condition of weaned Hawaiian monk seal pups associated with El Niño events: potential benefits to an endangered species. *Marine Mammal Science* 19:590–598.

Archembault, M-C., M. Bricelj, J. Grant, and D. M. Anderson. 2002. Mitigation of harmful algal blooms with clay: effects on juvenile *Mercenaria mercenaria* (Abstract). 10th International Conference on Harmful Algal Blooms, 21–25 October, St. Petersburg Beach, FL.

Arctic Climate Impact Assessment. 2004. *Impacts of a Warming Arctic. Arctic Climate Impact Assessment.* Cambridge University Press, Cambridge, U.K.

Arctic Monitoring and Assessment Programme. 1998. *Assessment Report: Arctic Pollution Issues.* AMAP, Oslo, Norway.

Arctic Monitoring and Assessment Programme. *Pollution 2002.* www.amap.no/.

Arctic Monitoring and Assessment Programme. 2002. Chapters on Heavy Metals, Organochlorines, and Human Health. In *AMAP Pollution 2002.* www.amap.no/.

Arnould, J. P. Y. 2002. Southern squid jig fishery. Seal interaction project: report on observations of interactions between fur seals and fishing vessels. Australian Fisheries Management Authority Report. 17 pp.

Arrigo, K. R., and G. L. van Dijken. 2004. Annual cycles of sea ice and phytoplankton in Cape Bathurst polynya, southeastern Beaufort Sea, Canadian Arctic. *Geophysical Research Letters* 31:L08304.

Ashford, J. R., P. S. Rubilar, and A. R. Martin. 1996. Interactions between cetaceans and longline fishery operations around South Georgia. *Marine Mammal Science* 12:452–457.

Atlantic Large Whale Take Reduction Team. 2003. Atlantic Large Whale Take Reduction Plan. www.nero.noaa.gov/whaletrp/. Accessed 8 July 2003.

Au, W. W. 1993. *The Sonar of Dolphins.* Springer-Verlag, New York, NY.

Au, W. W. L., and P. W. B. Moore. 1984. Receiving beam pattern and directivity indices of the Atlantic bottlenose dolphin. *Journal of the Acoustical Society of America* 75:255–262.

Au, W. W. L., D. A. Carder, R. H. Penner, and B. L. Scronce. 1985. Demonstration of adaptation in beluga whale echolocation signals. *Journal of the Acoustical Society of America* 77:726–730.

Au, W. W., and M. Green. 2000. Acoustic interaction of humpback whales and whale-watching boats. *Marine Environmental Research* 49(5):469–481.

Auster, P. J., and R. W. Langton. 1999. The effects of fishing on fish habitat. *American Fisheries Society Symposium* 22:150–187.

Aydin, K. Y. 2004. Age structure or functional response? Reconciling the energetics of surplus production between single-species models and ECOSIM. In *Ecosystem Approaches to Fisheries in the Southern Benguela* (L. J. Shannon, K. L. Cochrane, and S. C. Pillar, eds.). *African Journal of Marine Science* 26:289–302.

Aydin, K. Y., and N. Friday. 2001. The early development of Ecosim as a predictive multispecies fisheries management tool. Document SC/53/E3 presented to the International Whaling Commission Scientific Committee, July 2001.

Bache, S. J. 2001. A primer on take reduction planning under the Marine Mammal Protection Act. *Ocean and Coastal Management* 44:221–239.

Bacon, S., and D. J. T. Carter. 1993. A connection between mean wave height and atmospheric pressure gradient in the North Atlantic. *International Journal of Climatology* 13:423–436.

Baden, D. G. 1989. Brevetoxin analysis. In *Toxic Dinoflagellates and Marine Mammal Mortalities* (D. M. Anderson and A. W. White, eds.). Proceedings of an expert consultation held at the Woods Hole Oceanographic Institution, Woods Hole, MA. Technical Report WHOI-89-36.

Baden, D. G. 1996. Analysis of brevetoxins (red tide) in manatee tissues. Report No. MR148. Marine and Freshwater Biomedical Sciences Center. National Institute of Environmental Health Sciences, Rosenstiel School of Marine and Atmospheric Sciences, University of Miami, Miami, FL.

Baggeroer, A. B., T. G. Birdsall, C. Clark, J. A. Colosi, B. D. Cornuelle, D. Costa, B. D. Dushaw, M. Dzieciuch, A. M. G. Forbes, C. Hill, B. M. Howe, J. Marshall, D. Menemenlis, J. A. Mercer, K. Metzger, W. Munk, R. C. Spindel, D. Stammer, P. F. Worcester, and C. Wunsch. 1998. Ocean climate change: comparison of acoustic tomography, satellite altimetry, and modeling. *Science* 281(5381):1327–1332.

Bagnis R. S. 1987. Ciguatera fish poisoning: an objective witness of coral reef stress. Pp. 241–253 In *Human Impacts on Coral Reefs: Facts and Recommendations* (B. Salvat, ed.). Antenne Museum, French Polynesia.

Baker, A. 1999. Unusual mortality of the New Zealand sea lion *Phocarctos hookeri*, Auckland Islands, January–February 1998. In Report of a workshop held on 8–9 June 1998 and a contingency plan for future events. New Zealand Department of Conservation, Wellington.

Baker, J. R., A. Hall, L. Hiby, R. Munro, I. Robinson, H. M. Ross, and J. F Watkins. 1995. Isolation of salmonellae from seals from UK waters. *Veterinary Record* 136:471–472.

Bakun, A. 2001. "School-mix feedback": a different way to think about low frequency variability in large mobile fish populations. *Progress in Oceanography* 49:485–511.

Balch, G. 2002. Wildlife health research and the influence of contaminants in Nunavut. *Arctic Bulletin* 4(2):7.

Balcomb, K. C. I., and D. E. Claridge. 2001. A mass stranding of cetaceans caused by naval sonar in the Bahamas. *Bahamas Journal of Science* 2:2–12.

Banner, A., and M. Hyatt. 1973. Effects of noise on eggs and larvae of 2 estuarine fish. *Transactions of the American Fisheries Society* 102(1):134–136.

Bannister, J. L., C. M. Kemper, and R. M. Warneke. 1996. *The Action Plan for Australian Cetaceans.* Wildlife Australia Endangered Species Program Project Number 380, Canberra Printing Services, Canberra, 242 pp.

Bard, E. 2002. Climate shock: abrupt changes over millennial time scales. *Physics Today* 12:32–38.

Bargu, S., C. L. Powell, S. L. Coale, M. Busman, G. J. Doucette, and M. W. Silver. 2002. Krill: a potential vector for domoic acid in marine food webs. *Marine Ecology Progress Series* 237:209–216.

Barlough, J. E., E. S. Berry, E. A. Goodwin, R. F. Brown, R. L. Delong, and A. W. Smith. 1987. Antibodies to marine caliciviruses in the Steller sea lion (*Eumetopias jubatus schreber*). *Journal of Wildlife Diseases* 23:34–44.

Barlow, J., and G. A. Cameron. 2003. Field experiments show that acoustic pingers reduce marine mammal by-catch in the California drift gill net fishery. *Marine Mammal Science* 19:265–283.

Barlow, J., S. L. Swartz, T. C. Eagle, and P. R. Wade. 1995. U.S. marine mammal stock Assessments: guidelines for preparation, background, and a summary of 1995 assessments. NOAA Technical Memorandum NMFS-OPR-95-6, September 1995. 76 pp.

Barlow, J., and B. Taylor. 1998. Preliminary abundance of sperm whales in the northeastern temperate Pacific estimated from a combined visual and acoustic survey. International Whaling Commission Working Paper SC/50/CAWS20.

Barlow, K. E., I. L. Boyd, J. P. Croxall, K. Reid, I. J. Staniland, and A. S. Brierley. 2002. Are penguins and seals in competition for Antarctic krill at South Georgia? *Marine Biology (Berlin)* 140(2):205–213.

Barrett-Lennard, L. G. 2000. Population structure and mating patterns of killer whales (*Orcinus orca*) as revealed by DNA analysis. Ph.D. Thesis. University of British Columbia, Vancouver, BC, Canada.

Barros, N. 1993. Feeding ecology and foraging strategies of bottlenose dolphins on the central east coast of Florida. Ph.D. Dissertation. University of Miami, Coral Gables, FL. 328 pp.

Barros, N. B., and R. S. Wells. 1998. Prey and feeding patterns of bottlenose dolphins (*Tursiops truncatus*) resident in Sarasota Bay, Florida. *Journal of Mammalogy* 79:1045–1059.

Bartlett, M. L., and G. R. Wilson. 2002. Characteristics of small boat signatures. *Journal of the Acoustical Society of America* 112:2221.

Bauer, G. B., J. R. Mobley, and L. M. Herman. 1993. Responses of wintering humpback whales to vessel traffic. *Journal of the Acoustical Society of America* 94:1848.

Baum, J. K., R. A. Myers, D. G. Kehler, B. Worm, S. J. Harley, and P. A. Doherty. 2003. Collapse and conservation of shark populations in the northwest Atlantic. *Science* 299:389–392.

Bauman, R. A., N. Elsayed, J. M. Petras, and J. Widholm. 1997. Exposure to sublethal blast overpressure reduces the food intake and exercise performance of rats. *Toxicology* 121(1):65–79.

Baur, D. C., M. J. Bean, and M. L. Gosliner. 1999. The laws governing marine mammal conservation in the United States. Pp. 48–86 In *Conservation and Management of Marine Mammals* (J. R. Twiss, Jr., and R. R. Reeves, eds.). Smithsonian Institution Press, Washington, DC.

Bax, N. J. 1998. The significance and prediction of predation in marine fisheries. *ICES Journal of Marine Science* 55:997–1030.

Bearzi, G., R. R. Reeves, G. Notarbartolo di Sciara, E. Politi, A. Cañadas, A. Frantzis, and B. Mussi. 2003. Ecology, status and conservation of short-beaked common dolphins (*Delphinus delphis*) in the Mediterranean Sea. *Mammal Review* 33:224–252.

Becker, P. R., M. M. Krahn, E. A. Mackey, R. Demiralp, M. M. Schantz, M. S. Epstein, M. K. Donais, B. J. Porter, D. C. G.

Muir, and S. A. Wise. 2000. Concentration of polychlorinated biphenyls (PCBs), chlorinated pesticides and heavy metals and other elements in tissues of belugas, *Delphinapterus leucas,* from Cook Inlet, Alaska. *Marine Fisheries Review* 62(3):81–98.

Beckmen, K. B., J. E. Blake, G. M. Ylitalo, J. L. Stott, and T. M. O'Hara. 2003. Organochlorine contaminant exposure and associations with hematological and humoral immune functional assays with dam age as a factor in free-ranging northern fur seal pups (*Callorhinus ursinus*). *Marine Pollution Bulletin* 46(5):594–606.

Beckmen, K., L. J. Lowenstein, and F. Galey. 1995. Epizootic seizures of California sea lions. Proceedings of the 11th Biennial Conference of the Marine Mammal Society, Orlando, FL.

Beckmen, K. B., G. M. Ylitalo, R. G. Towell, M. M. Krahn, T. M. O'Hara, and J. E. Blake. 1999. Factors affecting organochlorine contaminant concentrations in milk and blood of northern fur seal (*Callorhinus ursinus*) dams and pups from St. George Island, Alaska. *Science of the Total Environment* 231:183.

Begley, J. 2004. *Gadget User Guide.* Available online: www.hafro.is/gadget.

Begley, J., and D. Howell. 2004. An Overview of GADGET, the Globally applicable Area-Disaggregated General Ecosystem Toolbox. ICES CM 2004/FF:13. 15 pp.

Benedetti-Cecchi, L. 2000. Variance in ecological consumer-resource interactions. *Nature* 407:370–374.

Bengtson, J. L., P. Boveng, U. Franzen, P. Have, M. P. Heide-Jorgensen, and T. J. Harkonen. 1991. Antibodies to canine distemper virus in Antarctic seals. *Marine Mammal Science* 7:85–87.

Bengtson, J. L., and R. M. Laws. 1985. Trends in crabeater seal age at maturity: an insight into Antarctic marine interactions. Pp. 669–675 In *Antarctic Nutrient Cycles and Foodwebs* (W. R. Siegfried, P. R. Condy, and R. M. Laws, eds.). Springer-Verlag, Berlin, Germany.

Beranek, L. L., and I. L. Ver. 1992. *Noise and Vibration Control Engineering.* Wiley and Sons, New York, NY.

Berger, J. O. 1985. *Statistical Decision Theory and Bayesian Analysis.* Springer Verlag, New York, NY.

Bergeron, E., L. Measures, and J. Huot. 1997. Lungworm (*Otostrongylus circumlitus*) infections in ringed seals (*Phoca hispida*) from eastern Arctic Canada. *Canadian Journal of Fisheries and Aquatic Sciences* 54:2443–2448.

Bergin, A. 1997. Albatross and longlining: managing seabird bycatch. *Marine Policy* 21:63–72.

Berkes, F. 2004. Rethinking community-based conservation. *Conservation Biology* 18(3):621–630.

Beverton, R. J. H. 1985. Analysis of marine mammal-fishery interaction. Pp. 34–38 In *Marine Mammals and Fisheries* (J. R. Beddington, R. J. H. Beverton, and D. M. Lavigne, eds.). Allen and Unwin, London, U.K.

Birkun A., Jr. 2002. Cetacean habitat loss and degradation in the Black Sea. In *Cetaceans of the Mediterranean and Black Seas: State of Knowledge and Conservation Strategies* (G. Notarbartolo di Sciara, ed.). Report to the ACCOBAMS Secretariat, Monaco, February 2002. Section 8, 19 pp.

Birkun, A., T. Kuiken, S. Krivokhizhin, D. M. Haines, A. D. M. E. Osterhaus, M. W. G. van de Bildt, C. R. Joiris, and U. Siebert. 1999. Epizootic of morbilliviral disease in common dolphins (*Delphinus delphis ponticus*) from the Black Sea. *Veterinary Record* 144:85–92.

Bisack, K., and G. DiNardo. 1992. Estimating total effort in the Gulf of Maine sink gillnet fishery. In Report of the Fall Stock Assessment Workshop (13th SAW). Northeast Fisheries Science Center Reference Document 92-02.

Bjerregaard, P., T. K. Young, and R. A. Hegele. 2003. Low incidence of cardiovascular disease among the Inuit: what is the evidence? *Atherosclerosis* 166:351–357.

Bjørge, A., T. Bekkby, V. Bakkestuen, and E. Framstad. 2002. Interactions between harbour seals, *Phoca vitulina,* and fisheries in complex coastal waters explored by combined Geographic Information System (GIS) and energetics modelling. *ICES Journal of Marine Science* 59:29–42.

Blane, J. M., and R. Jaakson. 1994. The impact of ecotourism boats on the St. Lawrence beluga whales. *Environmental Conservation* 21:267–269.

Blixenkrone-Møller, M., G. Bolt, T. D. Jensen, T. Harder, and V. Svanson. 1996. Comparative analysis of the attachment protein gene (H) of dolphin morbillivirus. *Virus Research* 34:291–304.

Blomkvist, G., A. Roos, S. Jensen, A. Bignert, and M. Olsson. 1992. Concentrations of DDT and PCB in seals from Swedish and Scottish waters. *Ambio* 21:539–545.

Blus, L. J., S. N. Wiemeyer, and C. J. Henny. 1996. Organochlorine pesticides. In *Noninfectious Diseases of Wildlife* (2nd Edition) (A. Fairbrother, L. N. Locke, and G. L. Hoff, eds.). Iowa State University Press, Ames, IA.

Boesch, D. F., D. M. Anderson, R. A. Horner, S. E. Shumway, P. A. Tester, and T. E. Whitledge. 1997. Harmful algal blooms in coastal waters: options for prevention, control and mitigation. NOAA Coastal Ocean Program Analysis Series No. 10. NOAA Coastal Ocean Office, Silver Spring, MD.

Bogstad, B., T. Hauge, and S. Mehl. 2000. Who eats whom in the Barents Sea? Pp. 98–119 In *Minke Whales, Harp and Hooded Seals: Major Predators in the North Atlantic Ecosystem* (G. A. Víkingsson and F. O. Kappel, eds.). NAMMCO Scientific Publication 2. The North Atlantic Marine Mammal Commission, Tromsø, Norway.

Bogstad, B., K. H. Hauge, and Ø. Ulltang. 1997. MULTSPEC: A multi-species model for fish and marine mammals in the Barents Sea. *Journal of Northwest Atlantic Fisheries Science* 22:317–341.

Booth, D. T., D. H. Clayton, and B. A. Block. 1993. Experimental demonstration of the energetic cost of parasitism in free ranging hosts. *Proceedings of the Royal Society of London. Series B. Biological Sciences* 253:125–129.

Bordelais, A. J., C. R. Tomas, J. Naar, J. Kubanek, and D. G. Baden. 2002. New fish killing alga in coastal Delaware produces neurotoxins. *Environmental Health Perspectives* 110:465–470.

Bordino, P., S. Kraus, D. Albareda, A. Fazio, A. Palmerio, M. Mendez, and S. Botta. 2002. Reducing incidental mortality of Franciscana dolphin *Pontoporia blainvillei* with acoustic warning devices attached to fishing nets. *Marine Mammal Science* 18:833–842.

Born, E. W., N. Levermann, S. Rysgaard, G. Ehlme, M. Sejr, and M. Acquarone. 2003. Underwater observations of foraging free-living Atlantic walruses (*Odobenus rosmarus rosmarus*) and estimates of their food consumption. *Polar Biology* 26(5):348–357.

Borrell, A., and A. Aguilar. 1990. Loss of organochlorine compounds in the tissues of a decomposing stranded dolphin. *Bulletin of Environmental Contamination and Toxicology* 45:46–53.

Bossart, G. D., D. G. Baden, R. Y. Ewing, B. Roberts, and S. D. Wright. 1998. Brevetoxicosis in manatees (*Trichechus manatus latirostris*) from the 1996 epizotic: gross, histologic and immunohistochemical features. *Toxicologic Patholology* 26:276–282.

Bowen, W. D., J. McMillan, and R. Mohn. 2003. Sustained exponential population growth of grey seals at Sable Island, Nova Scotia. *ICES Journal of Marine Science* 60:1265–1274.

Bowen, W. D., and D. B. Siniff. 1999. Distribution, population biology and feeding ecology of marine mammals. Pp. 423–484 In *Biology of Marine Mammals* (J. E. Reynolds, III, and S. A. Rommel, eds.). Smithsonian Institution Press, Washington, DC.

Bowles, A. E., M. Smultea, B. Wursig, D. P. DeMaster, and D. Palka. 1994. Relative abundance and behavior of marine mammals exposed to transmissions from the Heard Island Feasibility Test. *Journal of the Acoustical Society of America* 96(4):2469–2484.

Boyd, I. L. 1998. Time and energy constraints in pinniped lactation. *American Naturalist* 152:717–728.

Boyd, I. L. 2002. Estimating food consumption of marine predators: Antarctic fur seals and macaroni penguins. *Journal of Applied Ecology* 39(1):103–119.

Boyd, I. L., D. J. McCafferty, K. Reid, R. Taylor, and T. R. Walker. 1998. Dispersal of male and Antarctic fur seals (*Arctocephalus gazella*). *Canadian Journal of Fisheries and Aquatic Sciences* 55:845–852.

Bracewell, R. N. 2000. *The Fourier Transform and Its Applications* (3rd Edition). McGraw-Hill, Boston, MA.

Bradshaw, C. J. A., M. A. Hindell, N. J. Best, K. L. Phillips, G. Wilson, and P. D. Nichols. 2003. You are what you eat: describing the foraging ecology of southern elephant seals (*Mirounga leonina*) using blubber fatty acids. *Proceedings of the Royal Society of London. Series B. Biological Sciences* 270:1283–1292.

Bradshaw, C. J. A., M. A. Hindell, K. J. Michael, and M. D. Sumner. 2002. The optimal spatial scale for the analysis of elephant seal foraging as determined by geo-location in relation to sea surface temperatures. *ICES Journal of Marine Science* 59:770–781.

Braund, S., and E. L. Moorehead. 1995. Contemporary Alaska Eskimo bowhead whaling villages. In *Hunting the Largest Animals. Native Whaling in the Western Arctic and Subarctic* (A. P. McCartney, ed.). Canadian Circumpolar Institute, Alberta, Canada.

Breeding, J. E. J. 1993. Description of a noise model for shallow water: RANDI-III. *Journal of the Acoustical Society of America* 94:1920.

Brierly, A. S., P. G. Fernandes, M. A. Brandon, F. Armstrong, N. W. Millard, S. D. McPhail, P. Stevenson, M. Pebody, J. Perrett, M. Squires, D. G. Bone, and G. Griffiths. 2002. Antarctic krill under sea ice: elevated abundance in a narrow band just south of ice edge. *Science* 295:1890–1892.

Broecker, W. S., S. Sutherland, and T.-H. Peng. 1999. A possible 20th century slowdown of southern ocean deep water formation. *Science* 286:1132–1135.

Brower, C. D., A. Carpenter, M. L. Branigan, W. Calvert, T. Evans, A. S. Fishbach, J. A. Nagy, S. Schliebe, and I. Stirling. 2002. The polar bear management agreement for the southern Beaufort Sea: an evaluation of the first ten years of a unique conservation agreement. *Arctic* 55 (4):362–372.

Brownell, R. L., T. Yamada, J. G. Mead, and A. L. van Helden. 2004. Mass Strandings of Cuvier's Beaked Whales in Japan: U.S. Naval Acoustic Link? International Whaling Commission SC/56E37. 10 pp.

Brownstein, C. 2002. Restoring abundance in the seas. Terrain.org.www.terrain.org/articles/12/brownstein.htm. Accessed 8 July 2003.

Bryant, P. J., C. M. Lafferty, and S. K. Lafferty. 1984. Reoccupation of Laguna Guerrero Negro, Baja California, Mexico, by gray whales. Pp. 375–386 In *The Gray Whale* Eschrichtius robustus (M. L. Jones, ed.). Academic Press, Orlando, FL.

Buckingham, C. A., L. W. Lefebvre, J. M. Schaefer, and H. I. Kochman. 1999. Manatee response to boating activity in a thermal refuge. *Wildlife Society Bulletin* 27(2):514–522.

Bundy, A. 2001. Fishing on ecosystems: the interplay of fishing and predation in Newfoundland-Labrador. *Canadian Journal of Fisheries and Aquatic Sciences* 58:1153–1167.

Burek, K., R. Zarnke, T. Spraker, and A. Trites. 2001. Historical and current serology in Steller sea lions in Alaska. In *Proceedings of the 14th Biennial Conference on the Biology of Marine Mammals*. Vancouver, BC, Canada.

Burkholder, J. M. 1998. Implications of harmful microalgae and heterotrophic dinoflagellates in management of sutainable fisheries. *Ecological Applications* 8:S37–S62.

Burnham, K. P., and D. R. Anderson. 1998. *Model Selection and Inference: A Practical Information Theoretic Approach*. Springer, New York, NY.

Burnham, K. P., and D. R. Anderson. 2002. *Model Selection and Multimodel Inference: A Practical Information-Theoretic Approach* (2nd Edition). Springer-Verlag, New York, NY.

Burtenshaw, J. C., E. M. Oleson, J. A. Hildebrand, M. A. McDonald, R. K. Andrew, B. M. Howe, and J. A. Mercer. 2004. Acoustic and satellite remote sensing of blue whale seasonality and habitat in the northeast Pacific. *Deep-Sea Research II* 51:967–986.

Butterworth, D. S. 1986. Antarctic marine ecosystem management. *Polar Record* 23(142):37–47.

Butterworth, D. S. 1999. Do increasing marine mammal populations threaten national fisheries? Chapter 7 In *Issues Related to Indigenous Whaling, Tonga* (M. R. Freeman, ed.). World Council of Whalers.

Butterworth, D. S., D. C. Duffy, P. B. Best, and M. O. Bergh. 1988. On the scientific basis for reducing the South African seal population. *South African Journal of Science* 84:179–188.

Butterworth, D. S., and J. Harwood (eds.). 1991. *Report on the Benguela Ecology Programme Workshop on Seal-Fishery Biological-Interactions*. Reports of the Benguela Ecology Programme of South Africa 22. 65 pp.

Butterworth, D. S., and É. E. Plagányi. 2004. A brief introduction to some multi-species/ecosystem modelling approaches in the context of their possible application in the management of South African fisheries. Pp. 53–61 In *Ecosystem Approaches to Fisheries in the Southern Benguela* (L. J. Shannon, K. L. Cochrane, and S. C. Pillar, eds.). *African Journal of Marine Science* 26.

Butterworth, D. S., A. E. Punt, H. F. Geromont, H. Kato, and Y. Fujise. 1999. Inferences on the dynamics of Southern Hemisphere minke whales from ADAPT analyses of catch-at-age information. *Journal of Cetacean Research and Management* 1(1):11–32.

Butterworth, D. S., A. E. Punt, W. H. Oosthuizen, and P. A. Wickens. 1995. The effects of future consumption by the Cape fur seal on catches and catch rates of the Cape hakes. Part III:Modelling the dynamics of the Cape fur seal

Arctocephalus pusillus pusillus. South African Journal of Marine Science 16:161–183.

Butterworth, D. S., and R. B. Thomson. 1995. Possible effects of different levels of krill fishing on predators: some initial modelling attempts. *CCAMLR Science* 2:79–97.

Caldwell, M. C., and D. K. Caldwell. 1965. Individual whistle contours in bottlenosed dolphins (*Tursiops truncatus*). *Nature* 207:434–435.

Calkins, D. G. 1979. Marine Mammals of Lower Cook Inlet and the potential for impact from outer continental shelf oil and gas exploration, development, and transport. Pp. 171–263 In *Environmental Assessment of the Alaskan Continental Shelf: Final Reports of Principal Investigators,* Vol. 20. NTIS PB85-201226. NOAA, Juneau, AK.

Carey, P. W. 1992. Fish prey species of the New Zealand fur seal (*Arctocephalus forsteri,* Lesson). *New Zealand Journal of Ecology* 16(1):41–46.

Casey, J. M., and R. A. Myers. 1998. Near extinction of a large, widely distributed fish. *Science* 281:690–692.

Caswell, H., M. Fujiwara, and S. Brault. 1999. Declining survival probability threatens the North Atlantic right whale. *Proceedings of the National Academy of Sciences* 96:3308–3313.

Cato, D.H. 1998. Prediction of ambient noise in the ocean, the missing components. *Journal of the Acoustical Society of America* 103:2858.

Cato, D. H., and R. D. McCauley. 2002. Australian research in ambient sea noise. *Acoustic Australia* 30:13–20.

Cembella, A. D., N. I. Lewis, and M. A. Quilliam. 2000. The marine dinoflagellate *Alexandrium ostenfeldii* (Dinophyceae) as the causative organism of spirolide shellfish toxins. *Phycologia* 39(1):67–74.

Center for Coastal Studies. 2003. Whale rescue. www.coastalstudies.org/rescue/netw.htm. Accessed 8 July 2003.

Chang, E. F., and M. M. Merzenich. 2003. Environmental noise retards auditory cortical development. *Science* 300:498–502.

Chang, Y. S., D. Jeon, H. Lee, H. S. An, J. W. Seo, and Y. H. Youn. 2004. Interannual variability and lagged correlation during strong El Niño events in the Pacific Ocean. *Climate Research* 27:51–58.

Chavez, F. P., J. Ryan, S. E. Lluch-Cota, and M. Niquen. 2003. From anchovies to sardines and back: multidecadal change in the Pacific Ocean. *Science* 299:217–221.

Christensen, V. 1995. Ecosystem maturity: towards quantification. *Ecological Modeling* 77:3–32.

Christensen, V., and D. Pauly. 1992. ECOPATH II: A software for balancing steady-state models and calculating network characteristics. *Ecological Modeling* 61:169–185.

Christensen, V., and C. J. Walters. 2004. Ecopath with Ecosim: methods, capabilities and limitations. *Ecological Modeling* 172:109–139.

Christensen, V., C. J. Walters, and D. Pauly. 2000. *Ecopath with Ecosim: A User's Guide* (October 2000 Edition). Fisheries Centre, University of British Columbia, Vancouver, Canada and ICLARM, Penang, Malaysia. 130 pp.

Clapham, P. A. 2001. Why do baleen whale migrate? *Marine Mammal Science* 17:432–436.

Clapham, P. J., and R. L. Brownell, Jr. 1996. The potential for interspecific competition in baleen whales. *Reports of the International Whaling Commission* 46:361–367.

Clark, J. B., K. L. Russell, M. E. Knafelc, and C. C. Steevens. 1996. Assesment of vestibular function of divers exposed to high intensity low frequency underwater sound. *Undersea & Hyperbaric Medicine* 23 (Suppl.):33.

Clarke, A., and C. M. Harris. 2003. Polar marine ecosystems: major threats and future change. *Environmental Conservation* 30(1):1–25.

Cohen, P., C. F. B. Holmes, and Y. Tsukitani. 1990. Okadaic acid: a new probe for the study of cellular regulation. *Toxicology Information Briefs* 15:98–102.

Colborn, T., and M. J. Smolen. 1996. Epidemiological analysis of persistent organochlorine contaminants in cetaceans. *Reviews of Environmental Contamination and Toxicology* 146:92–172.

Cole, R., D. S. Lindsay, D. K. Howe, C. L. Roderick, J. P. Dubey, N. J. Thomas, and L. A. Baeten. 2000. Biological and molecular characterizations of *Toxoplasma gondii* strains obtained from southern sea otters (*Enhydra lutris nereis*). *Journal of Parasitology* 86:526–530.

Collins, J. W. 1886. United States Commission of Fish and Fisheries. Part XII: Report of the Commissioner for 1884. Government Printing Office, Washington, DC.

Coltman, D. W., K. Wilson, J. G. Pilkington, M. J. Stear, and J. M. Pemberton. 2001. A microsatellite polymorphism in the gamma interferon gene is associated with resistance to gastrointestinal nematodes in a naturally-parasitized population of Soay sheep. *Parasitology* 122:571–582.

Colwell, R. 1996. Global climate and infectious disease: the cholera paradigm. *Science* 274:2025–2031.

Comiso, J. C. 2002. A rapidly declining perennial sea ice cover in the Arctic. *Geophysical Research Letters* 29(20):1–4.

Comiso, J. C., and C. L. Parkinson. 2004. Satellite-observed changes in the Arctic. *Physics Today* 57(8):38–44.

Commander Naval Air Warfare Center. 1994. Marine Mammal Protection/Mitigation and Results Summary for the Shock Trial of the USS *John Paul Jones* (DDG 53). Naval Air Warfare Center, Weapons Division, 521 9th St., Point Mugu, CA 93042-5001. Prepared for the Assistant Administrator for Fisheries, National Oceanic and Atmospheric Administration, U.S. Department of Commerce, Silver Spring, MD.

Committee on the Alaska Groundfish Fishery and Steller Sea Lions. 2003. *Decline of the Steller Sea Lion in Alaskan Waters: Untangling Food Webs and Fishing Nets.* National Academies Press, Washington, DC. 204 pp.

Constable, A. J. 2001. The ecosystem approach to managing fisheries: achieving conservation objectives for predators of fished species. *CCAMLR Science* 8:37–64.

Cooke, J. G. 2002. Some aspects of the modelling of effects of changing cetacean abundance on fishery yields. International Whaling Commission document SC/J02/FW10. 5 pp.

Cooper, L. W., I. L. Larsen, T. M. O'Hara, S. Dolvin, V. Woshner, and G. Cota. 2000. Radionuclide contaminant burdens in arctic marine mammals harvested during subsistence hunting. *Arctic* 53(2):174–182.

Cornish, V. R., M. Payne, D. Palka, and F. Julian. 1998. The use of observer data to develop methods for predicting the success of marine mammal by-catch reduction strategies. 12th Biennial Conference of the Society for Marine Mammalogy, Monaco.

Cote, I. M., and J. D. Reynolds. 2002. Predictive ecology to the rescue. *Science* 289:1181–1182.

Coulson, T., E. J. Milner-Gulland, and T. Clutton-Brock. 2000. The relative roles of density and climatic variation on population dynamics and fecundity rates in three contrasting ungulate species. *Proceedings of the Royal Society of London. Series B. Biological Sciences* 267:1771–1779.

Cox, T. M., A. J. Reed, A. Solow, and N. Treganza. 2002. Will harbor porpoises, *Phocoena phocoena,* habituate to pingers? *Journal of Cetacean Research and Management* 3:81–86.

Craig, M. P., and T. J. Ragen. 1999. Body size, survival, and decline of juvenile Hawaiian monk seals *Monachus schauinslandi*. *Marine Mammal Science* 15:786–809.

Crespo, E. A., S. N. Pedraza, S. L. Dans, M. K. Alonso, L. M. Reyes, N. A. Garcia, M. Coscarella, and C. M. Schiavini. 1997. Direct and indirect effects of the high seas fisheries on the marine mammal populations in the northern and central Patagonian coast. *Journal of Northwest Atlantic Fisheries Science* 22:189–207.

Croll, D. A., B. R. Tershy, R. P. Hewitt, D. A. Demer, P. C. Fiedler, S. E. Smith, W. Armstrong, J. M. Popp, T. Kiekhefer, V. R. Lopez, J. Urban, and D. Gendron. 1998. An integrated approach to the foraging ecology of marine birds and mammals. *Deep-Sea Research II* 45:1353–1371.

Croxall, J. P., P. N. Trathan, and E. J. Murphy. 2002. Environmental change and Antarctic seabird populations. *Science* 297:1510–1514.

Crum, L. A., and Y. Mao. 1996. Acoustically enhanced bubble growth at low frequencies and its implications for human diver and marine mammal safety. *Journal of the Acoustical Society of America* 99:2898.

Cudahy, E., and W. T. Ellison. n.d. A review of the potential for *in vivo* tissue damage by exposure to underwater sound. Naval Submarine Medical Research Laboratory, Groton, CT.

Cudahy, E., E. Hanson and D. Fothergill 1999. Summary Report on the Bioeffects of Low Frequency Water Borne Sound. Naval Submarine Medical Research Laboratory, Groton, CT. 29 pp.

Culik, B. M., S. Koschinski, N. Treganza, and G. M. Ellis. 2001. Reactions of harbor porpoises *Phocoena phocoena* and herring *Clupea harengus* to acoustic alarms. *Marine Ecology Progress Series* 211:255–260.

Daan, N., H. Gislason, J. G. Pope and J. C. Rice. 2005. Changes in the North Sea fish community: evidence of indirect effects of fishing? *ICES Journal of Marine Science* 62:177–188.

D'Amico, A., and W. C. Verboom. 1998. Summary record of the SACLANTCEN Bioacoustics Panel, La Spiezia, 15–17 June. SACLANT Undersea Research Center.

Dahlheim, M. E. 1988. Killer whale (*Orcinus orca*) depredation on longline catches of sablefish (*Anoplopoma fimbria*) in Alaskan waters. NWAFC Processed Report 88-14. 31 pp.

Dailey, M. D. 2001. Parasitic diseases. In *CRC Handbook of Marine Mammal Medicine* (L. A. Dierauf and F. M. D. Gulland, eds.). CRC Press, Boca Raton, FL.

Dao, J. 2003. Waterways declared nearly clear of mines. *New York Times*, 28 March 2003.

Daszak, P., A. A. Cunningham, and A. D. Hyatt. 2000. Emerging infectious diseases of wildlife-threats to biodiversity and human health. *Science* 287:443–449.

Daszak, P., A. A. Cunningham, and A. D. Hyatt. 2001. Anthropogenic environmental change and the emergence of infectious diseases in wildlife. *Acta Tropica* 78:103–116.

Dayton, P. K. 1998. Reversal of the burden of proof in fisheries management. *Science* 279:821–822.

Dayton, P. K., M. J. Tegner, P. B. Edwards, and K. L. Riser. 1998. Sliding baselines, ghosts, and reduced expectations in kelp forest communities. *Ecological Applications* 8:309–322.

Dayton, P. K., S. Thrush, and F. C. Coleman. 2002. Ecological effects of fishing in marine ecosystems of the United States. Pew Oceans Commission, Arlington, VA.

de la Mare, W. K. 1997. Abrupt mid-twentieth-century decline in Antarctic sea-ice extent from whaling records. *Nature* 389:57–60.

de Wit, C. A., A. T. Fisk, K. E. Hobbs, D. C. G. Muir, G. W. Gabrielsen, R. Kallenborn, M. M. Krahn, R. J. Norstrom, and J. U. Skaare. 2004. *AMAP Assessment 2002: Persistent Organic Pollutants in the Arctic*. Arctic Monitoring and Assessment Programme (AMAP), Oslo, Norway. xvi +310 pp.

De Guise, S. 2002. Cellular immunology of cetaceans. In *Molecular and Cell Biology of Marine Mammals* (C. J. Pfeiffer, ed.). Krieger Publishing Company, Melbourne, FL.

De Guise, S., D. Martineau, P. Beland, and M. Fournier. 1998. Effects of in vitro exposure of beluga whale leukocytes to selected organochlorines. *Journal of Toxicology and Environmental Health* 55:479–493.

DeMaster, D. J., C. W. Fowler,. S. L. Perry, and M. E. Richlen. 2001. Predation and competition: the impact of fisheries on marine mammal populations over the next one hundred years. *Journal of Mammalogy* 82:641–651.

Derocher, A. E., N. J. Lunn, and I. Stirling. 2004. Polar bears in a warming climate. *Integrative and Comparative Biology* 44:163–176.

De Swart, R. L., P. S. Ross, E. J. Vedder, H. H. Timmerman, S. H. Heister-kamp, H. van Loveren, J. G. Vos, P. J. H. Reijnders, and A. D. M. E. Osterhaus. 1994. Impairment of immunological functions in harbor seals (*Phoca vitulina*) feeding on fish from polluted coastal waters. *Ambio* 23:155–159.

Dewailly, E., P. Ayotte, S. Bruneau, C. Laliberte, D. C. G. Muir, and R. J. Norstrom. 1993. Inuit exposure to organochlorines through the aquatic food chain in arctic Quebec. *Environmental Health Perspectives* 101:618–620.

Diachok, O. I., and R. S. Winokur. 1974. Spatial variability of underwater ambient noise at the Arctic ice-water boundary. *Journal of the Acoustical Society of America* 55(4):750–753.

Dickey, R. L., H. R. Granade, and F. D. McClure. 1994. Evaluation of a solid-phase immunobead assay for detection of ciguatera-related biotoxins in Caribbean finfish. *Memoirs of the Queensland Museum* 34:481–488.

Dizon, A. E., C. Lockyer, W. F. Perrin, D. P. DeMaster, and J. Sisson. 1991. Rethinking the stock concept: a phylogeographic approach. *Conservation Biology* 6:24–36.

Dokken, T. M., and K. H. Nisancioglu. 2004. Fresh angle on the polar seesaw. *Nature* 430:842–843.

Dolar, L. M. L., S. Leatherwood, C. L. Hill, and L. V. Aragones. 1994. Directed fisheries for cetaceans in the Philippines. *Reports of the International Whaling Commission* 44:439–450.

Domingo, M., L. Ferrer, M. Pumarola, A. Marco, J. Plana, S. Kennedy, M. McAliskey, and B. K. Rima. 1990. Morbillivirus in dolphins. *Nature* 348:21.

Donoghue, M., R. R. Reeves, and G. Stone. 2003. Report of the workshop on interactions between cetaceans and longline fisheries held in Apia, Samoa, November 2002. New England Aquarium Aquatic Forum Series Report 03-1. 44 pp.

Dortch, Q., R. Robichaux, S. Pool, D. Milted, G. Mire, N. Rabelais, T. M. Soniat, G. A. Fryxell, R. E. Turner, and M. L. Parsons. 1997. Abundance and vertical flux of Pseudonitzschia in the Gulf of Mexico. *Marine Ecology Progress Series* 146:249–264.

Doucette, G. J. R. M. Rolland, T. V. N. Cole, J. L. Martin, L. A. Hollen, and A. Anderson. 2002. Evidence for the occurrence of PSP toxins in North Atlantic right whales (*Eubalaena glacialis*) and their zooplankton prey in the Bay of Fundy, Canada(Abstract). 10th International Conference on Harmful Algal Blooms, 21–25 October, St. Petersburg Beach, FL.

Dragoset, W. 1984. A comprehensive method for evaluating the design of air guns and air gun arrays. *Geophysics: The Leading Edge of Exploration* 3(10):52–61.

Dragoset, W. 2000. Introduction to air guns and air-gun arrays. *Geophysics: The Leading Edge of Exploration* 19:892–897.

Dubrovsky, N. A., and S. V. Kosterin. 1993. Noise in the ocean caused by lightning strokes. Pp. 697–709 In *Natural Physical Sources of Underwater Sound* (B. R. Kerman, ed.). Kluwer Academic Publishers, Dordrecht.

Dunn, J. L., J. D. Buck, and T. R. Robeck. 2001. Bacterial diseases of cetaceans and pinnipeds. In *CRC Handbook of Marine Mammal Medicine* (L. A. Dierauf and F. M. D. Gulland, eds.). CRC Press, Boca Raton, FL.

Durbin, E., G. Teegarden, R. Campbell, A. Cembella, M. F. Baumgartner, and B. R. Mate. 2002. North Atlantic right whales, *Eubalaena glacialis,* exposed to paralytic shellfish poisoning (PSP) toxins via a zooplankton vector, *Calanus finmarchicus. Harmful Algae* 1:243–251.

Edds, P. L. 1988. Characteristics of finback *Balenoptera physalus* vocalizations in the St. Lawrence Estuary, Canada. *Bioacoustics* 1:131–150.

Edds-Walton, P. L. 1997. Acoustic communication signals of mysticete whales. *Bioacoustics* 8:47–60.

Efurd, D., G. Miller, D. Rokop, F. Roensch, M. Attrep, J. Thompson, W. Inkert, H. Poths, J. Banar, J. Musgrave, E. Rios, M. Fowler, R. Gritzo, J. Headstream, D. Dry, M. Hameedi, A. Robertson, N. Valette-Silver, S. Dolvin, L. Thorsteinson, T. O'Hara, and R. Olsen. 1996. Evaluation of the anthropogenic radionuclide concentrations in sediments and fauna collected in the Beaufort Sea and Northern Alaska. Los Alamos National Laboratory, Los Alamos, NM. Report No. LAMS-13115-MS UC-721.

Egeland, G. M., L. A. Feyk, and J. P. Middaugh. 1998. The use of traditional foods in a healthy diet in Alaska: risks in perspective. *State of Alaska Epidemiology Bulletin* 2:1–140.

Emery, L., M. Bradley, and T. Hall. 2001. Data Base Description (DBD) for the Historical Temporal Shipping Data Base (HITS), Version 4.0, PSI Technical Report TRS-301. Planning Systems Inc., Slidell, LA.

Epstein, P., B. Sherman, E. Spanger-Seigfried, A. Langston, S. Prasad, and B. McKay. 1998. Marine ecosystems: emerging diseases as indicators of change. Year of the Oceans Special Report. NOAA Office of Global Programs, Washington, DC.

Erbe, C. 2000. Detection of whale calls in noise: performance comparison between a beluga whale, human listeners, and a neural network. *Journal of the Acoustical Society of America* 108(1):297–303.

Erbe, C. 2002. Underwater noise of whale-watching boats and potential effects on killer whales (*Orcinus orca*), based on an acoustic impact model. *Marine Mammal Science* 18:394–418.

Erbe, C., and D. M. Farmer. 1998. Masked hearing thresholds of a beluga whale (*Delphinapterus leucas*) in icebreaker noise. *Deep-Sea Research Part II: Topical Studies in Oceanography* 45(7):1373–1388.

Erbe, C., and D. M. Farmer. 2000. Zones of impact around icebreakers affecting beluga whales in the Beaufort Sea. *Journal of the Acoustical Society of America* 108(3):1332–1340.

Erbe, C., A. R. King, M. Yedlin, and D. M. Farmer. 1999. Computer models for masked hearing experiments with beluga whales (*Delphinapterus leucas*). *Journal of the Acoustical Society of America* 105(5):2967–2978.

Erlich, M. A., and W. Lawson 1980. The incidence and significance of the Tullio phenomenon in man. *Otolaryngology—Head and Neck Surgery* 88(5):630–636.

Estes, J. A. 1990. Growth and equilibrium in sea otter populations. *Journal of Animal Ecology* 59:385–402.

Estes, J. A., and J. L. Bodkin. 2002. Otters. Pp. 842–858 In *Encyclopedia of Marine Mammals* (W. F. Perrin, B. W. Wursig, and J. G. M. Thewissen, eds.). Academic Press, San Diego, CA.

Estes, J. A., R. J. Jameson, and A. M. Johnson. 1981. Food selection and some foraging tactics of sea otters. Pp. 606–6541 In *Worldwide Furbearer Conference Proceedings,* Vol. 1 (J. A. Chapman and D. Pursley, eds.). University of Maryland Press, Baltimore, MD.

Estes, J. A., M. L. Riedman, M. M. Staedler, M. T. Tinker, and B. E. Lyon. 2003. Individual variation in prey selection by sea otters: patterns, causes and implications. *Journal of Animal Ecology* 72:144–155.

Estes, J. A., M. T. Tinker, T. M. Williams, and D. F. Doak. 1998. Killer whale predation on sea otters linking oceanic and nearshore ecosystems. *Science* 282:473–476.

European Commission. 2001. European transport policy for 2010: time to decide. Office for Official Publications for the European Communities, Luxembourg.

Evans, D. L., and G. R. England. 2001. Joint Interim Report, Bahamas Marine Mammal Stranding Event of 14–16 March 2000. Washington, DC, U.S. Department of Commerce and U.S. Navy, available at www.nmfs.noaa.gov/prof_res/overview/Interim_Bahamas_Report.pdf.

Fanshawe, S., G. R. Vanblaricom, and A. A. Shelly. 2003. Restored top carnivores as detriments to the performance of marine protected areas intended for fishery sustainability: a case study with red abalones and sea otters. *Conservation Biology* 17:273–283.

Fay, F. H., B. P. Kelley, P. H. Gehnrigh, J. L. Sease, and A. A. Hoover. 1984. Modern populations, migrations, demography, tropics, and historical status of the Pacific walrus. Pp. 231–376 In *Environmental Assessment of the Alaskan Continental Shelf: Final Reports of Principal Investigators* (Vol. 37). OCS Study MMS 86-0021. NTIS PB87-107546, NOAA, Anchorage, AK.

Feore, S. M., M. Bennett, J. Chantrey, T. Jone, D. Baxby, and M. Begon. 1997. The effect of cowpox virus infection on fecundity in bank voles and wood mice. *Proceedings of the Royal Society of London. Series B. Biological Sciences* 264:1457–1461.

Fernández, A. 2004. Pathological findings in stranded beaked whales during the naval military manoeuvers near the Canary Islands. Pp. 37–40 In *Proceedings of the Workshop on Active Sonar and Cetaceans* (P. G. H. Evans and L. A. Miller, eds.). European Cetacean Society.

Fertl, D., and S. Leatherwood. 1997. Cetacean interactions with trawls: a preliminary review. *Journal of the Northwest Atlantic Fisheries Science* 22:219–248.

Fiedler, P. C. 1992. Seasonal climatologies and variability of eastern tropical Pacific surface waters. U.S. Department of Commerce, NOAA Technical Report NMFS 109.

Fiedler, P. C. 2002. The annual cycle and biological effects of the Costa Rica Dome. *Deep-Sea Research I* 49:321–338.

Fiedler, P. C., and S. B. Reilly. 1994. Interannual variability of dolphin habitats in the eastern tropical Pacific II: effects on abundances estimated from tuna vessel sightings, 1975–1990. *Fishery Bulletin* 92:451–463.

Fiedler, P. C., S. B. Reilly, R. Hewitt, D. Demer, V. Philbrick, S. Smith, W. Armstrong, D. Croll, B. Tershy, and B. Mate. 1998. Blue whale habitat and prey in the Channel Islands. *Deep-Sea Research II* 45:1781–1801.

Finneran, J. J. 2003. Whole-lung resonance in a bottlenose dolphin (*Tursiops truncatus*) and white whale (*Delphinapterus leucas*). *Journal of the Acoustical Society of America* 114(1): 529–535.

Finneran, J. J., C. E. Schlundt, R. Dear, D. A. Carder, and S. H. Ridgway. 2002. Temporary shift in masked hearing thresholds in odontocetes after exposure to single underwater impulses from a seismic watergun. *Journal of the Acoustical Society of America* 111(6):2929–2940.

Fish, J. F., and J. S. Vania. 1971. Killer whale, *Orcinus orca,* sounds repel white whales, *Delphinapterus leucas. Fishery Bulletin* 69:531–535.

Fisheries Global Information System. 2003. Global fishing fleet statistics (1990–1995). www.fao.org/figis/servlet/FiRefServlet?ds=staticXML&xml=webapps/figis/wwwroot/fi/figis/index.xml&xsl=webapps/figis/staticXML/format/webpage.xsl. Accessed 18 February 2003.

Fisk, A. T., M. Holst, K. A. Hobson, J. Duffe, J. Moisey, and R. J. Norstrom. 2002. Persistent organochlorine contaminants and enantiomeric signatures of chiral pollutants in ringed seals (*Phoca hispida*) collected on the east and west side of the Northwater Polynya, Canadian Arctic. *Archives of Environmental Contamination and Toxicology* 42:118–126.

Fletcher, D., F. M. D. Gulland, M. Haulena, L. J. Lowenstine, and M. Dailey. 1998. Nematode-associated gastrointestinal perforations in stranded California sea lions (*Zalophus californianus*). In *Proceedings of the 29th International Association for Aquatic Animal Medicine Annual Conference.* San Diego, CA.

Fletcher, E. R., J T. Yelverton, and D. R. Richmond. 1976. The thoraco-abdominal system's response to underwater blast. Final Technical Report for ONR contract N00014-75-C-1079, Arlington, VA.

Flewelling, L. J., J. P. Abbott, D. G. Hammond, J. Landsberg, E. Haubold, T. Pulfer, P. Speiss, and M. S. Henry. 2004. Seagress as a route of brevetoxin exposure in the 2002 red tide-related manatee mortality event. *Harmful Algae* 3:205.

Flinn, R. D., A. W. Trites, E. J. Gregr, and R. I. Perry. 2002. Diets of fin, sei, and sperm whales in British Columbia: an analysis of commercial whaling records, 1963–1967. *Marine Mammal Science* 18:663–679.

Floeter, J., and A. Temming. 2003. Explaining diet composition of North Sea cod (*Gadus morhua*): prey size preference vs. prey availability. *Canadian Journal of Fisheries and Aquatic Sciences* 60:140–150.

Florida Fish and Wildlife Conservation Commission. 2003. Final Biological Status Review of the Florida Manatee (*Trichechus manatus latirostris*). Report by the Florida Fish and Wildlife Conservation Commission, Florida Marine Research Institute, St. Petersburg, FL.

Food and Agriculture Organization of the United Nations. 1995. Technical consultation on the precautionary approach to capture fisheries and species introductions. Food and Agriculture Organization of the United Nations, Fisheries Department, Rome, Italy. 54 pp.

Food and Agriculture Organization of the United Nations. 2000. *Fisheries Resources: Trends in Production, Utilization and Trade.* World Review of Fisheries and Aquaculture. United Nations (UN) Publication.

Food and Agriculture Organization of the United Nations. 2002. The state of world fisheries and aquaculture. Food and Agriculture Organization of the United Nations, Fisheries Department, Rome, Italy. 149 pp.

Food and Agriculture Organization of the United Nations. 2003. Deepwater fisheries: the challenge for sustainable fisheries at the final frontier. Background document: COFI 2003.

Food and Agriculture Organization of the United Nations Statistical Databases–FAOSTAT. 2002. FAOSTAT fisheries database. Total marine fish catches (1990–1999). apps.fao.org/page/collections?subset=fisheries. Accessed 16 July 2002.

Foote, A. D., R. W. Osborne, and A. R. Hoelzel. 2004. Environment: whale-call response to masking boat noise. *Nature* 428(6986):910.

Forbes, L. B., O. Nielsen, L. Measures, and D. R. Ewalt. 2000. Brucellosis in ringed seals and harp seals from Canada. *Journal of Wildlife Diseases* 36:595–598.

Forcada, J., P. S. Hammond, and A. Aguilar. 1999. Status of the Mediterranean monk seal *Monachus monachus* in the western Sahara and implications of a mass mortality event. *Marine Ecology Progress Series* 188:249–261.

Ford, J. B. K. 1991. Vocal traditions among resident killer whales (*Orcinus orca)* in coastal waters of British Columbia. *Canadian Journal of Zoology* 69:1454–1483.

Ford, J. K. B., G. M. Ellis, L. G. Barrett-Lennard, A. B. Morton, R. S. Palm, and K. C. Balcomb III. 1998. Dietary specialization in two sympatric populations of killer whales (*Orcinus orca*) in coastal British Columbia and adjacent waters. *Canadian Journal of Zoology* 76:1456–1471.

Fordyce, R. E., L. G. Barnes, and N. Miyazaki. 1994. General aspects of the evolutionary history of whales and dolphins. *The Island Arc* 3:373–391.

Fordyce, E., and C. de Muizon. 2001. Evolutionary history of cetaceans: a review. Pp. 169–233 In *Secondary Adaptation of Tetrapods to Life in Water* (J. M. Mazin and V. deBuffrenil, eds.). Verlag, Berlin, Germany.

Forney, K. A. 2000. Environmental models of cetacean abundance: reducing uncertainty in population trends. *Conservation Biology* 14(5):1271–1286.

Forney, K. A., and J. Barlow. 1998. Seasonal patterns in the abundance and distribution of California cetaceans, 1991–1992. *Marine Mammal Science* 14:460–489.

Forney, K. A., J. Barlow, M. M. Muto, M. Lowry, J. Baker, G. Cameron, J. Mobley, C. Stinchcomb, and J. Carretta. 2000. U.S. Pacific Marine Mammal Stock Assessments: 2000. NOAA-TM-NMFS-SWFSC-300.

Fowler, C. W. 1981. Density dependence as related to life history strategy. *Ecology* 62:602–610.

Fowler, C. W. 1990. Density-dependence in northern fur seals (*Callorhinus ursinus*). *Marine Mammal Science* 6:171–195.

Fowler, C. W. 2003. Tenets, principles, and criteria for management: the basis for systemic management. *Marine Fisheries Review* 65:1–55.

Fowler, C. W., and R. J. M. Crawford. 2004. Systemic management of fisheries in space and time: tradeoffs, complexity, ecosystems, sustainability. *Biosphere Conservation* 6(1):25–42.

Fowler, C. W., and L. Hobbs. 2002. Limits to natural variation: implications for systemic management. *Animal Biodiversity and Conservation* 25(2):7–45.

Frances, R. C., S. R. Hare, A. B. Hollowed, and W. S. Wooster. 1998. Effects of interdecadal climate variability on the oceanic ecosystems of the NE Pacific. *Fisheries Oceanography* 7(1):1–21.

Frantzis, A. 1998. Does acoustic testing strand whales? *Nature* 392(6671):29.

Frantzis, A. 2004. The first mass stranding that was associated with the use of active sonar (Kyparissiakos Gulf, Greece, 1996). Pp. 14–20 In *Proceedings of the Workshop on Active Sonar and Cetaceans* (P. G. H. Evans and L. A. Miller, eds.). European Cetacean Society.

Fraser, W. R., and E. E. Hofmann. 2003. A predator's perspective on causal links between climate change, physical forcing and ecosystem response. *Marine Ecology Progress Series* 265:1–15.

Freese, C. H. 2000. *The Consumptive Use of Wild Species in the Arctic: Challenges and Opportunities for Ecological Sustainability.* World Wildlife Fund, Toronto, Canada.

Freitas, L. 2004. The stranding of three Cuvier's beaked whales *Ziphius cavirostris* in Madeira Archipelago—May 2000. Pp. 28–32 In *Proceedings of the Workshop on Active Sonar and Cetaceans* (P. G. H. Evans and L. A. Miller, eds.). European Cetacean Society.

Frie, A. K., V. A. Potelov, M. C. S. Kingsley, and T. Haug. 2003. Trends in age-at-maturity and growth parameters of female Northeast Atlantic harp seals, *Pagophilus groenlandicus* (Erxleben, 1777). *ICES Journal of Marine Science* 60:1018–1032.

Furey A., C. Moroney, A. B. Magdelena, M. J. F. Saez, M. Lehane, and K. J. James. 2003. Geographical, temporal, and species variation of polyether toxins, azaspiracids, in shellfish. *Environmental Science & Technology* 37:3078–3084.

Furness, R. W. 2002. Management implications of interactions between fisheries and sandeel-dependent seabirds and seals in the North Sea. *ICES Journal of Marine Science* 59:261–269.

García-Rodríguez, F. J., and D. Aurioles-Gamboa. 2004. Spatial and temporal variation in the diet of the California sea lion (*Zalophus californianus*) in the Gulf of California, Mexico. *Fishery Bulletin* 102:47–62.

Gärdenfors, U., C. Hilton-Taylor, G. M. Mace, and J. P. Rodríguez. 2001. The application of IUCN red list criteria at regional levels. *Conservation Biology* 15:1206–1212.

Garner, M. M., D. M. Lambourn, S. J. Jeffries, P. B. Hall, J. C. Rhyan, D. R. Ewalt, L. M. Polzin, and N. F. Cheville. 1997. Evidence of Brucella infection in Parafilaroides lungworms in a Pacific harbor seal (*Phoca vitulina richardi*). *Journal of Veterinary Diagnostic Investigation* 9:298–303.

George, J. C., J. Bada, J. Zeh, L. Scott, S. E. Brown, T. O'Hara, and R. Suydam. 1999. Age and growth estimates of bowhead whales (*Balaena mysticetus*) via aspartic acid racemization. *Canadian Journal of Zoology* 77:571–580.

George, J. L., and D. E. H. Frear. 1996. Pesticides in the Antarctic. Pp. 155–167 In *Pesticides in the Environment and Their Effects on Wildlife* (N. W. Moore, ed.). *Journal of Applied Ecology 3* (Suppl.).

George-Nascimento, M., R. Bustamante, and C. Oyarzún. 1985. Feeding ecology of the South American sea lion *Otaria flavescens:* food contents and food selectivity. *Marine Ecology Progress Series* 21:135–143.

George-Nascimento, M., M. Lima, and E. Ortiz, E. 1992. A case of parasite-mediated competition? Phenotypic differentiation among hookworms *Uncinaria* spp. (Nematoda: Ancylostomatidae) in sympatric and allopatric populations of South American sea lions, *Otaria byronia,* and fur seals, *Arctocephalus australis* (Carnivora: Otariidae). *Marine Biology* 112:527–533.

Geraci, J. R. 1989. Clinical investigations of the 1987–1988 mass mortality of bottlenose dolphins along the U.S. central and south Atlantic coast. Final report to the U.S. Marine Mammal Commission, Washington, DC.

Geraci, J. R. 1990. Physiologic and toxic effects on cetaceans. In *Sea Mammals and Oil: Confronting the Risks* (J. R. Geraci and D. J. St. Aubin, eds.). Academic Press, San Diego, CA.

Geraci, J. R., D. M. Anderson, R. J. Timperi, D. J. St. Aubin, G. A. Early, J. H. Prescott, and C. A. Mayo. 1989. Humpback whales (*Megaptera novaeangeliae*) fatally poisoned by dinoflagellate toxin. *Canadian Journal of Fisheries and Aquatic Sciences* 46: 1895–1898.

Geraci, J. R., J. F. Fortin, D. J. St. Aubin, and B. D. Hicks. 1981. The seal louse, *Echinophthirius horridus,* an intermediate host of the seal heartworm, *Dipetalonema spirocauda* (Nematoda). *Canadian Journal of Zoology* 59:1457–1459.

Geraci, J. R., J. Harwood, and V. J. Lounsbury. 1999. Marine mammal die-offs: causes, investigations and issues. Pp. 367–395 In *Conservation and Management of Marine Mammals* (J. R. Twiss, Jr., and R. R. Reeves, eds.). Smithsonian Institution Press, Washington, DC.

Geraci, J. R., D. J. St. Aubin, I. K. Barker, R. G. Webster, V. S. Hinshaw, W. J. Bean, H. L. Ruhnke, J. Prescott, G. Early, A. S. Baker, S Madoff, and R. T. Schooley. 1982. Mass mortality of harbor seals: pneumonia associated with influenza virus. *Science* 215:1129–1131.

Gerber, L. R., W. S. Wooster, D. P. DeMaster, and G. R. Van Blaricom. 1999. Marine mammals: new objectives in U.S. fishery management. *Reviews in Fisheries Science* 7:23–38.

Gerken, L. 1986. *ASW versus Submarine Technology Battle.* American Scientific Corp., Chula Vista, CA.

Gerrodette, T., and D. P. DeMaster. 1990. Quantitative determination of optimum sustainable population level. *Marine Mammal Science* 6:1–16.

Giam, C. S., and L. E. Ray. 1987. *Pollutant Studies in Marine Animals.* CRC Press, Boca Raton, FL.

Gill, P. C., and D. Thiele, D. 1997. A winter sighting of killer whales (*Orcinus orca*) in Antarctic sea ice. *Polar Biology* 17(5):401–404.

Gilmartin, W. G. 1979. Fetal hepatoencephalopathy in a group on California sea lions. Proceedings of the 10th Annual International Association for Aquatic Animal Medicine Conference. St. Augustine, FL.

Gilmartin, W. G., and G. A. Antonelis. 1998. Recommended recovery actions for the Hawaiian monk seal population at Midway Island. NOAA Technical Memorandum. NMFS NOAA-NMFS-SWFSC-253.

Gilmartin, W. G., R. L. DeLong, L. A. Smith, L. A. Griner, and M. D. Dailey. 1980. An investigation into an unusual mortality event in the Hawaiian monk seal, *Monachus schauinslandi.* Proceedings of the Symposium on the Status of Resource Investigations in the Northwestern Hawaiian Islands. UNIHI–Sea Grant Report No. MR-80-04:32–41.

Gilmartin, W. G., R. L. DeLong, A. W. Smith, J. C. Sweeney, B. W. DeLappe, R. W. Riseborough, L. A. Griner, M. Dailey, and D. B. Peakall. 1976. Premature parturition in the California sea lion. *Journal of Wildlife Diseases* 12:104–115.

Gislason, H. 2003. Effects of fishing on non-target species. Pp. 255–274 In *Responsible Fisheries in the Marine Ecosystem* (M. Sinclair and G. Valdimarsson, eds.). U.N. Food and Agriculture Organization, Rome, Italy, and CABI Publishing, Wallingford, U.K.

Gjøsæter, H. 1998. The population biology and exploitation of capelin (*Mallotus villosus*) in the Barents Sea. *Sarsia* 83:453–496.

Gloersen, P., and W. B. White. 2001. Reestablishing the circumpolar wave in sea ice around Antarctica from one winter to the next. *Journal of Geophysical Research* 106 (C3):4391–4395.

Goldsworthy, S. D., X. He, G. N. Tuck, M. Lewis, and R. Williams. 2001. Trophic interactions between the Patagonian toothfish, its fishery, and seals and seabirds around Macquarie Island. *Marine Ecology Progress Series* 218:283–302.

González, L. M., A. Aguilar, L. F. López-Jurado and E. Grau. 1997. Status and distribution of the Mediterranean monk seal

Monachus monachus on the Cabo Blanco peninsula (Western Sahara – Mauritania) in 1993–1994. *Biological Conservation* 80:225–233.

Goodman, D. 2002a. Uncertainty, risk, and decision: the PVA example. Pp. 171–196 In *Incorporating Uncertainty into Fisheries Models* (J. M. Berkson, L. L. Kline, and D. J. Orth, eds). American Fisheries Society Symposium.

Goodman, D. 2002b. Predictive Bayesian PVA: A Logic for Listing Criteria, Delisting Criteria, and Recovery Plans. Pp. 447–469 In *Population Viability Analysis* (S. R. Beissinger and D. R. McCullough, eds.). University of Chicago Press, Chicago, IL.

Goodman, D., and S. Blacker. 1998. Site cleanup: An integrated approach for project optimization to minimize cost and control risk. Pp. 4329–4347 In *The Encyclopedia of Environmental Analysis and Remediation* (R. A. Meyers, ed.). Wiley, New York, NY.

Goodman, D., M. Mangel, G. Parkes, T. Quinn, V. Restrepo, T. Smith, and K. Stokes, with assistance from G. Thompson. 2002. Scientific review of the harvest strategy currently used in the BSAI and GOA groundfish fishery management plans. Draft report to the North Pacific Fishery Management Council (available at www.fakr.noaa.gov/npfmc/Reports/f40review1102.pdf).

Goold, J. C., and P. J. Fish. 1998. Broadband spectra of seismic survey air-gun emissions, with reference to dolphin auditory thresholds. *Journal of the Acoustical Society of America* 103(4): 2177–2184.

Gordon, C. J., B. Padnos, E. Smith, M.-Y. Bottein Dechraoui, R. Woofter, S. Dover, and J. S. Ramsdell. 2002. Oral exposure of brevetoxin in mice: examining the possible protective effects of cholestyramine (Abstract). 10th International Conference on Harmful Algal Blooms, 21–25 October, St. Petersburg Beach, FL.

Gordon, J. C. D., J. N. Matthews, S. Panigada, A. Gannier, F. J. Borsani, and G. Notarbartolo di Sciara. 2000. Distribution and relative abundance of striped dolphins in the Ligurian Sea Cetacean Sanctuary: results from an acoustic collaboration. *Journal of Cetacean Research and Management* 2:27–36.

Gosliner, M. L. 1999. The tuna-dolphin controversy. Pp. 120–155 In *Conservation and Management of Marine Mammals* (J. R. Twiss, Jr., and R. R. Reeves, eds.). Smithsonian Institution Press, Washington, DC.

Government of Japan. 2002. Report of the 2000 and 2001 Feasibility Study of the Japanese Whale Research Program under Special Permit in the western North Pacific. Phase II (JARPN II) (Y. Fujise, S. Kawahara, L. A. Pastene, and H. Hatanaka, eds.). Paper SC/54/017 submitted to the International Whaling Commission Scientific Committee, May 2002.

Grachev, M. A., V. P. Kumarev, L. V. Mamaev, V. L. Zorin, L. V. Baranova, N. N. Denikina, S. I. Belikov, E. A. Petrov, V. S. Kolesnik, R. S. Kolesnik, V. M. Dorofeev, A. M. Beim, V. N. Kudelin, F. G. Nagieva, and V. N. Sidorov. 1989. Distemper virus in Baikal seals. *Nature* 338:209.

Graham, N. E., and H. F. Diaz. 2001. Evidence for intensification of North Pacific winter cyclones since 1948. *Bulletin of the American Meteorological Society* 82:1869–1893.

Grattan, L. M., C. Parrott, R,C. Shoemaker, C. L. Kauffman, M. P. Wasserman, et al. 1998. Learning and memory difficulties after environmental exposure to waterways containing toxin-producing *Pfiesteria* and *Pfiersteia*-like dinoflagellates. *Lancet* 352:532–539.

Gray, L. M., and D. S. Greeley. 1980. Source level model for propeller blade rate radiation for the world's merchant fleet. *Journal of the Acoustical Society of America* 67(2):516–522.

Greene, C. R. J., and S. E. Moore. 1995. Man-made Noise. Pp. 101–158 In *Marine Mammals and Noise* (D. H. Thomson, ed.). Academic Press, San Diego, CA.

Gregr, E. J., and A. W. Trites. 2001. Predictions of critical habitat for five whale species in the waters of coastal British Columbia. *Canadian Journal of Fisheries and Aquatic Sciences* 58:1265–1285.

Grenfell, B. T., M. E. Lonergan, and J. Harwood. 1992. Quantitative investigations of the epidemiology of phocine distemper virus (PDV) in European common seal populations. *Science of the Total Environment* 115:15–29.

Griffin, E. 2003. The cost of U.S. cetacean by-catch reduction measures as a reason for supporting international action. Master's Project. Duke University, Durham, NC.

Gucu, A. C., G. Gucu, and H. Orek. 2004. Habitat use and preliminary demographic evaluation of the critically endangered Mediterranean monk seal (*Monachus monachus*) in the Cilician Basin (Eastern Mediterranean). *Biological Conservation* 116:417–431.

Guisan, A., T. C. Edwards, Jr., and T. Hastie. 2002. Generalized linear and generalized additive models in studies of species distributions: setting the scene. *Ecological Modelling* 157:89–100.

Gulland, F. M. D. 1992. The role of nematode parasites in Soay sheep (*Ovis aries* L) mortality during a population crash. *Parasitology* 105:493–503.

Gulland, F. 2000. Domoic acid toxicity in California sea lions (*Zalophus californianus*) stranded along the central California coast, May–October, 1998. Report to the National Marine Fisheries Service Working Group on Marine Mammal Unusual Mortality Events. U.S. Department of Commerce, NOAA Technical Memorandum NMFS-OPR-17.

Gulland, F. M. D., K. Beckmen, K. Burek, L. Lowenstine, L. Werner, T. Spraker, and E. Harris. 1997. *Otostrongylus circumlitus* infestation of northern elephant seals (*Mirounga angustirostris*) stranded in central California. *Marine Mammal Science* 13:446–459.

Gulland, F. M. D., M. Koski, L. J. Lowenstine, A. Colagross, L. Morgan, and T. Spraker. 1996. Leptospirosis in California sea lions (*Zalophus californianus*) stranded along the central California coast, 1981–1994. *Journal of Wildlife Diseases* 32:572–580.

Gulland, F. M. D., H. Perez-Cortes, J. Urban, L. Rojas-Bracho, G. Ylitalo, C. Kreuder, and T. Rowles. 2004. Eastern North Pacific gray whale (*Eschrichtius robustus*) unusual mortality event, 1999–2000: a compilation. NOAA Technical Memorandum.

Gunter, G., R. H. Williams, C. C. Davis, and F. G. Walton Smith, F. G. 1948. Catastrophic mass mortality of marine mammals and coincident phytoplankton bloom on the west coast of Florida, November 1946 to August 1947. *Ecological Monographs* 18:309–324.

Gupta, S., J. Swinton, and R. M. Anderson. 1994. Theoretical studies of the effects of heterogeneity in the parasite population on the transmission dynamics of malaria. *Proceedings of the Royal Society of London. Series B. Biological Sciences* 245:231–238.

Hales, S., P. Weinstein, and A. Woodward. 1999. Ciguatera (fish poisoning), El Nino, and Pacific sea surface temperatures. *Ecosystem Health* 5:20–25.

Hall, A. J., R. J. Law, D. E. Wells, J. Harwood, H. M. Ross, S. Kennedy, C. R. Allchin, L. A. Campbell, and P. P. Pomeroy. 1992. Organochlorine levels in common seals (*Phoca vitulina*) which were victims and survivors of the 1988 phocine distemper epizootic. *Science of the Total Environment* 115:145–162.

Hall, A. J., B. J. McConnell, and R. J. Barker. 2002. The effect of total immunoglobulin levels, mass and condition on the first-year survival of grey seal pups. *Functional Ecology* 16:462–474.

Hall, A. J., P. Pomeroy, N. Green, K. Jones, and J. Harwood. 1997. Infection, hematology and biochemistry in grey seal pups exposed to chlorinated biphenyls. *Marine Environmental Research* 43:81–98.

Hall, M. A. 1996. On by-catches. *Reviews in Fish Biology and Fisheries* 6:319–352.

Hall, M. A. 1998. An ecological view of the tuna-dolphin problem: impacts and trade-offs. *Reviews in Fish Biology and Fisheries* 8:1–34.

Hall, M. A., D. L. Alverson, and K. I. Metuzals. 2000. By-catch: problems and solutions. In *Seas at the Millennium: An Environmental Evaluation* (C. Sheppard, ed.). Elsevier Science, New York, NY.

Hallegraeff, G. M. 1993. A review of harmful algal blooms and their apparent global increase. *Phycologia* 32:79–99.

Hallegraeff, G. M., 1998. Transport of toxic dinoflagellates via ships' ballast water bioeconomic risk assessment and efficacy of possible ballast water management strategies. *Marine Ecology Progress Series* 168:297–309.

Hammill, M. O., C. Lydersen, K. M. Kovacs, and B. Sjare. 1997. Estimated fish consumption by hooded seals (*Cystophora cristata*) in the Gulf of St. Lawrence. *Journal of the Northwest Atlantic Fisheries Science* 22:249–257.

Hansen, B., S. Osterhus, D. Quadfasel, and W. Turrell. 2004. Already the day after tomorrow? *Science* 305:953–954.

Hansen, J. C. 1990. Human exposure to metals through consumption of marine foods: a case study of exceptionally high intake among Greenlanders. In *Heavy Metals in the Marine Environment* (R. W. Furness and D. S. Rainbow, eds.). CRC Press, Boca Raton, FL.

Hansen, J. C., A. Gilman, V. Klopov, and J. Ø. Odland (eds.). 2003. *AMAP Assessment 2002: Human Health in the Arctic.* Arctic Monitoring and Assessment Programme (AMAP), Oslo, Norway. xiv+137 pp.

Harbitz, A., and U. Lindstrøm. 2001. Stochastic spatial analysis of marine resources with application to minke whales (*Balaenoptera acutorostrata*) foraging: a synoptic case study from the southern Barents Sea. *Sarsia* 86:485–501.

Harcourt, R. G., C. J. A. Bradshaw, K. Dickson, and L. S. Davis. 2002. Foraging ecology of a generalist predator, the female New Zealand fur seal. *Marine Ecology Progress Series* 227:11–24.

Harder, T. C., T. Willhaus, W. Leibold, and B. Liess. 1992. Investigations on course and outcome of phocine distemper virus infection in harbor seals (*Phoca vitulina*) exposed to polychlorinated biphenyls. *Journal of Veterinary Medicine B* 39:19–31.

Hare, S. R., and N. J. Mantua. 2000. Empirical evidence for North Pacific regime shifts in 1977 and 1989. *Progress in Oceanography* 47:103–145.

Hartman, D. S. 1979. Ecology and behavior of the manatee (*Trichechus manatus*) in Florida. American Society of Mammalogists (Special Publication 5).

Harvell, C. D., K. Kim, J. M. Burkholder, R. R. Colwell, P. R. Epstein, D. J. Grimes, E. E. Hofmann, E. K. Lipp, A. D. M. E. Osterhaus, R. M. Overstreet, J. W. Porter, G. W. Smith, and G. R. Vasta. 1999. Emerging marine diseases: climate links and anthropogenic factors. *Science* 285:1505–1510.

Harwood, J. 2001. Marine mammals and their environment in the twenty-first century. *Journal of Mammalogy* 82(3):630–640.

Harwood, J., and J. P. Croxall. 1988. The assessment of competition between seals and commercial fisheries in the North Sea and the Antarctic. *Marine Mammal Science* 4:13–33.

Harwood, J., and A. Hall. 1990. Mass mortality in marine mammals: its implications for population dynamics and genetics. *Trends in Ecology and Evolution* 5:254–257.

Harwood, J., and I. McLaren. 2002. Modelling interactions between seals and fisheries: model structures, assumptions and data requirements. International Whaling Commission Document SC/J02/FW4.

Haury, L. R., J. A. McGowan, and P. H. Wiebe. 1978. Patterns and processes in the time-space scales of plankton distributions. In *Spatial Pattern in Plankton Communities* (J. H. Steele, ed.). Plenum Press, New York, NY.

Hayes M. L., J. Bonaventura, T. P. Mitchell, J. M. Prospero, E. A. Shinn, F. Van Dolah, and R. T. Barber. 2001. How are climate and marine biological outbreaks functionally linked? *Hydrobiologia.* 460:213–220.

Haynes, D., J. F. Muller, and M. S. McLachlan. 1999. Polychlorinated dibenzo-*p*-dioxins and dibenzofurans in Great Barrier Reef (Australia) dugongs (*Dugong dugon*). *Chemosphere* 38:255–262.

Haywood, A., L. MacKenzie, I. Garthwaite, and N. Towers. 1996. *Gymnodinium breve* look-alikes: three *Gymnodinium* isolates from New Zealand. Pp. 227–230 In *Harmful and Toxic Algal Blooms* (T. Yasumoto, Y. Oshima, and Y. Fukuyo, eds.). International Oceanographic Committee of UNESCO, Paris, France.

Heesterbeek, J. A., and M. G. Roberts. 1995. Threshold quantities for helminth infections. *Journal of Mathematical Biology* 33:415–434.

Heide-Jørgensen, M. P., T. Harkonen, and P. Aberg. 1992a. Long term effect of epizootic in harbor seals in the Kattegat-Skagerrak and adjacent areas. *Ambio* 21:511–516.

Heide-Jørgensen, M. P., T. Harkonen, R. Dietz, and P. M. Thompson. 1992b. Retrospective of the 1988 European seal epizootic. *Diseases of Aquatic Organisms* 13:37–62.

Heise, K., L. G. Barrett-Lennard, E. Saulitis, C. Matkin, and D. Bain. 2003. Examining the evidence for killer whale predation on Steller sea lions in British Columbia and Alaska. *Aquatic Mammals* 29(3):325–334.

Heithaus, M. R., and L. M. Dill. 2002. Feeding strategies and tactics. Pp. 412–422 In *Encyclopedia of Marine Mammals* (W. F. Perrin, B. Würsig, and J. G. M. Thewissen, eds.). Academic Press, San Diego, CA.

Hellou, J. 1996. Polycyclic aromatic hydrocarbons in marine mammals, finfish and mollusks. In *Environmental Contaminants in Wildlife: Interpreting Tissue Concentrations* (W. N. Beyer, G. H. Heinz, and A. W. Redmon-Norwood, eds.). Lewis Publishers, Boca Raton, FL.

Hernandez, M., I. Robinson, A. Aguilar, L. M. Gonzalez, L. F. Lopez-Jurado, M. I. Reyero, E. Cacho, J. Francom, V. Lopez-Rodas, and E. Costas. 1998. Did algal toxins cause monk seal mortality? *Nature* 393:28–29.

Hewitt, J. E., S. F. Thrush, P. Legendre, V. J. Cummings, and A. Norkko. 2002. Integrating heterogeneity across spatial scales: interactions between *Atrina zelandica* and benthic macrofauna. *Marine Ecology Progress Series* 239:115–128.

Heyning, J. E. 2003. Draft final report on unusual mortality event along the southern California coast 2003. Report to the NMFS Working Group on Marine Mammal Unusual Mortality Events.

Highsmith, R. C., and K. O. Coyle. 1992. Productivity of arctic amphipods relative to gray whale energy requirements. *Marine Ecology Progress Series* 83:141–150.

Hill, P. S., D. P. DeMaster, and R. J. Small. 1997. Alaska marine mammal stock assessments, 1996. U.S. Department of Commerce, NOAA Technical Memorandum NMFS-AFSC-78. 150 pp.

Hindell, M. A. 1991. Some life-history parameters of a declining population of southern elephant seals, *Mirounga leonina*. *Journal of Animal Ecology* 60:119–134.

Hing, M. C. 1998. A database for environmental contaminants in traditional foods in northern and arctic Canada: development and applications. *Food Additives and Contaminants* 15: 127–134.

Hodgkiss I. J., and K. C. Ho. 1997. Are changes in N:P ratios in coastal waters the key to increased red tide blooms? *Hydrobiologia* 852:141–147.

Hoegh-Guldberg, O. 1999. Climate change, coral bleaching and the future of the world's coral reefs. *Marine and Freshwater Research* 50:839–866.

Hoekstra, P. F., T. M. O'Hara, S. Pallent, K. R. Solomon, and D. C. G. Muir. 2002b. Bioaccumulation of organochlorine contaminants in bowhead whales (*Balaena mysticetus*) from Barrow, Alaska. *Archives of Environmental Contamination and Toxicology* 42:497–507.

Hoekstra, P. F., T. M. O'Hara, C. Teixeira, S. Backus, A. T. Fisk, and D. C. G. Muir. 2002a. Spatial trends and bioaccumulation of organochlorine pollutants in marine zooplankton from the Alaskan and Canadian Arctic. *Environmental Toxicology and Chemistry* 21:575–583.

Hoekstra, P. F., C. S. Wong, T. M. O'Hara, K. R. Solomon, S. Mabury, and D. C. G. Muir. 2002c. Enantiomer-specific accumulation of pcb atropisomers in the bowhead whale (*Balaena mysticetus*). *Environmental Science and Technology* 36:1419–1425.

Hokama, Y. 1985. A simplified enzyme immunoassay stick test for the detection of ciguatoxin and related polyethers from fish tissue. *Toxicon* 23:939–946.

Holden, A. V., and K. Marsden. 1967. Organochlorine pesticides in seals and porpoises. *Nature* 216:1274–1276.

Holt, R. D., and J. Pickering. 1985. Infectious disease and species coexistence: a model of Lotka-Volterra form. *American Naturalist* 126:196–211.

Holt, S. J., and L. M. Talbot. 1978. New principles for the conservation of wild living resources. *Wildlife Monographs* 59. 33 pp.

Hooker, S. K., I. L. Boyd, M. Jessopp, O. Cox, J. Blackwell, P. L. Boveng, and J. L. Bengtson. 2002. Monitoring the prey-field of marine predators: combining digital imaging with datalogging tags. *Marine Mammal Science* 18:680–697.

Hooker, S. K., and H. Whitehead. 2002. Click characteristics of northern bottlenose whales (*Hyperoodon ampullatus*). *Marine Mammal Science* 18:69–80.

Hooker, S. K., H. Whitehead, and S. Gowans. 1999. Marine protected area design and the spatial and temporal distribution of cetaceans in a submarine canyon. *Conservation Biology* 13:592–602.

Hoover, A. A. 1988. Steller sea lion. Pp. 159–193 In *Selected Marine Mammals of Alaska: Species Accounts with Research and Management Recommendations* (J. W. Lentfer, ed.). Marine Mammal Commission, Washington, DC.

Houser, D. S., R. Howard, and S. Ridgway. 2001. Can diving-induced tissue nitrogen supersaturation increase the chance of acoustically driven bubble growth in marine mammals? *Journal of Theoretical Biology* 213(2):183–195.

Howe, B. M. 1996. Acoustic Thermometry of Ocean Climate (ATOC): Pioneer Seamount Source Installation. Technical Memo, Applied Physics Laboratory, University of Washington, Seattle, WA.

Hückstädt, L. A., and T. Antezana. 2003. Behaviour of the southern sea lion (*Otaria flavescens*) and consumption of the catch during purse-seining for jack mackerel (*Trachurus symmetricus*) off central Chile. *ICES Journal of Marine Science* 60:1003–1011.

Hudson, P. 1986. The effect of a parasitic nematode on the breeding production of red grouse. *Journal of Animal Ecology* 55:85–92.

Hudson, P. J., A. Rizzoli, B. T. Grenfell, H. Heesterbeek, and A. P. Dobson (eds.). 2002. *Ecology of Wildlife Diseases*. Oxford University Press, New York, NY.

Hughes, T. P. 1994. Catastrophies, phase shifts, and large-scale degradation of a Caribbean coral reef. *Science* 265:1547–1551.

Hume, F., M. A. Hindell, D. Pemberton, and R. Gales. 2004. Spatial and temporal variation in the diet of a high trophic level predator, the Australian fur seal (*Arctocephalus pusillus doriferus*). *Marine Biology* 144:407–415.

Hunt, G. L., and P. J. Stabeno. 2002. Climate change and the control of energy flow in the southeastern Bering Sea. *Progress in Oceanography* 55(1–2):5–22.

Hunt, G. L., P. Stabeno, G. Walters, E. Sinclair, R. D. Brodeur, J. M. Napp, and N. A. Bond. 2002. Climate change and control of the southeastern Bering Sea pelagic ecosystem. *Deep Sea Research II* 49(6):5821–5854.

Hunter, J. E., P. J. Duignan, C. Dupont, L. Fray, S. G. Fenwick, and A. Murray. 1998. First report of potentially zoonotic tuberculosis in fur seals in New Zealand. *New Zealand Medical Journal* 11:130–131.

Huntington, H. P. 1992. The Alaska Eskimo Whaling Commission and other cooperative marine mammal management organizations in northern Alaska. *Polar Record* 28:119–126.

Huntington, H. 2000. Traditional knowledge of the ecology of belugas, *Delphinapterus leucas,* in Cook Inlet, Alaska. *Marine Fisheries Review* 62(3):134–140.

Intergovernmental Panel on Climate Change. 2001. *Climate Change 2001: The Scientific Basis* (J. T. Houghton, Y. Ding, D. J. Griggs, M. Noguer, P. J. van der Linden, X. Dai, K. Maskell, and C. A. Johnson, eds.). Cambridge University Press, Cambridge, U.K.

International Committee for the Recovery of the Vaquita. 1999. Report of the second meeting of the International Committee for the Recovery of the Vaquita (CIRVA). Ensenada, Baja California, Mexico, 7–11 February.

International Convention for the Regulation of Whaling. http://environment.harvard.edu/HERO/wrapper/pageid=/guides/intenvpol/indexes/treaties/ICRW.html.

International Whaling Commission. 1994. Gillnets and cetaceans. In *Report of the Workshop on Mortality of Cetaceans in Passive Fishing Nets and Traps* (W. F. Perrin, G. P. Donovan, and J. Barlow, eds.). Report of the International Whaling Commission, Special Issue 15, Cambridge, U.K.

International Whaling Commission. 2000. Report of the subcommittee on small cetaceans. *Journal of Cetacean Research and Management* 2 (Suppl.):235–263.

International Whaling Commission. 2001. Report of the Scientific Committee. *Journal of Cetacean Research and Management* 3 (Suppl.):1–76.

International Whaling Commission. 2002. Report of the standing subcommittee on small cetaceans. *Journal of Cetacean Research and Management* 4 (Suppl.):325–338.

International Whaling Commission. 2004a. Report of the Modeling Workshop on Cetacean-Fishery Competition, La Jolla, CA, June 2002. *Journal of Cetacean Research and Management* 6 (Suppl.):413–426.

International Whaling Commission. 2004b. Report of the workshop to design simulation based performance testing for evaluating methods used to identify population structure from genetic data. *Journal of Cetacean Research and Management* 6 (Suppl.):469–485.

International Whaling Commission. 2004c. Report of the Standing Working Group on Environmental Concerns: Mini-Symposium on Acoustics Report of the Scientific Committee Meeting, Annex K, Cambridge, U.K.

International Whaling Commission. 2005. Report of the Standing Working Group on Scientific Permit Proposals, 2004. *Journal of Cetacean Research and Management* 7 (Suppl.): 343–353.

International Whaling Commission. www.iwcoffice.org.

International Whaling Commission Scientific Committee (SC). 2002. *Journal of Cetacean Research and Management* 4 (Suppl.). Cambridge, U.K.

Inter-Organization Programme for the Sound Management of Chemicals United Nations Environment Programme. 2002. *PCB Transformers and Capacitors: From Management to Reclassification and Disposal.* IOMC United Nations Environment Programme, First Issue, May 2002.

Iverson, F., J. Truelove, E. Nera, L. Tryphonas, J. Campbell, and E. Lok. 1989. Domoic acid poisoning and mussel-associated intoxication: preliminary investigations into the response of mice and rats to toxic mussel extract. *Food Chemistry and Toxicology* 27:377–384.

Jackson, J. B. C. et al. 2001. Historical overfishing and the recent collapse of coastal ecosystems. *Science* 293:629–638.

Jackson, J. B. C., M. X. Kirby, W. H. Berger, K. A. Bjorndal, L. W. Botsford, B. J. Bourque, R. H. Bradbury, R. Cooke, J. Erlandson, J. A. Estes, T. P. Hughes, S. Kidwell, C. B. Lange, H. S. Lenihan, J. M. Pandolfi, C. H. Peterson, R. S. Steneck, M. J. Tegner, and R. R. Warner. 2001. Historical overfishing and the recent collapse of coastal ecosystems. *Science* 293:629–637.

Jackson, R., and R. Kopke. 1998. The effects of underwater high intensity low frequency sound on vestibular function. Naval Submarine Medical Research Laboratory, Groton, CT.

James, K. J., C. Moroney, C. Roden, M. Satake, T. Yasumoto, M. Lehane, and A. Furey. 2003. Ubiquitous benign alga emerges as cause of shellfish contamination responsible for the human toxic syndrome, azaspiracid poisoning. *Toxicon* 41:145–151.

Janik, V. M. 2000. Source levels and the estimated active space of bottlenose dolphin *(Tursiops truncatus)* whistles in the Moray Firth, Scotland. *Journal of Comparative Physiology A-Sensory Neural & Behavioral Physiology* 186(7–8):673–680.

Jefferson, T. A., and B. E. Curry. 1994. Review and evaluation of potential acoustic methods of reducing or eliminating marine mammal-fishery interactions. Marine Mammal Research Program, Texas A&M University, for the U.S. Marine Mammal Commission, Washington, DC.

Jensen, M. N. 2003. Consensus on ecological impacts remains elusive. *Science* 299:38.

Jensen, T., M. van de Bildt, H. H. Dietz, T. H. Andersen, A. S. Hammer, T. Kuiken, and A. Osterhaus. 2002. Another phocine distemper outbreak in Europe. *Science* 297:209.

Jenssen, B. M., J. U. Skaare, M. Ekker, D. Vongraven, and S. H. Lorentsen. 1996. Organochlorine compounds in blubber, liver and brain in neonatal grey seal pups. *Chemosphere* 32:2115–2125.

Jepson, P. D., M. Arbelo, R. Deaville, I. A. P. Patterson, P. Castro, J. R. Baker, E. Degollada, H. M. Ross, P. Herráez, A. M. Pocknell, F. Rodríguez, F. E. Howie, A. Espinosa, R. J. Reid, J. R. Jaber, V. Martin, A. A. Cunningham, and A. Fernández. 2003. Gas-bubble lesions in stranded cetaceans. *Nature* 425:575–576.

Jepson, P. D., P. M. Bennett, C. R. Allchin, R. J. Law, T. Kuiken, J. R. Baker, E. Rogan, and J. K. Kirkwood. 1999. Investigating potential associations between chronic exposure to polychlorinated biphenyls and infectious disease mortality in harbour porpoises from England and Wales. *Science of the Total Environment* 243/244:339–348.

Jepson, P. D., S. Brew, A. P. MacMillan, J. R. Baker, J. Barnett, J. K. Kirkwood, T. Kuiken, I. R. Robinson, and V. R. Simpson. 1997. Antibodies to *Brucella* in marine mammals around the coast of England and Wales. *Veterinary Record* 141:513–515.

Jimenez, O., B. Jimenez, and M. J. Gonzalez. 2000. Isomer-specific polychlorinated biphenyl determination in cetaceans from the Mediterranean Sea: enantioselective occurrence of chiral polychlorinated biphenyl congeners. *Environmental Toxicology and Chemistry* 19:2653–2660.

Johannessen, O. M., L. Bengtsson, M. W. Miles, S. I. Kuzmina, V. A. Semenov, G. V. Alekseev, A. P. Nagurnyi, V. F. Zakharov, L. P. Bobylev, L. H. Pettersson, K. Hasselmann, and H. P. Cattle. 2004. Arctic climate change: observed and modeled temperature and sea-ice variability. *Tellus* 56A:328–341.

Johnson, J. 2002. Final Overseas Environmental Impact Statement and Environmental Impact Statement for Surveillance towed Array Sensor System Low Frequency Active (SURTASS LFA) Sonar, Vols. 1 and 2.

Johnson, M., P. T. Madsen, W. M. X. Zimmer, N. A. d. Soto, and P. L. Tyack. 2004. Beaked whales echolocate on prey. *Biology Letters* DOI: 10.1098/rsbl.2004.0208.

Johnson, S. P., S. Nolan, and F. M. D. Gulland. 1998. Antimicrobial susceptibility of bacteria isolated from pinnipeds stranded in central and northern California. *Journal of Zoo and Wildlife Medicine* 29:288–294.

Johnston, D. W. 2002. The effect of acoustic harassment devices on harbour porpoises (*Phocoena phocoena*) in the Bay of Fundy, Canada. *Biological Conservation* 108(1):113–118.

Jones, M. L., and S. L. Swartz. 1984. Demography and phenology of gray whales and evaluation of whale-watching activities in Laguna San Ignacio, Baja California sur, Mexico. Pp. 309–374 In *The Gray Whale* (Eschrichtius robustus) (M. L. Jones, ed.) Academic Press, Orlando, FL.

Joseph, J. 1994. The tuna-dolphin controversy in the eastern Pacific Ocean: biological, economic and political impacts. *Ocean Development and International Law* 25:1–30.

Joyner, C. C., and Z. Tyler. 2000. Marine conservation versus international free trade: reconciling dolphins with tuna and sea turtles with shrimp. *Ocean Development and International Law* 31:127–150.

Kaly, U. L., and G. P. Jones. 1994. Test of the effect of disturbance on ciguatera in Tuvalu. *Memoirs of the Queensland Museum* 34:523–532

Kamminga, C. 1988. Echolocation signal types of odontocetes. Pp. 9–22 In *Animal Sonar Processes and Performance* (P. W. B. Moore, ed). Plenum Press, New York, NY.

Kamrin, M. A., and R. K. Ringer. 1996. Toxicological implication of PCB residues in mammals. In *Environmental Contaminants in Wildlife: Interpreting Tissue Concentrations* (W. N. Beyer, G. H. Heinz, and A. W. Redmon-Norwood, eds). Lewis Publishers, Boca Raton, FL.

Kasamatsu, F., G. G. Joyce, P. Ensor, and J. Mermoz. 1996. Current occurrence of baleen whales in Antarctic waters. *Report of the International Whaling Commission* 46:293–304.

Kaschner, K. 2004. Modelling and mapping of resource overlap between marine mammals and fisheries on a global scale. Ph.D. Thesis, MMRU, Fisheries Centre, Department of Zoology, University of British Columbia, Vancouver, Canada.

Kaschner, K., R. Watson, V. Christensen, A. W. Trites, and D. Pauly. 2001. Modeling and mapping trophic overlap between marine mammals and commercial fisheries in the North Atlantic. Pp. 35–45 In *Fisheries Impacts on North Atlantic Ecosystems: Catch, Effort and National/regional Datasets* (D. Zeller and R. D. Pauly Watson, eds.). Fisheries Centre Research Reports.

Kastak, D., and R. J. Schusterman. 1998. Low-frequency amphibious hearing in pinnipeds: methods, measurements, noise, and ecology. *Journal of the Acoustical Society of America* 103(4):2216–2228.

Kasuya, T. 1985. Fishery-dolphin conflict in the Iki Island area of Japan. Pp. 354–364 In *Marine Mammals and Fisheries* (J. R. Beddington, R. J. H. Beverton, and D. M. Lavigne, eds.). Allen and Unwin, London, U.K.

Keeling, C. W. 1989. Records of carbon dioxide in the atmosphere. In *Aspects of Climate Variability in the Pacific and Western Americas* (D. H. Peterson, ed.). Geophysical Monographs. American Geophysical Union, Number 55.

Kennedy, S., T. Kuiken, P. D. Jepson, R. Deaville, M. Forsyth, T. Barrett, M. W. G. van de Bildt, A. D. M. E. Osterhaus, T. Eybatov, C. Duck, A. Kydyrmanov, I. Mitrofanov, and S. Wilson. 2000. Mass die-off of Caspian seals caused by canine distemper virus. *Emerging Infectious Diseases* 6:637–639.

Kennedy, S., J. A. Smyth, P. F. Cush, S. J. McCullough, G. M. Allan, and S. McQuaid. 1988. Viral distemper now found in porpoises. *Nature* 336:21.

Kenney, R. D., C. A. Mayo, and H. E. Winn. 2002. Migration and foraging strategies at varying spatial scales in western North Atlantic right whales. *Journal of Cetacean Research and Management,* Special Issue 2:251–260.

Kenney, R. D., P. M. Payne, D. W. Heinemann, and H. E. Winn. 1996. Shifts in Northeast Shelf cetacean distributions relative to trends in Gulf of Maine/Georges Bank finfish abundance. Pp. 169–196 In *The Northeast Shelf Ecosystem: Assessment, Sustainability and Management* (K. Sherman, N. A. Jaworski, and T. J. Smayda, eds.). Blackwell Science, Oxford, U.K.

Kenney, R. D., H. E. Winn, and M. C. Macaulay. 1995. Cetaceans in the Great South Channel, 1979–1989: right whale (*Eubalaena glacialis*). *Continental Shelf Research* 15(4/5):385–414.

Ketten, D. R., J. Lien, and S. Todd. 1993. Blast injury in humpback whale ears: evidence and implications. *Journal of the Acoustical Society of America* 94:1849–1850.

Ketten, D. R., T. Rowles, S. Cramer, J. O'Malley, J. Arruda, and P. G. H. Evans. 2004. Cranial trauma in beaked whales. Pp. 21–27 In *Proceedings of the Workshop on Active Sonar and Cetaceans* (P. G. H. Evans and L. A. Miller, eds.). European Cetacean Society.

Khan, S., M. S. Ahmed, O. Arakawa, and Y. Onoue. 1995a. Properties of neurotoxins separated from harmful red tide organism *Chattonella marina. Israeli Journal of Aquaculture* 47:137–140.

Khan, S., M. Haque, O. Arakawa, and Y. Onoue. 1995b. Toxin profiles and ichthyotoxicity of three phytoflagellates. *Bangladesh Journal of Fisheries* 15:73–81.

Kimm-Brinson, K. L., P. D. Moeller, M. Barbier, J. H. Glasgow, J. M. Burkholder, and J. S. Ramsdell. 2001. Identification of a P2X7 receptor in GH4C1 rat pituitary cells: a potential target for a bioactive substance produced by *Pfiesteria piscicida. Environmental Health Perspectives* 109:457–462.

Kinloch, D., H. Kuhnlein, and D. C. G. Muir. 1992. Inuit foods and diet: a preliminary assessment of benefits and risks. *Science of the Total Environment* 122:247–278.

Kishiro, T., and T. Kasuya. 1993. Review of Japanese dolphin drive fisheries and their status. *Reports of the International Whaling Commission* 43:439–452.

Knudsen, V. O., R. S. Alford, and J. W. Emling. 1948. Underwater ambient noise. *Journal of Marine Research* 7:410–429.

Knutson, T. R., T. L. Delworth, K. W. Dixon, and R. J. Stouffer, R. J. 1999. Model assessment of regional surface temperature trends. *Journal of Geophysical Research* 104 (D24):30,981–30,996.

Koeman, J. H., and H. van Genderen. 1966. Some preliminary notes on residues of chlorinated hydrocarbon insecticides in birds and mammals in the Netherlands. In *Pesticides in the Environment and Their Effects on Wildlife* (N. W. Moore, ed.). *Journal of Applied Ecology* 3 (Suppl.):99–1069.

Koen-Alonso, M., E. A. Crespo, S. N. Pedraza, N. A. García, and M. Coscarella. 2000. Food habits of the South American sea lion, *Otaria flavescens,* off Patagonia, Argentina. *Fishery Bulletin* 98:250–263.

Koen-Alonso, M., and P. Yodzis. 2005. Multispecies modelling of some components of the marine community of northern and central Patagonia, Argentina. *Canadian Journal of Fisheries and Aquatic Sciences* 62:1490–1512.

Kohler, S. T., and C. C. Kohler. 1992. Dead bleached coral provides new surfaces for dinoflagellates implicated in ciguatera fish poisonings. *Environmental Biology of Fishes* 35:413–416.

Koslow, J. A., G. W., Boehlert, J. D. M. Gordon, R. L. Haedrich, P. Lorance, and P. Parine. 2000. Continental slope and deep-sea fisheries: implications for a fragile ecosystem. *ICES Journal of Marine Science* 57:548–557.

Krahn, M. M., P. R. Becker, K. L. Tilbury, and J. E. Stein. 1997. Organochlorine contaminants in blubber of four seal species: integrating biomonitoring and specimen banking. *Chemosphere* 10:2109–2121.

Krahn, M. M., D. G. Burrows, J. E. Stein, P. R. Becker, M. M. Schantz, D. C. G. Muir, T. M. O'Hara, and T. K. Rowles. 1999. White whales (*Delphinapterus leucas*) from three Alaskan stocks: concentrations and patterns of persistent organochlorine contaminants in blubber. *Journal of Cetacean Research and Management* 1:239–249.

Krahn, M. M., M. J. Ford, W. F. Perrin, P. R. Wade, R. P. Angliss, M. B. Hanson, B. L. Taylor, G. M. Ylitalo, M. E. Dahlheim, J. E. Stein, and R. S. Waples. 2004. Status review of southern resident killer whales (*Orcinus orca*) under the Endangered Species Act. U.S. Department of Commerce, NOAA Technical Memorandum NMFS-NWFSC-62. 95 pp.

Krahn, M. M., P. R. Wade, S. T. Kalinowski, M. E. Dahlheim, B. L. Taylor, M. B. Hanson, G. M. Ylitalo, R. P. Angliss, J. E. Stein, and R. S. Waples. 2002. Status review of southern resident killer whales (*Orcinus orca*) under the Endangered Species Act. U.S. Department of Commerce, NOAA Technical Memorandum NMFS-NWFSC-54. 133 pp.

Krahn, M. M., G. M. Ylitalo, D. G. Burrows, J. Calambokidis, S. E. Moore, M. Gosho, P. Gearin, P. D. Plesha, R. L. Brownell, Jr., S. A. Blokhin, K. L. Tilbury, T. K. Rowles, and J. E. Stein. 2001. Organochlorine contaminant concentrations and lipid profiles in eastern North Pacific gray whales (*Eschrichtius robustus*). *Journal of Cetacean Research and Management* 3:19–29.

Kraus, S. D., P. K. Hamilton, R. D. Kenney, A. R. Knowlton, and C. K. Slay. 2001. Reproductive parameters of the North Atlantic right whale. *Journal of Cetacean Research and Management* 2:321–336.

Kraus, S., A. J. Read, A. Solow, K. Baldwin, T. Spradlin, E. Anderson, and J. Williamson. 1997. Acoustic alarms reduce porpoise mortality. *Nature* 388:525.

Krecek, R. C., F. S. Malan, C. E. Rupprecht, and J. E. Childs. 1987. Nematode parasites from Burchell's zebras in South Africa. *Journal of Wildlife Diseases* 23:404–411.

Kucklick, J. R., S. K. Sivertsen, M. Sanders, and G. I. Scott. 1997. Factors influencing polycyclic aromatic hydrocarbon distributions in South Carolina estuarine sediments. *Journal of Experimental Marine Biology and Ecology* 213:13–29.

Kucklick, J. R., W. D. J. Struntz, P. R. Becker, G. W. York, T. M. O'Hara, and J. E. Bohonowych. 2001. Persistent organochlorine pollutants in ringed seals and polar bears collected from northern Alaska. *Science of the Total Environment* 287:45–59.

Kuiken, T., P. M. Bennett, C. R. Allchin, J. K. Kirkwood, J. R. Baker, C. H. Lockyer, M. J. Walton, and M. C. Sheldrick. 1994. PCB's, cause of death and body condition in harbor porpoises (*Phocoena phocoena*) from British waters. *Aquatic Toxicology* 28:13–28.

Kuiken, T., V. R. Simpson, C. R. Allchin, P. M. Bennett, G. A. Codd, E. A. Harris, G. J. Howes, S. Kennedy, J. K. Kirkwook, R. J. Law, N. R. Merrett and S. Phillips. 1994. Mass mortality of common dolphins (*Delphinus delphis*) in south west England due to incidental capture in fishing gear. *Veterinary Record* 134:81–89.

Kvitek, R. R., and M. K. Beitler. 1991. Relative insensitivity of butter clam neurons to saxitoxin: a pre-adaptation for sequestering paralytic shellfish poisoning toxins as a chemical defense. *Marine Ecology Progress Series* 69:47–54.

Kvitek, R. R., A. R. DeGrange, and M. K. Beitler, M. K. 1991. Paralytic shellfish poisoning toxins mediate feeding behavior in otters. *Limnology and Oceanography* 36:393–404.

Lagardere, J. P. 1982. Effects of noise on growth and reproductive of *Crangon crangon* in rearing tanks. *Marine Biology* (Berlin) 71(2):177–186.

Laist, D. W., J. M. Coe, and K. J. O'Hara. 1999. Marine debris pollution. Pp. 342–366 In *Conservation and Management of Marine Mammals* (J. R. Twiss, Jr., and R. R. Reeves, eds.). Smithsonian Institution Press, Washington, DC.

Lam, C. W. Y., and K. C. Ho. 1989. Red tides in Tolo Harbor, Hong Kong. Pp. 49–52 In *Red Tides: Biology, Environmental Science and Toxicology* (T. Okaichi, D. M. Anderson, and T. Nemoto, eds.). Elsevier Science, New York, NY.

Lambertson, R. H. 1986. Disease of the common fin whale (*Balaenoptera physalus*): Crassicaudosis of the urinary system. *Journal of Mammalogy* 67:353–366.

Lambertson, R. H. 1992. Crassicaudosis, a parasitic disease threatening the health and population recovery of large baleen whales. In *Scientific and Technical Review.* International Office of Epizzotics, Paris, France.

Landsberg, J. H. 2002. The effects of harmful algal blooms on aquatic organisms. *Reviews in Fishery Science* 10:113–390.

Landsberg, J. H., G. H. Balazs, K. A. Steidinger, D. G. Baden, T. M. Work, and D. J. Russel. 1999. The potential role of natural tumor promoters in turtle fibropapillomatosis. *Journal of Aquatic Animal Health* 11:19–210.

Landsberg, J. H., S. Hall, J. N. Johannessen, K. D. White, S. M. Conrad, L. J. Flewelling, R. W. Dickey, F. M. Van Dolah, M. A. Quilliam, T. A. Leighfield, Z. Yinglin, C. G. Beaudry, W. R. Richardson, K. Hayes, L. Baird, R. A. Benner, P. L. Rogers, J. Abbott, D. Tremain, D. Heil, R. Hammond, D. Bodager, G. McRae, C. M. Stephenson, T. Cody, P. S. Scott, W. S. Arnold, H. Schurz-Rogers, A. J. Haywood, and K. A. Steidinger. 2002. Pufferfish poisoning: widespread implications of saxitoxin in Florida (Abstract). 10th International Conference on Harmful Algae, St. Petersburg Beach, Florida, 21–25 October.

Landsberg, J. H., and K. A. Steidinger. 1998. A historical review of *Gymnodinium breve* red tides implicated in mass mortalities of the manatee (*Trichechus manatus latirostris*) in Florida, USA. Pp. 97–100 In *Harmful Algae* (B. Reguera, J. Blanco, M. L. Fernandez, and T. Wyatt, eds.). Xunta de Galicia and Intergovernmental Oceanographic Commission of UNESCO.

LaPointe, J. M., F. M. Gulland, D. M. Haines, B. C. Barr, and P. J. Duignan. 1999. Placentitis due to Coxiella burnetti in a Pacific harbor seal (*Phoca vitulina richardsi*). *Journal of Veterinary Diagnostic Investigation* 11:541–543.

Larsen, F. 1997. The effects of acoustic alarms on the by-catch of harbour porpoises in bottom set gill nets. Danish Institute for Fisheries Research Report No. 44.

Larsen, F., O. R. Eigaard, and J. Tougaard. 2002. Reduction of harbour porpoise by-catch in the North Sea by high density gillnets. Paper SC/54/SM30 submitted to the Scientific Committee of the International Whaling Commission, Shimonoseki, Japan.

Laurer, H. L., A. N. Ritting, A. B. Russ, F. M. Bareyre, R. Raghupathi, and K. E. Saatman. 2002. Effects of underwater sound exposure on neurological function and brain histology. *Ultrasound in Medicine & Biology* 28(7):965–973.

Law, R. J. 1996. Metals in marine mammals. In *Environmental Contaminants in Wildlife: Interpreting Tissue Concentrations* (W. N. Beyer, G. H. Heinz, and A. W. Redmon-Norwood, eds.). Lewis Publishers, Boca Raton, FL.

Laws, R. M. 1977. The significance of vertebrates in the Antarctic marine ecosystem. In *Adaptations within Antarctic Ecosystems, Third Symposium on Antarctic Biology* (G. A. Llano, ed.). Scientific Committee for Antarctic Research.

Laws, R. M., and R. F. J. Taylor. 1957. A mass mortality of crabeater seals *Lobodon carconiphagus* (Gray). *Proceedings of the Zoological Society of London. Series B. Biological Sciences* 129: 315–325.

Layne, J. N. 1965. Observations on marine mammals in Florida waters. *Bulletin of the Florida State Museum* 9:131–181.

Leaper, R., O. Chappell, and J. Gordon. 1992. The development of practical techniques for surveying sperm whale populations acoustically. *Reports of the International Whaling Commission* 42:549–560.

Leatherwood, S., and R. Reeves. 1989. Marine mammal research and conservation in Sri Lanka, 1985–1986. Marine Mammal Technical Report 1, United Nations Environment Programme.

LeBoeuf, B. J., H. Perez-Cortes, J. Urban, B. R. Mate, and U. Ollervides. 2000. High gray whale mortality and low recruitment in 1999: potential causes and implications. *Journal of Cetacean Research and Management* 2:85–99.

LeDuc, R. J., W. L. Perryman, J. W. Gilpatrick, Jr., J. Hyde, C. Stinchcomb, J. V. Caretta, and R. L. Brownell, Jr. 2001. A note on recent surveys for right whales in the southeastern Bering Sea. *Journal of Cetacean Research and Management* (Special Issue) 2:287–289.

Lee, M. (ed). 2000. *Seafood Lover's Almanac.* National Audubon Society's Living Oceans Program, Islip, NY.

Lefebvre, K. A., S. Bargu, T. Lieckhefer, and M. W. Silver. 2002. From sanddabs to blue whales: the pervasiveness of domoic acid. *Toxicon* 40:971–977.

Lehmann, A., J. McC. Overton, and M. P. Austin. 2002. Regression models for spatial prediction: their role for biodiversity and conservation. *Biodiversity and Conservation* 11:2085–2092.

Lemly, A. D. 1996. Selenium in aquatic organisms. In *Environmental Contaminants in Wildlife: Interpreting Tissue Concentrations* (W. N. Beyer, G. H. Heinz, and A. W. Redmon-Norwood, eds.). Lewis Publishers, Boca Raton, FL.

Lesage, V., C. Barrette, M. C. S. Kingsley, and B. Sjare. 1999. The effect of vessel noise on the vocal behavior of belugas in the St. Lawrence River Estuary. *Marine Mammal Science* 15:65–84.

Letcher, R. J. 1996. The ecological and analytical chemistry of chlorinated hydrocarbon contaminants and methyl sulfonyl-containing metabolites in the polar bear *Ursus maritimus* food chain. Ph.D. Thesis, Carleton University, Ottawa, Canada.

Letcher, R. J., R. J. Norstrom, S. Line, M. A. Ramsay, and S. M. Bandiera. 1996. Immunoquantitation and microsomal monooxygenase activities of hepatic cytochromes p4501a and p4502b and chlorinated hydrocarbon contaminant levels in polar bear *Ursus maritimus. Toxicology and Applied Pharmacology* 137:127–140.

Letcher, R. J., R. J. Norstrom, and D. C. G. Muir. 1998. Biotransformation versus bioaccumulation: sources of methyl sulfone pcb and 4,4-dde metabolites in the polar bear food chain. *Environmental Science and Technology* 32:1656–1661.

Levin, R. E. 1992. Paralytic shellfish toxins: their origins, characteristics, and methods of detection—a review. *Journal of Food Biochemistry* 15:405–417.

Levin, S. A. 1992. The problem of pattern and scale in ecology. *Ecology* 73(6):1943–1967.

Lindholm, J. B., P. J. Auster, M. Ruth, and L. Kaufman. 2001. Modelling the effects of fishing and implications for the design of marine protected areas: juvenile fish responses to variations in seafloor habitat. *Conservation Biology* 15(2): 424–437.

Lindstrøm, U., and T. Haug. 2001. Feeding strategy and prey selectivity in common minke whales (*Balaenoptera acutorostrata*) foraging in the southern Barents Sea during early summer. *Journal of Cetacean Research and Management* 3:239–249.

Lindstrøm, U., T. Haug, and I. Røttingen. 2002. Predation on herring, *Clupea harengus,* by minke whales, *Balaenoptera acutorostrata,* in the Barents Sea. *ICES Journal of Marine Science* 59:58–70.

Lipscomb, T. P., F. Y. Schulman, D. Moffett, and S. Kennedy. 1994. Morbilliviral disease in Atlantic bottlenose dolphins (*Tursiops truncatus*) from the 1987–1988 epizootic. *Journal of Wildlife Diseases* 30:567–571.

Liu, J., D. G. Martinson, X. Yuan, and D. Rind. 2002a. Evaluating Antarctic sea ice variability and its teleconnections in global climate models. *International Journal of Climatology* 22:885–900.

Liu, J., X. Yuan, D. Rind, and D. G. Martinson. 2002b. Mechanism study of the ENSO and southern high latitude climate teleconnections. *Geophysical Research Letters* 20 (14):24–28.

Livingston, P. A., and S. Tjelmeland. 2000. Fisheries in Boreal ecosystems. *ICES Journal of Marine Science* 57:619–627.

Loeb, V., V. Siegel, O. Holm-Hansen, R. Hewitt, W. Fraser, W. Trivelpiece, and S. Trivelpiece. 1997. Effects of sea-ice extent and krill or salp dominance on the Antarctic food web. *Nature* 387:897–900.

Lopez, A., G. J. Pierce, M. B. Santos, J. Gracia, and A. Guerra. 2003. Fishery by-catches of marine mammals in Galician waters: results from onboard observations and an interview survey of fishermen. *Biological Conservation* 111:25–40.

Loughlin, T. R. (ed.). 1994. *Marine Mammals and the Exxon Valdez.* Academic Press, San Diego, CA.

Loughlin, T. R., W. J. Ingraham, Jr., N. Baba, and B. W. Robson. 1999. Use of a surface-current model and satellite telemetry to assess marine mammal movements in the Bering Sea. In *The Bering Sea: Physical, Chemical and Biological Dynamics* (T. R. Loughlin and K Ohtani, eds.). University of Alaska Sea Grant Program, Fairbanks, AK.

Loughlin, T. R., A. S. Perlov, and V. A. Vladimirov. 1992. Range-wide survey and estimation of total number of Steller sea lions in 1989. *Marine Mammal Science* 8:220–239.

Loughlin, T. R., and A. E. York. 2000. An accounting of the sources of Steller sea lion, *Eumetopias jubatus,* mortality. *Marine Fisheries Review* 62:40–45.

Loughrey, A. G. 1959. Preliminary investigations of the Atlantic walrus, *Odobenus rosmarus rosmarus* (Linnaeus). *Canadian Wildlife Service Wildlife Management Bulletin Series 1,* Number 14:123.

Ludwig, D., R. Hilborn, and C. Walters. 1993. Uncertainty, resource exploitation, and conservation: lessons from history. *Science* 260:17–36.

Lyons, C., M. J. Welsh, J. Thorsen, K. Ronald, and B. K. Rima. 1993. Canine distemper isolated from a captive seal. *Veterinary Record* 132:487–488.

Lyons, E. T., R. L. DeLong, S. R. Melin, and S. C. Tolliver. 1997. Uncinariasis in northern fur seal and California sea lion pups from California. *Journal of Wildlife Diseases* 33:848–852.

MacKenzie, B. R., J. Alheit, D. J. Conley, P. Holm, and C. C. Kinze. 2002. Ecological hypotheses for a historical reconstruction of upper trophic level biomass in the Baltic Sea and Skagerrak. *Canadian Journal of Fisheries and Aquatic Sciences* 59:173–190.

Mackinson, S., J. L. Blanchard, J. K. Pinnegar, and R. Scott. 2003. Consequences of alternative functional response formulations in models exploring whale-fishery interactions. *Marine Mammal Science* 19:661–681.

Mackintosh, N. A. 1946. The natural history of whalebone whales. *Biological Reviews of the Cambridge Philosophical Society* 21:60–74.

Madsen, P. T., and B. Mohl. 2000. Sperm whales (*Physeter catodon* L. 1758) do not react to sounds from detonators. *Journal of the Acoustical Society of America* 107(1):668–671.

Magnússon, K. G. 1995. An overview of the multispecies VPA: theory and applications. *Reviews in Fish Biology and Fisheries* 5(2):195–212.

Malakoff, D. 2002. Seismology: Suit ties whale deaths to research cruise. *Science* 298(5594):722–723.

Malme, C. I., P. R. Miles, C. W. Clark, P. Tyack, and J. E. Bird. 1984. Investigations on the potential effects of underwater noise from petroleum industry activities on migrating gray whale behavior. Phase II: January 1984 migration. BBN Laboratories Inc., Cambridge, MA for U.S. Minerals Management Service, Washington, DC.

Malme, C. I., P. R. Miles, P. Tyack, C. W. Clark, and J. E. Bird. 1985. Investigation of the potential effects of underwater noise from petroleum industry activities on feeding humpback whale behavior. BBN Laboratories Inc., Cambridge, MA for U.S. Minerals Management Service, Anchorage, AK.

Mamaev, L. V., N. N. Denikina, S. I. Belikov, V. E. Volchkov, I. K. Visser, M. Fleming, C. Kai, T. C. Harder, B. Liess, and A. D. Osterhaus. 1995. Characterisation of morbilliviruses isolated from Lake Baikal seals (*Phoca sibirica*). *Veterinary Microbiology* 44:2–4, 251–259.

Mangel, M., and R. J. Hofman. 1999. Ecosystems: patterns, processes and paradigms. Pp. 87–98 In *Conservation and Management of Marine Mammals* (J. R. Twiss, Jr., and R. R. Reeves, eds.). Smithsonian Institution Press, Washington, DC.

Mangel, M., and P. V. Switzer. 1998. A model at the level of the foraging trip for the indirect effects of krill (*Euphausia superba*) fisheries on krill predators. *Ecological Modeling* 105:235–256.

Mangel, M., L. M. Talbot, G. K. Meffe, M. T. Agardy, D. L. Alverson, J. Barlow, D. B. Botkin, G. Budowski, T. Clark, J. Cooke, R. H. Crozier, P. K. Dayton, D. L. Elder. C. W. Fowler, S. Funtowicz, J. Giske, R. J. Hofman, S. J. Holt, S. R. Kellert, L. A. Kimball, D. Ludwig, K. Magnusson, B. S. Malayang III, C. Mann, E. A. Norse, S. P. Northridge, W. F. Perrin, C. Perrings, R. M. Peterman, G. B. Rabb, H. A. Regier, J. E. Reynolds III, K. Sherman, M. P. Sissenwine, T. D. Smith, A. Starfield, R. J. Taylor, M. F. Tillman, C. Toft, J. R. Twiss, Jr., J. Wilen, and T. P. Young. 1996. Principles for the conservation of wild living resources. *Ecological Applications* 6:338–362.

Mantua, N. J., S. R. Hare, Y. Zhang, J. M. Wallace, and R. C. Francis. 1997. A Pacific interdecadal climate oscillation with impacts on salmon production. *Bulletin of the American Meteorology Society* 78:1069–1081.

Marchant, J. 2001. Are pollock a red herring in Alaska's fishing debate? *New Scientist* 170:18–21.

Marine Mammal Commission. 2000. Annual Report to Congress for Calendar Year 1999. Washington, DC.

Marine Mammal Commission. 2001. Hawaiian monk seal (*Monachus schauinslandii*). Pp. 54–70 In *Species of Special Concern, Annual Report to Congress, 2000.* Marine Mammal Commission, Bethesda, MD.

Marine Mammal Commission. 2002. Vaquita (*Phocoena sinus*). Pp. 70–72 In *Species of Special Concern, Annual Report to Congress, 2001.* Marine Mammal Commission, Bethesda, MD.

Marine Mammal Commission. 2003. *Annual Report to Congress, 2002.* Marine Mammal Commission, Bethesda, MD.

Marine Mammal Commission. 2004. *Annual Report to Congress, 2003.* Marine Mammal Commission, Bethesda, MD.

Marine Stewardship Council. 2003. The Marine Stewardship Council. www.msc.org/. Accessed on 8 July 2003.

Markgraf, V., and H. F. Diaz. 2000. *El Niño and the Southern Oscillation: Multiscale Variability and Global and Regional Impacts.* Cambridge University Press, Cambridge, U.K.

Marr, J. C., A. E. Jackson, and J. L. McLachlan. 1992. Occurrence of Prorocentrum lima, a DSP-toxin producing species from the Altantic coast of Canada. *Journal of Applied Phycology* 4:17–24.

Martien, K. K., and B. L. Taylor. 2003. The limitations of hypothesis-testing as a means of delineating demographically independent units. *Journal of Cetacean Research and Management* 5(3):213–218.

Martin, J. S., P. H. Rogers, E. Cudahy, and E. Hanson. 2000. Low frequency response of the submerged human lung. *Journal of the Acoustical Society of America* 107:2813.

Martín, V., A. Servidio, and S. García. 2004. Mass strandings of beaked whales in the Canary Islands. Pp. 33–36 In *Proceedings of the Workshop on Active Sonar and Cetaceans* (P. G. H. Evans and L. A. Miller, eds.). European Cetacean Society.

Mase, B., W. Jones, R. Ewing, G. Bossart, F. Van Dolah, T. Leighfield, M. Busman, J. Litz, B. Roberts, and T. L. Rowles. 2000. Epizootic of bottlenose dolphins in the Florida panhandle: 1999–2000. *Proceedings of the American Association of. Zoo Veterinarians and the International Association of Aquatic Animal Medicine Joint Conference,* New Orleans, LA.

Mate, B. R., and J. T. Harvey (eds.). 1987. Acoustical Deterrents in Marine Mammal Conflicts with Fisheries. Oregon Sea Grant Publication ORESU-W-86-001. Sea Grant Communications, Oregon State University, Corvallis, OR. 116 pp.

Mate, B., W. Jones, R. Ewing, G. Bossart, F. Van Dolah, T. Leighfield, M. Busman, J. Litz, B. Roberts, and T. Rowles. 2000. Epizootic of bottlenose dolphins in the Florida panhandle: 1999–2000. Proceedings of the American Association of Zoologists, American Association of Zoo Veterinarians, and International Association of Aquatic Animal Medicine Joint Conference, New Orleans, LA.

Matkin, C. O., and E. L. Saulitis. 1994. Killer Whale (*Orcinus Orca*) biology and management in Alaska. Report to the Marine Mammal Commission. 46 pp.

Matthews, J. N., L. E. Rendell, J. C. D. Gordon, and D. W. Macdonald. 1999. A review of frequency and time parameters of cetacean tonal calls. *Bioacoustics* 10:47–71.

Maury, O., D. Gascuel, F. Marsac, A. Fonteneau, and A. DeRosa. 2001. Hierarchical interpretation of nonlinear relationships linking yellowfin tuna (*Thunnus albacares*) distribution to the environment in the Atlantic Ocean. *Canadian Journal of Fisheries and Aquatic Sciences* 58:458–469.

Maybaum, H. L. 1993. Responses of humpback whales to sonar sounds. *Journal of the Acoustical Society of America* 94:1848–1849.

Mayer, K. A., M. D. Dailey, and M. D. Miller. 2003. Helminth parasites of the southern sea otter (*Enhydra lutris nereis*) in central California: abundance, distribution and pathology. *Diseases of Aquatic Organisms* 53:77–88.

Mazzuca, L. L. 2001. Potential Effects of Low Frequency Sound (LFS) from Commercial Vessels on Large Whales. Master of Marine Affairs, University of Washington, Seattle, WA.

McCarthy, E., and J. H. Miller. 2002. Is anthropogenic ambient noise in the ocean increasing? *Journal of the Acoustical Society of America* 112(5):2262.

McConnaughey, R. A., K. L. Mier, and C. B. Dew. 2000. An examination of chronic trawling effects on soft-bottom benthos of the eastern Bering Sea. *ICES Journal of Marine Science* 57:1377–1388.

McDonald, M. A., J. Calambokidis, A. M. Teranishi, and J. A. Hildebrand. 2001. The acoustic calls of blue whales off California with gender data. *Journal of the Acoustical Society of America* 109(4):1728–1735.

McKitrick, R. 2003. Emission scenarios and recent global warming projections. *Climate Change* 3:14–16.

McLachlan, M. S., D. Haynes, and J. F. Muller. 2001. PCDDs in the water/sediment-seagrass-dugong (*Dugong dugon*) food chain on the Great Barrier Reef (Australia). *Environmental Pollution* 113:129–134.

McLean, J. L. 1989. Indo-Pacific red tides 1985–1988. *Marine Pollution Bulletin* 20:304–310.

McMinn, A., G. M. Hallegraeff, P. Thompson, A. V. Jenkinson, and H. Heinjis. 1997. Cyst and radionuclide evidence for the recent introduction of the toxic dinoflagellate *Gymnodinium catenatum* into Tasmanian waters. *Marine Ecology Progress Series* 161:165–172.

Measures, L. N., and M. Olson. 1999. Giardiasis in pinnipeds from eastern Canada. *Journal of Wildlife Diseases* 35:779–782.

Medina-Vogel, G., C. Delgado-Rodriguez, P. Alvarez, E. Ricardo, V. Bartheld, and L. Jose. 2004. Feeding ecology of the marine otter (*Lutra felina*) in a rocky seashore of the south of Chile. *Marine Mammal Science* 20:134–144.

Melvin, E. F., and J. K. Parrish (eds.). 2001. Seabird By-catch: Trends, Roadblocks and Solutions. University of Alaska Sea Grant, AK-SC-01-01, Fairbanks, AK. 206 pp

Melvin, E. F., J. K. Parrish, and L. L. Conquest. 1999. Novel tools to reduce seabird by-catch in coastal gillnet fisheries. *Conservation Biology* 13:1–12.

Merrick, R. L., R. Brown, D. G. Calkins, and T. R. Loughlin. 1995. A comparison of Steller sea lion, *Eumetopias jubatus*, pup masses between rookeries with increasing and decreasing populations. *Fishery Bulletin* 93:753–758.

Messaudi, I., J. Guevara Patino, R. Dyall, J. LeMaoult, and J. Nikolich-Zugich. 2002. Direct link between MHC polymorphism, T cell avidity and diversity in immune defense. *Science* 298:1797–1800.

Mikaelian, I., J. Boisclair, J. P. Dubey, S. Kennedy, and D. Martineau. 2000. Toxoplasmosis in beluga whales (*Delphinapterus leucas*) from the St Lawrence Estuary: two case reports and a serological survey. *Journal of Comparative Pathology* 122:73–76.

Miksis, J. L., M. D. Grund, D. P. Nowacek, A. R. Solow, R. C. Connor, and P. L. Tyack. 2001. Cardiac responses to acoustic playback experiments in the captive bottlenose dolphin (*Tursiops truncatus*). *Journal of Comparative Psychology* 115(3): 227–232.

Miller, M. A., I. A. Gardner, C. Kreuder, D. M. Paradies, K. R. Worcester, D. A. Jessup, E. Dodd, M. D. Harris, J. A. Ames, A. E. Packham, and P. A. Conrad. 2002. Coastal freshwater runoff is a risk factor for *Toxoplasma gondii* infection of southern sea otters (*Enhydra lutris nereis*). *International Journal for Parasitology* 32:997–1006.

Miller, P. J., N. Biassoni, A. Samuels, and P. L. Tyack. 2000. Whale songs lengthen in response to sonar. *Nature* 405(6789):903.

Miller, W. G., L. G. Adams, T. A. Ficht, N. Cheville, J. P. Payeur, D. R. Harley, C. House, and S. H. Ridgway. 1999. Brucella-induced abortions and infection in bottlenose dolphins (*Tursiops truncatus*). *Journal of Zoo and Wildlife Medicine* 30:100–110.

Milne, A. R. 1967. Sound propagation and ambient noise under sea ice. Pp. 120–138 In *Underwater Acoustics* (M. Albers, ed.). Plenum Press, New York, NY.

Minobe, S. 2002. Interannual to interdecadal changes in the Bering Sea and concurrent 1998/99 changes over the North Pacific. *Progress in Oceanography* 55(1–2):45–64.

Mohl, B., M. Wahlberg, P. T. Madsen, L. A. Miller, and A. Surlykke. 2000. Sperm whale clicks: directionality and source level revisited. *Journal of the Acoustical Society of America* 107(1): 638–648.

Mohn, R., and W. Bowen. 1996. Grey seal predation on the eastern Scotian Shelf: modelling the impact on Atlantic cod. *Canadian Journal of Fisheries and Aquatic Sciences* 53:2722–2738.

Moore, S. E. 2000. Variability in cetacean habitat selection in the Alaskan Arctic. *Arctic* 53(4):432–447.

Moore, S. E., J. M. Grebmeier, and J. R. Davies. 2003. Gray whale distribution relative to forage habitat in the northern Bering Sea: current conditions and retrospective summary. *Canadian Journal of Zoology* 81(10):734–742.

Moore, S. E., J. Urban, W. L. Perryman, F. Gulland, H. Perez-Cortes, P. R. Wade, L. Rojas-Bracho, and T. Rowles. 2001. Are gray whales hitting "K" hard? *Marine Mammal Science* 17:954–958.

Moore, S. E., J. M. Waite, N. A. Friday, and T. Honkalehto. 2002a. Cetacean distribution and relative abundance on the central-eastern and the southeastern Bering Sea shelf with reference to oceanographic domains. *Progress in Oceanography* 55:249–261.

Moore, S. E., W. E. Watkins, M. A. Daher, J. R. Davies, and M. E. Dahlheim. 2002b. Blue whale habitat associations in the Northwest Pacific: analysis of remotely-sensed data using a Geographic Information System. *Oceanography* 15(3):20–25.

Mori, M., and D. S. Butterworth. 2004. Consideration of multi-species interaction in the Antarctic: an initial model of the minke whale–blue whale– krill interaction. Pp. 235–259 In *Ecosystem Approaches to Fisheries in the Southern Benguela* (L. J. Shannon, K. L. Cochrane, and S. C. Pillar, eds.). *African Journal of Marine Science* 26.

Moritz, C. 1994. Defining "evolutionary significant units" for conservation. *Trends in Ecology and Evolution* 9:373–375.

Mörner, J., R. Bos, and M. Fredrix. 2002. Reducing and eliminating the use of persistent organic pesticides. In *Guidance on Alternative Strategies for Sustainable Pest and Vector Management*. The Inter-Organization Programme for the Sound Management of Chemicals (IOMC), Geneva, Switzerland.

Morton, A. B., and H. K. Symonds. 2002. Displacement of *Orcinus orca* (L.) by high amplitude sound in British Columbia, Canada. *ICES Journal of Marine Science* 59:71–80.

Mukerjee, M. 1998. Stalking the wild dugong. *Scientific American* 279(3):20–21.

Murase, H., K. Matsuoka, T. Ichii, and S. Nishiwaki. 2002. Relationship between the distribution of euphausiids and baleen whales in the Antarctic (35°E–145°W). *Polar Biology* 25: 135–145.

Murphy, E. 1995. Spatial structure of the Southern Ocean ecosystem: predator-prey linkages in Southern Ocean food webs. *Journal of Animal Ecology* 64:333–347.

Murray, D. L., J. R. Cary, and L. B. Keith. 1997. Interactive effects of sublethal nematodes and nutritional status on snowshoe hares vulnerability to predation. *Journal of Animal Ecology* 66:250–264.

Murray, K. T., A. J. Read, and A. R. Solow. 2000. The use of time/area closures to reduce by-catches of harbour porpoises: lessons from the Gulf of Maine sink gillnet fishery. *Journal of Cetacean Research and Management* 2:135–141.

Myers, R. A., and B. Worm. 2003. Rapid worldwide depletion of predatory fish communities. *Nature* 423:280–283.

Nachtigall, P. E., D. W. Lemonds, and H. L. Roitblat. 2000. Psychoacoustic studies of dolphin and whale hearing. *Hearing by Whales and Dolphins* 12:330–363.

Nachtigall, P. E., J. Lien, W. W. L. Au, and A. J. Read (eds.). 1995. *Harbour Porpoises: Laboratory Studies to Reduce By-catch*. De Spil Publishers, Woerden, The Netherlands.

National Marine Fisheries Service. 2000. Endangered Species Act Section 7: Consultation, biological opinion and incidental take statement on authorization of Bering Sea / Aleutian Islands groundfish fisheries based on the Fishery Management Plan for the Bering Sea / Aleutian Islands Groundfish and authorization of Gulf of Alaska groundfish fisheries based on the Fishery Management Plan for Groundfish of the Gulf of Alaska. Department of Commerce, National Oceanic and Atmospheric Administration, National Marine Fisheries Service, Juneau, AK.

National Marine Manufacturers Association. 2003. Recreational Boating Statistical Abstract. www.nmma.org/facts/boatingstats/2003/index.asp.

National Oceanic and Atmospheric Administration. 1998. *Managing the Nation's By-catch*. U.S. Department of Commerce, NOAA, Silver Spring, MD.

National Oceanic and Atmospheric Administration. 2000. *U.S. Atlantic and Gulf of Mexico Marine Mammal Stock Assessments, 2000*. NOAA-TM-NMFS-NE-162.

National Oceanic and Atmospheric Administration. 2002. *Fisheries of the United States, 2001*. U.S. Department of Commerce, NOAA, Silver Spring, MD.

National Oceanic and Atmospheric Administration Fisheries. 2003a. National By-Catch Strategy. www.nmfs.noaa.gov/bycatch.htm. Accessed on 8 July 2003.

National Oceanic and Atmospheric Administration Fisheries. 2003b. Marine mammal stock assessment reports. www.nmfs.noaa.gov/prot_res/PR2/Stock_Assessment_Program/sars.html. Accessed on 8 July 2003.

National Research Council. 1990. *Decline of the Sea Turtles: Causes and Prevention*. National Academy Press. Washington, DC.

National Research Council. 1994. *Low-Frequency Sound and Marine Mammals: Current Knowledge and Research Needs*. National Academy Press, Washington, DC.

National Research Council. 2000a. *Marine Mammals and Low-Frequency Sound*. National Academy Press, Washington, DC.

National Research Council. 2000b. *Clean Coastal Waters: Understanding and Reducing the Effects of Nutrient Pollution*. National Academy Press, Washington, DC.

National Research Council. 2002. *Abrupt Climate Change: Inevitable Surprises*. National Academy Press, Washington, DC.

National Research Council. 2003a. *Decline of the Steller Sea Lion in Alaskan Waters: Untangling Food Webs and Fishing Nets*. National Academy Press, Washington, DC.

National Research Council. 2003b. *Potential Impacts of Ambient Noise in the Ocean on Marine Mammals*. National Academy Press, Washington, DC.

Nelson R. J., G. E. Demas, S. L. Klein, and L. J. Kriegsfeld. 2002. *Seasonal Patterns of Stress, Immune Function and Disease*. Cambridge University Press, Cambridge, U.K.

Nerini, M. 1984. A review of gray whale feeding ecology. In *The Gray Whale, Eschrichtius robustus* (M. L. Jones, S. L. Swartz, and S. Leatherwood, eds.). Academic Press, Inc., San Diego, CA.

Nicol, S., T. Pauly, N. L. Bindoff, S. Wright, D. Thiele, G. W. Hosie, P. G. Strutton, and E. Woehler. 2000. Ocean circulation off east Antarctic affects ecosystem structure and sea-ice extent. *Nature* 406:504–507.

Nicol, S., and G. Robertson. 2003. Ecological consequences of Southern Ocean harvesting. Pp. 48–61 In *Marine Mammals: Fisheries, Tourism and Management Issues* (N. Gales, M. Hindell, and R. Kirkwood, eds.). CSIRO Publishing, Melbourne, Australia.

Nielsen, O., K. Nielson, and R. E. Stewart. 1996. Serological evidence of *Brucella* spp. exposure in Atlantic walruses (*Odobenus rosmarus rosmarus*) and ringed seals (*Phoca hispida*). *Arctic* 49:383–386.

Nieukirk, S. L., K. M. Stafford, D. K. Mellinger, R. P. Dziak, and C. G. Fox. 2004. Low-frequency whale and seismic airgun sounds recorded in the mid-Atlantic Ocean. *Journal of the Acoustical Society of America* 115(4):1832–1843.

Niimi, A. J. 1996. PCBs in aquatic organisms. In *Environmental Contaminants in Wildlife: Interpreting Tissue Concentrations* (W. N. Beyer, G. H. Heinz, and A. W. Redmon-Norwood, eds.). Lewis Publishers, Boca Raton, FL.

Nilssen, K. T., T. Haug, and V. Potelov. 1997. Seasonal variation in body condition of adult Barents Sea harp seals (*Phoca groenlandica*). *Journal of Northwest Atlantic Fisheries Science* 22:17–25.

Nilssen, K. T., O.-P. Pedersen, L. P. Folkow, and T. Haug. 2000. Food consumption estimates of Barents Sea harp seals. Pp. 9–27 In *Minke Whales, Harp and Hooded Seals: Major Predators in the North Atlantic Ecosystem* (G. A. Vikingsson and F. O. Kapel, eds.). NAMMCO Scientific Publication 2. The North Atlantic Marine Mammal Commission, Tromsø, Norway.

Nixon, S. W. 1995. Coastal marine eutrophication: a definition, social causes and future concerns. *Ophelia* 41:199–220.

Norris, S. 2003. Neutral theory: a new, unified model for ecology. *BioScience* 53(2):124–129.

Norris, K. S., G. W. Harvey, L. A. Burzell, and D. K. Kartha. 1972. Sound production in the freshwater porpoise *Sotalia cf. fluviatilis* Gervais and Deville and *Inia geoffrensis* Blainville in the Rio Negro Brazil. *Investigations on Cetacea* 4:251–262.

North Atlantic Marine Mammal Commission. 1998. Report of the Scientific Committee Working Group on the Role of Minke Whales, Harp Seals and Hooded Seals in the North Atlantic Ecosystems. Pp. 125–146 In *NAMMCO Annual Report 1997*. The North Atlantic Marine Mammal Commission, Tromsø, Norway.

North Atlantic Marine Mammal Commission. 2001. Report of the Scientific Committee Working Group on the Economic Aspects of Marine Mammal–Fisheries Interactions. Pp. 169–202 In *NAMMCO Annual Report 2000*. The North Atlantic Marine Mammal Commission, Tromsø, Norway.

North Atlantic Marine Mammal Commission. 2002. Report of the Scientific Committee. In *NAMMCO Annual Report 2001*. The North Atlantic Marine Mammal Commission, Tromsø, Norway.

North Atlantic Marine Mammal Commission. 2003. Report of the Scientific Committee. In *NAMMCO Annual Report 2002*. The North Atlantic Marine Mammal Commission, Tromsø, Norway.

Northridge, S. P. 1984. World review of interactions between marine mammals and fisheries. FAO Fisheries Technical Paper 251. 190 pp.

Northridge, S. P. 1991a. Driftnet fisheries and their impacts on non-target species: a worldwide review. FAO Fisheries Technical Paper 320.

Northridge, S. P. 1991b. An updated world review of interactions between marine mammals and fisheries. FAO Fisheries Technical Paper 251, Suppl. 1. 58 pp.

Northridge, S. P. 2003. Investigations into cetacean by-catch in a pelagic trawl fishery in the English Channel: preliminary results. Paper SC/55/SM26 presented to the Scientific Committee of the International Whaling Commission, Berlin, Germany.

Northridge, S. P., and R. J. Hofman. 1999. Marine mammal interactions with fisheries. Pp. 99–119 In *Conservation and Management of Marine Mammals* (J. R. Twiss and R. R. Reeves, eds.). Smithsonian Institution Press, Washington, DC.

Northridge, S., D. Vernicos, and D. Raitsos-Exarchopolous. 2003. Net depredation by bottlenose dolphins in the Aegean: first attempts to quantify and to minimize the problem. Paper SC/55/SM25 presented to the Scientific Committee of the International Whaling Commission, Berlin, Germany.

Notarbartolo di Sciara, G., A. Aguilar, G. Bearzi, A. Birkun, Jr., and A. Frantzis. 2002. Overview of known or presumed impacts on the different species of cetaceans in the Mediterranean and Black Seas. In *Cetaceans of the Mediterranean and Black Seas: State of Knowledge and Conservation Strategies* (G. Notarbartolo di Sciara, ed). Report to the ACCOBAMS Secretariat, Monaco, February.

Nowacek, S. M., R. S. Wells, and A. R. Solow. 2001. Short-term effects of boat traffic on bottlenose dolphins, *Tursiops truncatus,* in Sarasota Bay, Florida. *Marine Mammal Science* 17(4): 673–688.

Nystuen, J. A., and D. M. Farmer. 1987. The influence of wind on the underwater sound generated by light rain. *Journal of the Acoustical Society of America* 82:270–274.

O'Corry-Crowe, G. M., K. K. Martien, and B. L. Taylor. 2003. The analysis of population genetic structure in Alaskan harbor seals, *Phoca vitulina,* as a framework for the identification of management stocks. Administrative Report LJ-03-08. National Marine Fisheries Service, Southwest Fisheries Science Center, La Jolla, CA.

O'Hara, T. M., M. M. Krahn, D. Boyd, P. R. Becker, and L. M. Philo. 1999. Organochlorine contaminant levels in Eskimo harvested bowhead whales of arctic Alaska. *Journal of Wildlife Diseases* 35:741–752.

O'Hara, T. M., and T. J. O'Shea. 2001. Toxicology. In *CRC Handbook of Marine Mammal Medicine,* 2nd Edition (L. A. Dierauf and F. M. D. Gulland, eds.). CRC Press, Boca Raton, FL.

Ólafsdóttir, D., and E. Hauksson. 1997. Anisakid (Nematoda) infestations in Icelandic grey seals (*Halichoerus grypus Fabr.*). *Journal of the Northwest Atlantic Fisheries Science* 22:259–269.

Olesiuk, P. F., L. M. Nichol, M. J. Sowden, and J. K. B. Ford. 2002. Effect of the sound generated by an acoustic harassment device on the relative abundance and distribution of harbor porpoises (*Phocoena phocoena*) in retreat passage, British Columbia. *Marine Mammal Science* 18:843–862.

Olsen, O. W., and E. T. Lyons. 1965. Life cycle of *Uncinaria lucasi* Stiles, 1901 (Nematoda Ancylostomatidae) of fur seals, *Callorhinus ursinus* Linn., on the Pribilof Islands, Alaska. *Journal of Parasitology* 51:689–700.

Onderka, D. K. 1989. Prevalence and pathology of nematode infections in the lungs of ringed seals (*Phoca hispida*) of the western Arctic of Canada. *Journal of Wildlife Diseases* 25: 218–224.

O'Shea, T. J. 1999. Environmental contaminants in marine mammals. Pp. 485–563 In *Biology of Marine Mammals* (J. E. Reynolds, III, and S. A. Rommel, eds.). Smithsonian Institution Press, Washington, DC.

O'Shea, T. J. 2003. Toxicology of sirenians. In *Toxicology of Marine Mammals* (J. G. Vos, G. D. Bossart, M. Fournier, and T. J. O'Shea, eds.). Taylor & Francis Publishers, London, U.K.

O'Shea, T. J., and A. Aguilar. 2001. Cetacea and sirenia. In *Ecotoxicology of Wild Mammals* (R. F. Shore and B. A. Rattner, eds.). John Wiley and Sons, Ltd., West Sussex, U.K.

O'Shea, T. J., and R. L. Brownell. 1994. Organochlorine and metal contaminants in baleen whales: a review and evaluation of conservation implications. *Science of the Total Environment* 154:179–200.

O'Shea, T., and R. L. Brownell. 1998. California sea lion (*Zalophus californianus*) populations and DDT contamination. *Marine Pollution Bulletin* 36:159–164.

O'Shea, T. J., J. F. Moore, and H. I. Kochman. 1984. Contaminant concentrations in manatees in Florida. *Journal of Wildlife Management* 48:741–748.

O'Shea, T. J., G. B. Rathbun, R. K. Bonde, C. D. Buergelt, and D. K. Odell. 1991. An epizootic of Florida manatees associated with a dinoflagellate bloom. *Marine Mammal Science* 7:165–179.

O'Shea, T. J., R. R. Reeves, and A. K. Long (eds.). 1999. *Marine Mammals and Persistent Ocean Contaminants: Proceedings of the Marine Mammal Commission Workshop Keystone, Colorado, 12–15 October 1998.* Marine Mammal Commission, Bethesda, MD.

O'Shea, T. J., and S. Tanabe. 2003. Persistent ocean contaminants and marine mammals: a retrospective overview. In *Toxicology of Marine Mammals* (J. G. Vos, G. D. Bossart, M. Fournier, and T. J. O'Shea, eds). Taylor & Francis Publishers, London, U.K.

Osterberg, C., W. Pearcy, and N. Kujala. 1964. Gamma emitters in a fin whale. *Nature* 204:1006–1007.

Osterhaus, A., J. Groen, H. Niesters, M. van de Bildt, B. Martina, L. Vedder, J. Vos, H. van Egmond, B. A. Sidi, and M. E. O. Barham. 1997. Morbillivirus in monk seal mortality. *Nature* 388:838–839.

Osterhaus, A. D. M. E., G. F. Rimmelzwaan, B. E. Martina, T. M. Bestebroer, and R. A. Fouchier. 2000. Influenza B virus in seals. *Science* 288:1051–1053.

Osterhaus, A., M. Van De Bildt, L. Vedder, B. Martina, H. Niesters, J. Vos, H. Van Egmond, D. Liem, R. Baumann, E. Androukaki, S. Kotomatas, A. Komnenou, B. A. Sidi, A. B. Jiddou, and M. E. O. Barham. 1998. Monk seal mortality: virus or toxin? *Vaccine* 16:979–981.

Osterhaus, A. D. M. E., and E. J. Vedder. 1988. Identification of virus causing recent seal deaths. *Nature* 335:20.

Osterhaus, A. D. M. E., H. Yang, H. E. M. Spijkers, J. Groen, J. S. Teppema, and G. Van Steenis. 1985. The isolation and partial characterization of a highly pathogenic herpesvirus from the harbor seal (*Phoca vitulina*). *Archives of Virology* 86:239–251.

Overholtz, W. J., S. A. Murawski, and K. L. Foster. 1991. Impact of predatory fish, marine mammals, and seabirds on the pelagic fish ecosystem of the northeastern USA. *ICES Marine Science Symposium* 193:198–208.

Overland, J. E., and J. M. Adams. 2001. On the temporal character and regionality of the Arctic Oscillation. *Geophysical Research Letters* 28:2811–2814.

Overland, J. E., M. C. Spillane, and N. N. Soreide. 2004. Integrated analysis of physical and biological pan-Arctic change. *Climatic Change* 63(3):291–322.

Overpeck, J., K. Hughen, D. Hardy, R. Bradley, R. Case, M. Douglas, B. Finney, K. Gajewski, G. Jacoby, A. Jennings, S. Lamoureux, A. Lasca, G. MacDonald, J. Joore, M. Retelle, S. Smith, A. Wolfe, and G. Zieleinski. 1997. Arctic environmental change of the last four centuries. *Science* 278: 1251–1256.

Pacific Offshore Cetaceans Take Reduction Plan. 2003. www.nmfs.noaa.gov/prot_res/PR2/Fisheries_Interactions/TRT.htm#PacificOffshoreTRT. Accessed on 8 July 2003.

Pan, Y., M. L. Parsons, M. Busman, P. Moeller, Q. Dortch, C. Powell, G. A. Fyyxell, and G. J. Doucette. 2001. Pseudo-nitzschia pseudodelicatisimma: a confirmed domoic acid producer from the northern Gulf of Mexico. *Marine Ecology Progress Series* 220:83–92.

Parkinson, C. L. 2001. Trends in the length of the Southern Ocean sea ice season 1979–1999. *Annals of Glaciology* 34:435–440.

Parkinson, C. L., and D. J. Cavalieri. 2002. A 21-year record of Arctic sea-ice extents and their regional, seasonal and monthly variability and trends. *Annals of Glaciology* 34:441–446.

Parrish, F. A., M. P. Craig, T. J. Ragen, G. J. Marshall, and B. M. Buhleier. 2000. Identifying diurnal foraging habitat of endangered Hawaiian monk seals using a seal-mounted video camera. *Marine Mammal Science* 16:392–412.

Parsons, M. L., Q. Dortch, and R. E. Turner. 2002. Sedimentological evidence of an increase in *Pseudo-nitzschia* (Bacillariophyceae) abundance in response to coastal eutrophication. *Limnology and Oceanography* 47:551–558.

Pauly, D. 1995. Anecdotes and the shifting baseline syndrome of fisheries. *Trends in Ecology and Evolution* 10:430.

Pauly, D., V. Christensen, J. Dalsgaard, R. Froese, and F. Torres, Jr. 1998. Fishing down the marine food webs. *Science* 279: 860–863.

Pauly, D., V. Christensen, S. Guenette, T. J. Pitcher, U. R. Sumaila, C. J. Walters, R. Watson, and D. Zeller. 2002. Towards sustainability in world fisheries. *Nature* 418:689–695.

Payne, P. M., and D. C. Schneider. 1984. Yearly changes in abundance of harbor seals, *Phoca vitulina,* at a winter haulout site in Massachussetts. *Fishery Bulletin* 82:440–442.

Payne, R., and D. Webb. 1971. Orientation by means of long range acoustic signaling in baleen whales. *Annals of the New York Academy of Sciences* 188:110–141.

Pearl, H. W. 1999. Coastal eutrophication and harmful algal blooms: the importance of atmospheric deposition and groundwater as "new" nitrogen and other nutrient sources. *Limnology and Oceanography* 42:1154–1165.

Peddemors, V. M. 1999. Delphinids of southern Africa: a review of their distribution, status and life history. *Journal of Cetacean Research and Management* 1(2):157–165.

Peng, Y.-G., and J. S. Ramsdell. 1996. Brain fos is a sensitive biomarker for the lowest observed neuroexcitatory effects of domoic acid in mice. *Fundamental and Applied Toxicology* 31:162–168.

Perl, T. M., L. Bedard, T. Kosatsky, J. C. Hockin, E. C. D. Todd, and R. S. Remic. 1990. An outbreak of encephalopathy caused by eating mussels contaminated with domoic acid. *New England Journal of Medicine* 322:1775–1780.

Perrin, W. F. 1990. Subspecies of *Stenella longirostris* (Mammalia: Cetacea: Delphinidae). *Proceeding of the Biological Society of Washington* 103(2):453–463.

Perrin, W. F., and J. E. Powers. 1980. Role of a nematode in natural mortality of spotted dolphins. *Journal of Wildlife Management* 44:960–963.

Perrin, W. F., B. Wursig, and J. G. M. Thewissen. 2002. *Encyclopedia of Marine Mammals.* Academic Press, San Diego, CA. 1414 pp.

Perryman, W. L., M. A. Donahue, P. C. Perkins and S. B. Reilly. 2002. Gray whale calf production 1994–2000: are observed fluctuations related to changes in seasonal ice cover? *Marine Mammal Science* 18:121–144.

Pew Oceans Commission. 2003. *America's Living Oceans: Charting a Course for Sea Change.* Summary Report. Pew Oceans Commission, Arlington, VA.

Pfeiffer, C. J. (ed.). 2002. *Molecular and Cell Biology of Marine Mammals.* Krieger Publishing Company, Malabar, FL.

Philander, S. G. H. 1989. *El Niño, La Niña and the Southern Oscillation.* Academic Press, Inc., San Diego, CA.

Pierce, R. H., M. S. Henry, P. C. Blum, J. Lyons, T.-S. Cheng, D. Yazzie, et al. 2003. Brevetoxin concentrations in marine aerosol: human exposure levels during a *Karenia brevis* harmful algal bloom. *Bulletin of Environmental Contamination and Toxicology* 70:161–165.

Pistorius, P. A., M. N. Bester, and S. P. Kirkman. 1999. Survivorship of a declining population of southern elephant seals, *Mirounga leonina,* in relation to age, sex and cohort. *Oecologia* 121:201–211.

Pitcher, K. W. 1989. Harbor seal trend count surveys in southern Alaska, 1988. Final report contract MM4465852-1 submitted to U.S. Marine Mammal Commission, Washington, DC. 15 pp.

Pitcher, K. W. 1990. Major decline in number of harbor seals, *Phoca vitulina richardsi,* on Tugidak Island, Gulf of Alaska. *Marine Mammal Science* 6:121–134.

Pitman, R. L., L. T. Ballance, S. L. Mesnick, and S. J. Chivers. 2001. Killer whale predation on sperm whales: observations and implications. *Marine Mammal Science* 17:494–507.

Pitman, R. L., and P. Ensor. 2003. Three forms of killer whale (*Orcinus orca*) in the Antarctic. *Journal of Cetacean Research and Management* 5:131–139.

Plagányi, É. E., and D. S. Butterworth. 2002. Competition with fisheries. Pp. 268–273 In *Encyclopedia of Marine Mammals* (W. F. Perrin, B. Würsig, and J. G. M. Thewissen, eds.). Academic Press, San Diego, CA.

Plagányi, É. E., and D. S. Butterworth. 2004. A critical look at the potential of Ecopath with Ecosim to assist in practical fisheries management. Pp. 261–287 In *Ecosystem Approaches to Fisheries in the Southern Benguela* (L. J. Shannon, K. L. Cochrane, and S. C. Pillar, eds). *African Journal of Marine Science* 26.

Plakas S. M., K. R. el-Said, E. L. Jester, H. R. Granade, S. M. Musser, and R. W. Dickey. 2002. Confirmation of brevetoxin metabolism in the eastern oyster (*Crassostrea virginica*) by controlled exposures to pure toxins and to *Karenia brevis* cultures. *Toxicon* 40(6):721–729.

Podolska, M., and J. Horbowy. 2003. Infection of Baltic herring (*Clupea harengus membras*) with *Anisakis simplex* larvae, 1992–1999: a statistical analysis using generalized linear models. *ICES Journal of Marine Science* 60:85–93.

Poli, M. A., S. M. Musser, R. W. Dickey, P. P. Eilers, and S. Hall. 2000. Neurotoxic shellfish poisoning and brevetoxin metabolites: a case study. *Toxicon* 38:981–993.

Politi, E., and G. Bearzi. 2004. Evidence of decline for a coastal common dolphin community in the eastern Ionian Sea. *European Research on Cetaceans* 15:449–452.

Polovina, J. J. 1984. Model of a coral reef ecosystem. Part I. The ECOPATH model and its application to French Frigate Shoals. *Coral Reefs* 3:1–11.

Pope, J. G. 1991. The ICES multi-species assessment group: evolution, insights and future problems. *ICES Marine Science Symposium* 193:23–33.

Potter, J. R. 1994. ATOC: Sound policy or enviro-vandalism? Aspects of a modern media fueled policy issue. *Journal of Environment and Development* 3:47–76.

Punt, A. E. 1994. Data analysis and modelling of the seal-hake biological interaction off the South African West Coast. Working Paper, Report submitted to the Sea Fisheries Research Institute, Cape Town, South Africa.

Punt, A. E., and D. S. Butterworth. 1995. The effects of future consumption by the Cape fur seal on catches and catch rates of the Cape hakes. Part IV: Modelling the biological interaction between Cape fur seals *Arctocephalus pusillus pusillus* and the Cape hakes *Merluccius capensis* and *M. paradoxus*. *South African Journal of Marine Science* 16:255–285.

Quilliam, M., D. Wechsler, S. Marcus, B. Ruck, M. Weckell, and T. Hawryluk. 2002. Detection and identification of paralytic shellfish poisoning toxins in Florida pufferfish responsible for incidents of neurologic illness (Abstract). 10th International Conference on Harmful Algae, St. Petersburg Beach, FL, 21–25 October.

Rabalais, N. N., R. E. Turner, D. Justin, Q. Dortch, and W. J. Wiseman, Jr. 1999. Characterization of hypoxia. Topic 1: Report for the integrated assessment on hypoxia in the Gulf of Mexico. U.S. Department of Commerce, National Oceanic and Atmospheric Administration, Coastal Ocean Program, Decision Analysis Series No. 15.

Raga, J. A., F. J. Aznar, J. A., Balbuena, and M. Fernández. 2002. Parasites. Pp. 867–876 In *Encyclopedia of Marine Mammals* (W. F. Perrin, B. Wursig, and J. G. M. Thewissen, eds.). Academic Press, San Diego, CA.

Ragen, T. J., and D. M. Lavigne. 1999. The Hawaiian monk seal: biology of an endangered species. Pp. 224–245 In *Conservation and Management of Marine Mammals* (J. R. Twiss, Jr., and R R. Reeves, eds.). Smithsonian Institution Press, Washington, DC.

Ragen, T. J., and R. R. Reeves. 2004. Consultation on Future Directions in Marine Mammal Research: Executive Summary. Marine Mammal Commission, Bethesda, MD. 21 pp.

Rainbow, P. S. 1996. Heavy metals in aquatic invertebrates. In *Environmental Contaminants in Wildlife: Interpreting Tissue Concentrations* (W. N. Beyer, G. H. Heinz, and A. W. Redmon-Norwood, eds.). Lewis Publishers, Boca Raton, FL.

Ralls, K., D. P. DeMaster, and J. A. Estes. 1996. Developing a criterion for delisting the southern sea otter under the U.S. Endangered Species Act. *Conservation Biology* 10: 1528–1537.

Rathbun, G. B., J. P. Reid, R. K. Bonde, and J. A. Powell. 1995. Reproduction in free-ranging Florida manatees. Pp. 135–156 In *Population Biology of the Florida Manatee* (T. J. O'Shea, B. B. Ackerman, and H. F. Percival, eds.). U.S. Department of the Interior, National Biological Service, Information and Technology Report 1.

Read, A. J., P. Drinker, and S. P. Northridge. In review. By-catches of marine mammals in U.S. fisheries and a first attempt to estimate the magnitude of global marine mammal by-catch. *Conservation Biology.*

Read, A. J., and A. A. Rosenberg (convenors). 2002. Draft International Strategy for Reducing Incidental Mortality of Cetaceans in Fisheries. http://cetaceanbycatch.org/intlstrategy.cfm. Accessed on 8 July 2003.

Read, A. J., K. Van Waerebeek, J. C. Reyes, J. S. McKinnon, and L. C. Lehman. 1988. The exploitation of small cetaceans in coastal Peru. *Biological Conservation* 46:53–70.

Read, A. J., and P. R. Wade. 2000. Status of marine mammals in the United States. *Conservation Biology* 14:929–940.

Read, A. J., D. M. Waples, K. W. Urian, and D. Swanner. 2003. Fine-scale behaviour of bottlenose dolphins around gillnets. *Proceedings of the Royal Society of London. Series B. Biological Sciences* (Suppl.) DOI 10.1098/rsbl.2003.0021.

Reddy, M. L., L. A. Dierauf, and F. M. D. Gulland. 2001. Marine mammals as sentinels of ocean health. Pp. 471–520 In *CRC Handbook of Marine Mammal Medicine,* 2nd Edition (L. A. Dierauf and F. M. D. Gulland, eds.). CRC Press, Boca Raton, FL.

Reddy, M. L., and S. H. Ridgway. 2003. Opportunities for environmental contaminant research: what we can learn from marine mammals in human care. In *Toxicology of Marine Mammals* (J. G. Vos, G. D. Bossart, M. Fournier, and T. J. O'Shea, eds.). Taylor & Francis, London, U.K.

Reeves, R. R. (ed.). 2002a. Report of a workshop to develop a research plan on chemical contaminants and health status of southern sea otters. The Otter Project, Marina, CA.

Reeves, R. R. 2002b. The origins and character of 'aboriginal subsistence' whaling: a global view. *Mammal Review* 32 (2):71–106.

Reeves, R. R., W. F. Perrin, B. L. Taylor, C. S. Baker, and S. L. Mesnick. 2004. Report of the workshop on shortcomings of cetacean taxonomy in relation to needs of conservation and management, April 30–May 2, La Jolla, CA. NOAA Technical Memorandum NMFS. NOAA-TM-NMFS-SWFSC-363. 94 pp.

Reeves, R. R., and T. J. Ragen. 2004. Future directions in marine mammal research: report of the Marine Mammal Commission Consultation, August 3–7, 2003. Marine Mammal Commission, Bethesda, MD. 79 pp. + xiv.

Reeves, R. R., A. J. Read, and G. Notarbartolo di Sciara (eds). 2001. Report of the workshop on interactions between dolphins and fisheries in the Mediterranean: evaluation of mitigation alternatives. Istituto Centrale per la Ricerca Applicata al Mare, Rome, Italy.

Reeves, R. R., B. D. Smith, E. Crespo, G. Notarbartolo di Sciara, and the Cetacean Specialist Group. 2003. *Dolphins, Whales, and Porpoises: 2003–2010. Conservation Action Plan for the World's Cetaceans.* Species Survival Commission, IUCN. Gland, Switzerland.

Reeves, R. R., B. S. Stewart, P. J. Clapham, and J. A. Powell. 2002a. *Sea Mammals of the World.* A&C Black Publishers, London, U.K. 527 pp.

Reeves, R. R., B. S. Stewart, P. J. Clapham, and J. A. Powell. 2002b. *National Audubon Society Guide to Marine Mammals of the World.* Alfred A. Knopf, New York, NY.

Reich, S., Jimenez, B., Marsili, L., Hernandez, L. M., Schurig, V. and M. J. Gonzalez. 1999. Congener specific determination and enantiomeric ratios of chiral polychlorinated biphenyls in striped dolphins (*Stenella coeruleoalba*) from the Mediterranean Sea. *Environmental Science and Technology* 33:1787–1793.

Reidarson, T. H., J. F. McBain, L. M. Dalton, and M. G. Rinaldi. 2001. Mycotic diseases. In *CRC Handbook of Marine Mammal Medicine* (L. A. Dierauf and F. M. D. Gulland, eds.). CRC Press, Boca Raton, FL.

Reijnders, P. J. H. 1986. Reproductive failure in common seals feeding on fish from polluted coastal waters. *Nature* 324: 456–457.

Reijnders, P. J. H., A. Aguilar, and G. P. Donovan (eds.). 1999. *Chemical Pollutants and Cetaceans.* Report of the workshop on chemical pollution and cetaceans. *Journal of Cetacean Research and Management* (Special Issue 1). International Whaling Commission, Cambridge, U.K.

Reijnders, P. J. H., E. H. Ries, S. Tougaard, N. Nørgaard, G. Heidemann, J. Schwarz, E. Vareschi, and I. M. Traut. 1997. Population development of harbour seals *Phoca vitulina* in the Wadden Sea after the 1988 virus epizootic. *Journal of Sea Research* 38:161–168.

Reilly, S. B. 1990. Seasonal changes in distribution and habitat differences among dolphins in the eastern tropical Pacific. *Marine Ecology Progress Series* 66(1–2):1–11.

Reilly, S. B., and P. C. Fiedler. 1994. Interannual variability of dolphin habitats in the eastern tropical Pacific. Part I: Research vessel surveys, 1986–1990. *Fishery Bulletin* 92:434–450.

Reilly, S. B., and V. G. Thayer. 1990. Blue whale (*Balaenoptera musculus*) distribution in the eastern tropical Pacific. *Marine Mammal Science* 6:265–277.

Rendell, L. E., and H. Whitehead. 2003. Vocal clans in sperm whales (*Physeter macrocephalus*). *Proceedings of the Royal Society of London. Series B. Biological Sciences* 270(1512):225–231.

Resendes A. R, S. Almeria, J. P. Dubey, E. Obon, C. Juan-Salles, E. Degollada, F. Alegre, O. Cabezon, S. Pont, and M. Domingo. 2002. Disseminated toxoplasmosis in a Mediterranean pregnant Risso's dolphin (*Grampus griseus*) with transplacental fetal infection. *Journal of Parasitology* 88:1029–1032.

Resolve. 1999. The National Marine Fisheries Service take reduction team negotiation process evaluation. Unpub. Report. National Marine Fisheries Service Contract No. 50-DGNF-5-00164. Prepared by RESOLVE, Inc.

Reynolds, J. E. III. 1999. Efforts to conserve the manatees. In *Conservation and Management of Marine Mammals* (J. R. Twiss Jr. and R. R. Reevees, eds.) Smithsonian Institution Press, Washington DC.

Reynolds, J. E., III, D. L. Wetzel, and T. M. O'Hara. In press. Omega-3 fatty acids and omega-6 fatty acids in blubber of bowhead whales (*Balaena mysticetus*): health implications for subsistence-level cultures. *Arctic.*

Reynolds, J. E., and J. R. Wilcox. 1986. Distribution and abundance of the West Indian manatee, *Trichechus manatus,* around selected Florida power plants following winter cold fronts: 1984–1985. *Biological Conservation* 38:103–113.

Rice, D. W. 1998. *Marine Mammals of the World: Systematics and Distribution.* The Society for Marine Mammalogy (Special Publication 4), Lawrence, KS.

Rice, J. C. 2000. Evaluating fishery impacts using metrics of community structure. *ICES Journal of Marine Science* 57:682–688.

Rice, J. C., N. Daan, J. G. Pope, and H. Gislason. 1991. The stability of estimates of suitabilities in MSVPA over four years of data from predator stomachs. *ICES Marine Science Symposium* 193:34–45.

Richardson, W. J., C. R. J. Greene, C. I. Malme, and D. H. Thomson. 1995. *Marine Mammals and Noise.* Academic Press, San Diego, CA.

Richardson, W. J., and C. I. Malme. 1993. Man-made noise and behavioral responses. In *The Bowhead Whale* (J. J. Burns, et al., eds.). Society for Marine Mammology, Lawrence, KS.

Ridgway, S., D. A. Carder, R. R. Smith, T. Kamolnick, C. E. Schlundt, and W. Elsberry. 1997. Behavioral responses and temporary shift in masked hearing threshold of bottlenose dolphins *Tursiops truncatus,* to 1-second tones of 141 to 201 dB *re:* 1 μPa. NRAD, RDT&RE Div., Naval Command, Control & Ocean Surveillance Center, San Diego, CA. Technical Report 1751.

Ridgway, S. H., and J. G. McCormick, J. G. 1971. Anesthesia of the porpoise. Pp. 394–403 In *Textbook of Veterinary Anesthesia* (L. R. Soma, ed.). Williams and Wilkinson, Baltimore, MD.

Riedlinger, D., and F. Berkes. 2001. Contributions of traditional knowledge to understanding climate change in the Canadian Arctic. *Polar Record* 37 (203):315–328.

Rigor, I. G., and J. M. Wallace. 2004. Variations in the age of Arctic sea-ice and summer sea-ice extent. *Geophysical Research Letters* 31(9):L09401.

Rind, D., M. Chandler, J. Lerner, D. G. Martinson, and X. Yuan. 2001. Climate response to basin-specific changes in latitudinal temperature gradients and implications for sea ice variability. *Journal of Geophysical Research* 106: 20,161–20,173.

Rindorf, A., H. Gislason, and H. Lewy. 1998. Does the diet of cod and whiting reflect the species composition estimated from trawl surveys? ICES C. M. 1998/CC:5.

Robson, B. W., M. E. Goebel, J. D. Baker, R. R. Ream, T. R. Loughlin, R. C. Francis, G. A. Antonelis, and D. P. Costa. 2004. Separation of foraging habitat among breeding sites of a colonial marine predator, the northern fur seal (*Callorhinus ursinus*). *Canadian Journal of Zoology* 82:20–29.

Rolland, R. 2000. A review of chemically-induced alterations in thyroid and vitamin a status from field studies of wildlife and fish. *Journal of Wildlife Diseases* 36(4):615–635.

Romano, T. A., M. J. Keogh, C. Kelly, P. Feng, L. Berk, C. E. Schlundt, D. A. Carder, and J. J. Finneran. 2004. Anthropogenic sound and marine mammal health: measures of the nervous and immune systems before and after intense sound exposure. *Canadian Journal of Fisheries and Aquatic Sciences* 61:1124–1134.

Root, T. L., J. R. Price, K. R. Hall, S. H. Schneider, C. Rosenzweig, and J. A. Pounds. 2003. Fingerprints of global warming on wild animals and plants. *Nature* 421:57–60.

Rosen, D. A. S., and A. W. Trites. 2000. Pollock and the decline of Steller sea lions: testing the junk-food hypothesis. *Canadian Journal of Zoology* 78:1243–1250.

Rosenbaum, H. C., R. L. Brownell, Jr., M. W. Brown, C. Schaeff, V. Portway, B. N. White, S. Malik, L. A. Pastene, N. J. Patenaude, C. S. Baker, M. Goto, P. B. Best, P. J. Clapham, P. Hamilton, M. Moore, R. Payne, V. Rowntree, C. T. Tynan, J. L. Bannister and R. DeSalle. 2000. Worldwide genetic differentiation of *Eubalaena:* questioning the number of right whale species. *Molecular Ecology* 9:1793–1802.

Ross, D. G. 1987. *Mechanics of Underwater Noise.* Peninsula Publishing, Los Altos, CA.

Ross, D. G. 1993. On ocean underwater ambient noise. *Acoustics Bulletin* January / February:5–8.

Ross, P. S., and S. De Guise. 1998. Environmental contaminants and marine mammal health: research applications. *Canadian Technical Report of Fisheries and Aquatic Sciences* 2255:29.

Ross, P. S., G. M. Ellis, M. G. Ikonomou Barret, L. G. Lennard, and R. F. Addison. 2000. High PCB concentrations in free-ranging pacific killer whales, *Orcinus orca:* effects of age, sex, and dietary preference. *Marine Pollution Bulletin* 406:504–515.

Ross, P. S., and G. M. Troisi. 2001. Pinnipedia. In *Ecotoxicology of Wild Mammals* (R. F. Shore and B. A. Rattner, eds.). John Wiley & Sons, Ltd., West Sussex, U.K.

Rossby, T., D. Dorson, and J. Fontaine. 1986. The RAFOS system. *Journal of Atmospheric and Oceanic Technology* 3:672–679.

Rothman, K. J. 1986. *Modern Epidemiology.* Little Brown and Co., Boston, MA.

Rothrock, D. A., Y. Yu, and G. Maykut. 1999. Thinning of the Arctic sea-ice cover. *Geophysical Research Letters* 26:3469–3472.

Roussel, E. 2002. Disturbance to Mediterranean cetaceans caused by noise. In *Cetaceans of the Mediterranean and Black Seas: State of Knowledge and Conservation Strategies* (G. Notarbartolo di Sciara, ed.). Report to the ACCOBAMS (Agreement on the Conservation of Cetaceans in the Black and Mediterranean Sea) Secretariat, Section 13, 18 pp., Monaco, February.

Ruff, T. A. 1989. Ciguatera in the Pacific: a link with military activities. *Lancet* 1(8631):201–205.

Rugh, D. J., R. C. Hobbs, J. A. Lerczak, and J. M. Breiwick. 2004. Estimates of abundance of the eastern North Pacific stock of gray whales 1997 to 2002. *Journal of Cetacean Research and Management.*

Rugh, D. J., M. M. Muto, S. E. Moore, and D. P. DeMaster. 1999. Status review of the eastern North Pacific stock of gray whales. NOAA Technical Memorandum NMFS-AFSC-103.

Ruiz, G. M., J. T. Carleton, and A. H. Hines. 1997. Global invasions of marine and estuarine habitats by non-indiginous species: mechanisms, extent, and consequences. *American Zoologist* 37:621–632.

Ruiz, G. M., T. K. Rawlings, F. C. Dobbs, L. A. Drake, T. Mullady, A. Huq, and A. A. Colwell. 2000. Global spread of microorganisms by ships. *Nature* 408:49.

Safe, S. 1990. Polychlorinated biphenyls (PCBs), dibenzo-*p*-dioxins (PCDDs), dibenzofurans (PCDFs), and related compounds: environmental and mechanistic considerations which support the development of toxic equivalency factors (TEFs). *Critical Reviews in Toxicology* 21:51–88.

Sandegren, F. E., E. W. Chu, and J. E. Vandevere. 1973. Maternal behavior in the California sea otter. *Journal of Mammalogy* 54(3):668–679.

Save Wave. 2003. Save Wave—Dolphin Saver. www.savewave.net/. Accessed on 8 July 2003.

Schmidt, V. 2004. Seismic contractors realign equipment for industry's needs. *Offshore* 64:36–44.

Scholin, C. A., F. Gulland, G. J. Doucette, S. Benson, M. Busman, F. P. Chavez, J. Cordaro, R. DeLong, A. de Vogelaere, J. Harvey, M. Haulena, K. Lefebvre, T. Limscomb, S. Luscatoff, L. J. Lowenstine, R. Marin, P. Miller, W. A. McLellan, P. D. R. Moeller, C. Powell, T. Rowles, P. Silvagni, M. Silver, T. Spraker, V. Trainer, and F. M. Van Dolah. 2000. Mortality of sea lions along the central California coast linked to a toxic diatom bloom. *Nature* 403:80–84.

Schreiner, A. E., C. G. Fox, and R. P. Dziak. 1995. Spectra and magnitudes of t-waves from the 1993 earthquake swarm on the juan de fuca ridge. *Geophysical Research Letters* 22(2): 139–142.

Schreiner, H. F. J. 1990. The Randi-pe noise model. *Proceedings of the IEEE Oceans '90:* 576–577.

Schweder, T., G. S. Hagen, and E. Hatlebakk. 2000. Direct and indirect effects of minke whale abundance on cod and herring fisheries: a scenario experiment for the Greater Barents Sea. Pp. 120–132 In *Minke Whales, Harp and Hooded Seals: Major Predators in the North Atlantic Ecosystem* (G. A. Vikingsson and F. O. Kapel, eds.). NAMMCO Scientific Publication 2. The North Atlantic Marine Mammal Commission, Tromsø, Norway.

Schwing, F. B., T. Muurphree, and P. M. Green. 2002. The northern oscillation index (NOI): a new climate index for the northeast Pacific. *Progress in Oceanography* 53:115–139.

Scott, M. E. 1988. The impact of infection and disease on animal populations: implications for conservation biology. *Conservation Biology* 2:40–56.

Secchi, E. R., and T. Vaske, Jr. 1998. Killer whale (*Orcinus orca*) sightings and depredation on tuna and swordfish longline catches in southern Brazil. *Aquatic Mammals* 24:17–122.

Sellner, K. G., G. D. Doucette, and G. Kirkpatrick. 2003. Harmful algal blooms: causes, impacts, and detection. *Journal of Industrial Microbiology and Biotechnology* 30:383–406.

Serreze, M. C., J. A. Maslanik, T. A. Scambos, F. Fetterer, J. Stroeve, K. Knowles, C. Fowler, S. Drobot, R. G. Barry, and T. M. Haran. 2003. A record minimum Arctic sea ice extent and area in 2002. *Geophysical Research Letters* 30(3):10-1–10-4.

Seth, V., and A. Boetra. 1986. Malnutrition and the immune response. *Indian Pediatrics* 23:277–295.

Shaughnessy, P. D., and A. I. L. Payne. 1979. Incidental mortality of Cape fur seals during trawl fishing activities in South African waters. *Fishery Bulletin of South Africa* 12:20–25.

Shaughnessy P. D., S. K. Troy, R. Kirkwood, and A. O. Nicholls. 2000. Australian fur seals at Seal Rocks, Victoria: pup abundance by mark recapture estimation shows continued increase. *Wildlife Research* 27:629–633.

Shaw, D., and A. P. Dobson. 1995. Patterns of macroparasite abundance and aggregation in wildlife populations: a quantitative review. *Parasitology* 111:S111–133.

Shelden, K. E. W., S. E. Moore, J. M. Waite, P. R. Wade, and D. J. Rugh. 2005. Historic and current habitat use by North Pacific right whales, *Eubalaena japonica,* in the Bering Sea and Gulf of Alaska. *Mammal Review* 35(2):129–214.

Simmonds, M. P., and P. A. Johnston. 1994. Whale meat: a safe and healthy food? *British Food Journal* 96:26–31.

Simmonds, M. P., P. A. Johnston, M. C. French, R. Reeve, and J. D. Hutchinson. 1994. Organochlorines and mercury in pilot whale blubber consumed by Faroe Islanders. *Science of the Total Environment* 149:97–111.

Simmonds, M. P., and L. F. Lopez-Jurado. 1991. Whales and the military. *Nature* 351:448.

Širovic, A., J. A. Hildebrand, S. M. Wiggins, M. A. McDonald, S. E. Moore, and D. Thiele. 2004. Seasonality of blue and fin whale calls and the influence of sea ice in the western Antarctic Peninsula. *Deep Sea Research II* 51:2327–2344.

Sladen, W. J. L., C. M. Menzie, and W. L. Reichel. 1966. DDT residues in Adelie penguins and a crabeater seal from Antarctica. *Nature* 210:670–673.

Slate, J., L. E. B. Kruuk, T. C. Marshall, J. M. Pemberton, and T. H. Clutton-Brock. 2000. Inbreeding depression influences lifetime breeding success in a wild population of red deer (*Cervus elaphus*). *Proceedings of the Royal Society of London. Series B. Biological Sciences* 267:1657–1662.

Small, R. J., and D. P. DeMaster. 1995. Alaska marine mammal stock assessments, 1995. U.S. Department of Commerce, NOAA Technical Memorandum NMFS-AFSC-57. 93 pp.

Smayda, T. J. 1990. Novel and nuisance phytoplankton blooms in the sea: evidence for a global epidemic. Pp. 29–40 In *Toxic Marine Phytoplankton* (E. Graneli, B. Sundstrom, L. Edler, and D. M. Anderson, eds.). Elsevier Science, New York, NY.

Smith, J. A., K. Wilson, J. G. Pilkingon, and J. M. Pemberton. 1999. Heritable variation in resistance to gastro-intestinal nematodes in an unmanaged mammal population. *Proceedings of the Royal Society of London. Series B. Biological Sciences* 266: 1283–1290.

Sparre, P. 1991. Introduction to multispecies virtual population analysis. *ICES Marine Science Symposium* 193:12–21.

Spiess, F. N., J. Northrup, and E. W. Werner. 1968. Locations and enumeration of underwater explosions in the North Pacific. *Journal of the Acoustical Society of America* 43(3):640–641.

Springer, A. M. 1998. Is it all climate change? Why marine bird and mammal populations fluctuate in the North Pacific. In *Biotic Impacts of Extratropical Climate Variability in the Pacific. 'Aha Huliko'a Proceedings Hawaiian Winter Workshop* (G. Holloway, P. Muller, and D. Hender, eds.). University of Hawaii Press, Honolulu.

Springer, A. M., J. F. Piatt, V. P. Shuntov, G. B. Van Vliet, V. L. Vladimirov, A. E. Kuzin, and A. S. Perlov. 1999. Marine birds and mammals of the Pacific subarctic gyres. *Progress in Oceanography* 43:443–487.

St. Aubin, D., and L. Dierauf. 2001. Stress and marine mammals. In *CRC Handbook of Marine Mammal Medicine* (L. A. Dierauf and F. M. D. Gulland, eds.). CRC Press, Boca Raton, FL.

Stafford, K. M., D. R. Bohnenstiehl, M. Tolstoy, E. Chapp, D. K. Mellinger, and S. E. Moore. 2004. Antarctic-type blue whale calls recorded at low latitudes in the Indian and eastern Pacific Oceans. *Deep Sea Research I* 51(10):1337–1346.

Stafford, K. M., S. L. Nieukirk, and C. G. Fox. 2001. Geographic and seasonal variation of blue whale calls in the North Pacific. *Journal of Cetacean Research and Management* 3:65–76.

Stamper, M. A., F. M. D. Gulland, and T. Spraker. 1998. Leptospirosis in rehabilitated Pacific harbor seals from California. *Journal of Wildlife Diseases* 34:407–410.

Stefansson, G., and O. K. Palsson. 1998. The framework for multispecies modelling of Arcto-boreal systems. *Reviews in Fish Biology and Fisheries* 8:101–104.

Stefansson, G., J. Sigurjónsson, and G. A. Víkingsson. 1997. On dynamic interactions between some fish resources and cetaceans off Iceland based on a simulation model. *Journal of the Northwest Atlantic Fisheries Science* 22:357–370.

Steller Sea Lions. www.fakr.noaa.gov/protectedresources/stellers.htm.

Stenson, G. B., M. O. Hammill, M. C. S. Kingsley, B. Sjare, W. G. Warren, and R. A. Myers. 2002. Is there evidence of increased pup production in northwest Atlantic harp seals, *Pagophilus groenlandicus? ICES Journal of Marine Science* 59:81–92.

Stenson, G. B., M. O. Hammill, and J. W. Lawson. 1997. Predation by harp seals in Atlantic Canada: preliminary consumption estimates for Arctic cod, capelin and Atlantic cod. *Journal of the Northwest Atlantic Fisheries Science.* 22:137–154.

Stenvers, O., J. Plotz, and H. Ludwig. 1992. Antarctic seals carry antibodies against seal herpesviruses. *Archives of Virology* 123:421–424.

Stern, H. L., and M. P. Heide-Jørgensen. 2003. Trends and variability of sea ice in Baffin Bay and Davis Strait, 1953–2001. *Polar Research* 22 (1):11–18.

Stevens, C. C., K. L. Russell, M. E. Knafelc, P. F. Smith, E. W. Hopkins, and J. B. Clark. 1999. Noise-induced neurologic disturbances in divers exposed to intense water-borne sound: two case reports. *Undersea & Hyperbaric Medicine* 26(4):261–265.

Stewart, B. S., and R. L. DeLong. 1995. Double migrations of the northern elephant seal, *Mirounga angustirostris. Journal of Mammalogy* 76:196–205.

Stewart, B. S., W. E. Evans, and F. T. Awbrey. 1982. Effects of man-made waterborne noise on behavior of belukha whales (*Delphinapterus leucas*) in Bristol Bay, Alaska. Hubbs/Sea World Research Institute for the U.S. National Oceanic and Atmospheric Administration, Juneau, AK, San Diego, CA. HSWRI Technical Report 82-145.

Stirling, I. 2002. Polar bears and seals in the eastern Beaufort Sea and Amundsen Gulf: a synthesis of population trends and ecological relationships over three decades. *Arctic* 55 (Suppl. 1): 59–76.

Stirling, I., and A. E. Derocher. 1993. Possible impacts of climatic warming on polar bears. *Arctic* 46:240–245.

Stoskopf, M. K., S. Willens, and J. F. McBain. 2001. Pharmaceuticals and formularies. Pp. 703–727 In *CRC Handbook of Marine Mammal Medicine* (L. A. Dierauf and F. M. D. Gulland, eds.). CRC Press, Boca Raton, FL.

Stuen, S., P. Have, A. D. M. E. Osterhaus, J. M. Arnemo, and A. Moustgaard. 1994. Serological investigations of virus infections in harp seals (*Phoca groenlandica*) and hooded seals (*Cystophora cristata*). *Veterinary Record* 134:502–503.

Stumph, R. P., M. Culver, E. Truby, and M. Sorocco. 2002. A prototype system for monitoring and forecasting of harmful algal blooms in the Gulf of Mexico (Abstract). 10th International Conference on Harmful Algal Blooms, 21–25 October, St. Petersburg Beach, FL.

Suganuma, M., H. Fujiki, H. Suguri, S. Yoshizawa, M. Hirota, M. Nakayasu, M. Ojika, K. Wakamatsu, K. Yamada, and T. Sugimura. 1988. Okadaic acid: An additional non-phorbol-12-tetradecanoate-13-acetate-type tumor promoter. *Proceedings of the National Academy of Sciences* 85:1768–1771.

Tamura, T. 2003. Regional assessments of prey consumption and competition by marine cetaceans in the world. Pp. 143–170 In *Responsible Fisheries in the Marine Ecosystem* (M. Sinclair and G. Valdimarsson, eds.). U.N. Food and Agriculture Organization, Rome, Italy, and CABI Publishing, Wallingford, U.K.

Tamura, T., and Y. Fujise. 2002. Geographical and seasonal changes of the prey species of minke whale in the Northwestern Pacific. *ICES Journal of Marine Science* 59:516–528.

Tasker, M. L., M. C. J. Kees-Camphuysen, J. Cooper, S. Garthe, W. A. Montevecchi, and S. J. Blaber. 2000. The impacts of fishing on marine birds. *ICES Journal of Marine Science* 57: 531–547.

Taubenberger, J. K., M. Tsai, A. E. Krafft, J. H. Lichy, A. H. Reid, F. Y. Schulman, and T. P. Lipscomb. 1996. Two morbilliviruses implicated in bottlenose dolphin epizootics. *Emerging Infectious Diseases* 2:213–261.

Taylor, B. 1993. "Best" abundance estimates and best management: why they are not the same. U.S. Department of Commerce, NOAA Technical Memorandum NMFS-SWFSC-188.

Taylor, B. L. 1997. Defining "population" to meet management objectives for marine mammals. Pp. 49–65 In *Molecular Genetics of Marine Mammals* (A. E. Dizon, S. J. Chivers, and W. F. Perrin, eds.). *Marine Mammal Science* (Special Publication 33).

Taylor, B., J. Barlow, R. Pitman, L. Ballance, T. Klinger, D. DeMaster, J. Hildebrand, J. Urban, D. Palacios, and J. Mead. 2004. A call for research to assess risk of acoustic impact on beaked whale populations. International Whaling Commission SC/56/E36.

Taylor, B. L., and A. E. Dizon. 1996. The need to estimate power to link genetics and demography for conservation. *Conservation Biology* 2:661–664.

Taylor, B. L., and A. E. Dizon. 1999. First policy then science: why a management unit based solely on genetic criteria can't work. *Molecular Ecology* 8:S11–S16.

Taylor, B. L., and P. Wade. 2000. "Best" abundance estimates and best management: why they are not the same. In *Quantitative Methods for Conservation Biology* (S. Ferson and M. Burgman, eds.). Springer-Verlag, New York, NY.

Taylor, B. L., P. R. Wade, D. P. DeMaster, and J. Barlow. 2000. Incorporating uncertainty into management models for marine mammals. *Conservation Biology* 14:1243–1252.

Tegner, M. J., and P. K. Dayton. 1997. Shifting baselines and the problem of reduced expectations in nearshore fisheries. Pp. 110–128 In *California and the World Ocean '97* (O. T. Magoon, H. Converse, B. Baird, and M. Miller-Henson, eds.). American Society of Chemical Engineers, Reston, VA.

Tegner, M. J., P. L. Haaker, K. L. Riser, and L. I. Vilchis. 2001. Climate variability, kelp forests, and the Southern California red abalone fishery. *Journal of Shellfish Research* 20:755–763.

ter Harr, G., S. Daniels, K. C. Eastaugh, and C. R. Hill. 1982. Ultrasonically induced cavitation in vivo. *British Journal of Cancer* 45 (Suppl. V):151–155.

Tester, P. A., and K. A. Steidinger. 1997. *Gymnodinium* breve red tide blooms: initiation, transport and consequences of surface circulation. *Limnology and Oceanography* 42:1039–1051.

Tester, P. A., R. P. Stumpf, F. M. Vukovich, P. K. Fowler, and J. T. Turner. 1991. An expatriate red tide bloom: transport, distribution, and persistance. *Limnology and Oceanography* 36: 1053–1061.

Thiele, D., E. T. Chester, and P. C. Gill. 2000. Cetacean distribution off eastern Antarctica (80–150°E) during the Austral summer of 1995/96. *Deep-Sea Research II* 47:2543–2572.

Thiele, D., E. T. Chester, S. E. Moore, A. Širovic, J. A. Hildebrand, and A. S. Friedlaender. 2004. Seasonal variability in whale encounters in the western Antarctic Peninsula. *Deep-Sea Research II.*

Thiele, D., and P. C. Gill. 1999. Cetacean observations during a winter voyage into Antarctic sea ice south of Australia. *Antarctic Science* 11(1):48–53.

Thomas, J. A., and V. B. Kuechle. 1982. Quantitative analysis of Weddell seal (*Leptonychotes weddelli*) underwater vocalizations at McMurdo Sound, Antarctica. *Journal of the Acoustical Society of America* 72(6):1730–1738.

Thomas, J. A., and C. W. Turl. 1990. Echolocation characteristics and range detection threshold of a false killer whale *Pseudorca crassidens.* Pp. 321–334 In *Sensory Abilities of Cetaceans: Laboratory and Field Evidence* (R. A. Kastelein, ed.). Plenum Press, New York, NY.

Thomas, N. J., and R. A. Cole. 1996. The risk of disease and threats to the wild population. *Endangered Species Update* 13:23–27.

Thompson, D. W. J., and J. M. Wallace. 1998. The Arctic Oscillation signature in the wintertime geopotential height and temperature fields. *Geophysical Research Letters* 25:1297–1300.

Thompson, D. W. J., J. M. Wallace, and G. C. Hegerl. 2000. Annular modes in the extratropical circulation. Part II: Trends. *Journal of Climate* 13:1018–1036.

Thompson, P., and A. Hall. 1993. Seals and epizootics: what factors might effect the severity of mass mortalities? *Mammal Review* 23:147–152.

Thompson, P. M., D. J. Tollit, H. M. Corpe, R. J. Reid, and H. M. Ross. 1997. Changes in haematological parameters in relation to prey switching in a wild population of harbour seals. *Functional Ecology* 11(6):743–750.

Thomson, R. B., D. S. Butterworth, I. L. Boyd, and J. P. Croxall. 2000. Modeling the consequences of Antarctic krill harvesting on Antarctic fur seals. *Ecological Applications* 10:1806–1819.

Thomson, R. B., D. S. Butterworth, and H. Kato. 1999. Has the age at transition of southern hemisphere minke whales declined over recent decades? *Marine Mammal Science* 15: 661–682.

Thorne, E. T., and E. S. Williams. 1988. Disease and endangered species: the black-footed ferret as a recent example. *Conservation Biology* 2:66–74.

Tilbury, K. L., N. G. Adams, C. A. Krone, P. Meador, G. Early, and U. Varanasi. 1999. Organochlorines in stranded pilot whales (*Globicephala melaena*) from the coast of Massachusetts. *Archives of Environmental Contamination and Toxicology* 37:125–134.

Tilbury, K. L., J. Stein, C. A. Krone, R. L. Brownell, Jr., S. A. Blokhin, J. L. Bolton, and D. W. Ernest. 2002. Chemical contaminants in juvenile gray whales (*Eschrichtius robustus*) from a subsistence harvest in arctic feeding grounds. *Chemosphere* 47:555–564.

Tilbury, K. L., J. E. Stein, P. Meador, C. A. Krone, and S. L. Chan. 1997. Chemical contaminants in harbor porpoise (*Phocoena phocoena*) from the North Atlantic coast: tissue concentrations and intra- and inter-organ distribution. *Chemosphere* 34 (9/10):2159–2181.

Tilt, W. C. 1985. *Whales and Whalewatching in North America with Special Emphasis on the Issue of Harassment.* Yale School of Forestry and Environmental Studies, New Haven, CT.

Tjelmeland, S., and U. Lindstrøm. 2005. An ecosystem element added to the assessment of Norwegian spring-spawning herring: implementing predation by minke whales. *ICES Journal of Marine Science* 62:285–294.

Todd, S., P. Stevick, J. Lien, F. Marques, and D. R. Ketten. 1996. Behavioral effects of exposure to underwater explosions in humpback whales (*Megaptera novaeangliae*). *Canadian Journal of Zoology* 74:1661–1672.

Tolstoy, M., J. B. Diebold, S. C. Webb, D. R. Bohnenstiehl, E. Chapp, R. C. Holmes, and M. Rawson. 2004. Broadband calibration of R/V *Ewing* seismic sources. *Geophysical Research Letters* 31:L14310.

Townsend, C. H. 1935. The distribution of certain whales as shown by logbook records of American whaleships. *Zoologica* 19:1–50.

Trites, A. W., V. Christensen, and D. Pauly. 1997. Competition between fisheries and marine mammals for prey and primary production in the Pacific Ocean. *Journal of the Northwest Atlantic Fisheries Science* 22:173–187.

Trites, A. W., and C. P. Donnelly. 2003. The decline of Steller sea lions *Eumetopias jubatus* in Alaska: a review of the nutritional stress hypothesis. *Mammal Review* 33:3–28.

Trites, A. W., P. A. Livingston, S. Mackinson, M. C. Vasconcellos, A. M. Springer, and D. Pauly. 1999. Ecosystem change and the decline of marine mammals in the eastern Bering Sea: testing the ecosystem shift and commercial whaling hypotheses. *Fisheries Centre Research Reports,* University of British Columbia. Vol. 7. 106 pp.

Trivelpiece, W. Z., and S. Trivelpiece. 1998. The impact of global warming on Antarctica's krill-dependent predator populations. *Ecology* 84:213–227.

Truelove, J., and F. Iverson. 1994. Serum domoic acid clearance and clinical observations in the cynomolgus monkey and Sprague-Dawley rat following i.v. dose. *Bulletin of Environmental Contamination and Toxicology* 52:479–486.

Tubaro A., S. Sosa, M. Carbonatto, G. Altinier, F. Vita, M. Melato, M. Satake, T. Yasumoto, C. Vanderlip, and D. Sakumoto.

2003. Oral and intraperitoneal acute toxicity studies of yessotoxin and homoyessotoxins in mice. *Toxicon* 41:783–792.

Tyack, P. L. 2000. Functional aspects of cetacean communication. Pp. 270–307 In *Cetacean Societies: Field Studies of Dolphins and Whales* (J. Mann, ed.). University of Chicago Press, Chicago, IL.

Tyack, P. L., and C. W. Clark. 1998. Quick-look report: playback of low-frequency sound to gray whales migrating past the central California coast. Unpub. report.

Tynan, C. T., and D. P. DeMaster. 1997. Observations and predictions of Arctic climatic change: potential effects on marine mammals. *Arctic* 50(4):308–322.

United Nations Environment Programme. 1999a. Chemicals. *Inventory of Information Sources on Chemicals Persistent Organic Pollutants.* UNEP Chemicals, November.

United Nations Environment Programme. 1999b. *Report of the Scientific Advisory Committee of the Marine Mammals Action Plan.* United Nations Environment Programme [UNEP].

United Nations Environment Programme. Chemicals. 2002a. *Master List of Actions on the Reduction and/or Elimination of the Releases of Persistent Organic Pollutants,* 3rd Edition. Geneva, Switzerland: UNEP Chemicals, December 2000.

United Nations Environment Programme. 2002b. *Dugong status report and action plans for countries and territories.* United Nations Environment Programme, Division of Early Warning and Assessment.

Urick, R. J. 1975. *Principles of Underwater Sound.* McGraw-Hill, New York, NY.

U.S. Commission on Ocean Policy. 2004. *An Ocean Blueprint for the 21st Century.* U.S. Commission on Ocean Policy, Washington, DC.

U.S. Fish and Wildlife Service. 1996. Policy regarding the recognition of distinct vertebrate population segments under the Endangered Species Act. *Federal Register* 61:4722.

U.S. Fish and Wildlife Service. 2001. Florida manatee recovery plan (*Trichechus manatus latirostris*): Third revision. Southeast Region, U.S. Fish and Wildlife Service, Atlanta, GA.

U.S. Maritime Administration. 2003. www.marad.dot.gov/Marad_Statistics/index.html.

Van Bressem, M. F., K. Van Waerebeek, and J. A. Raga. 1999. A review of virus infections of cetaceans and the potential impact of morbilliviruses, poxviruses and papillomavirusesin host population dynamics. *Diseases of Aquatic Organisms* 38:53–65.

Vanderlip C., and D. Sakumoto. Unpubl. Ciguatera assessment at Midway Atoll. Student Skill Project Report to the University of Hawaii Marine Option Program, January 1993.

Van Dolah, F. M. 2000. Marine algal toxins: origins, health effects, and their increased occurrence. *Environmental Health Perspectives* 108S1:133–141.

Van Dolah, F. M., G. J. Doucette, F. M. D. Gulland, T. L. Rowles, and G. D. Bossart. 2003. Impacts of algal toxins on marine mammals. In *Toxicology of Marine Mammals* (J. G. Vos, G. D. Bossart, M. Fournier, and T. J. O'Shea, eds.). Taylor & Francis Publishers, London, U.K.

Vedros, N. A., A. W. Smith, J. Schoenwald, G. Migaki, and R. C. Hubbard. 1971. Leptospirosis epizootic among California sea lions. *Science* 172:1250.

Verbrugge, L. A., and J. P. Middaugh. 2004. Use of traditional foods in a healthy diet in Alaska: risks in perspective. 2nd Edition: Vol. 1. Polychlorinated biphenyls (PCBs) and related compounds. *State of Alaska Epidemiology Bulletin* 8:1–62.

Vinnikov, K. V., A. Robock, R. J. Stouffer, J. E. Walsh, C. L. Parkinson, D. J. Cavalieri, J. F. B. Mitchell, D. Garrett, and V. E. Zakharov. 1999. Global warming and northern hemisphere ice extent. *Science* 286:1934–1937.

Vinther, M. 1999. By-catches of harbour porpoises (*Phocoena phocoena* L.) in Danish set-net fisheries. *Journal of Cetacean Research and Management* 1:123–135.

Visser, I. K. G., J. S. Teppema, and A. D. M. E. Osterhaus. 1991. Virus infections of seals and other pinnipeds. *Reviews in Medical Microbiology* 2:105–114.

Vitousek, P. N., H. A. Nooney, J. Lubchenco, and J. M. Melillo, J. M. 1997. Human domination of Earth's ecosystems. *Science* 277:494–499.

Vos, J. G., G. D. Bossart, M. Fournier, and T. J. O'Shea (eds.). 2003a. *Toxicology of Marine Mammals.* Taylor & Francis Publishers, London, U.K.

Vos, J. G., P. S. Ross, R. L. DeSwart, H. van Loveren, and A. D. M. E. Osterhaus. 2003b. The effects of chemical contaminants on immune functions in harbor seals: results of a semifield study. In *Toxicology of Marine Mammals* (J. G. Vos, G. D. Bossart, M. Fournier, and T. J. O'Shea, eds.). Taylor & Francis Publishers, London, U.K.

Wade, P. R. 1994. Managing populations under the Marine Mammal Protection Act of 1994: a strategy for selecting values for N_{min}, the minimum abundance estimate, and F_r, the recovery factor. U.S. Department of Commerce, NOAA, National Marine Fisheries Service, Southwest Fisheries Science Center, Administrative Report L-94–19. 26 pp.

Wade, P. R. 1998. Calculating limits to the allowable human-caused mortality of cetaceans and pinnipeds. *Marine Mammal Science* 14:1–37.

Wade, P. R. 2002. Population dynamics. Pp. 974–979 In *Encyclopedia of Marine Mammals* (W. F. Perrin, B. Wursig, and J. G. M. Thewissen, eds.). Academic Press, San Diego, CA.

Wade, P. R., and R. Angliss. 1997. Guidelines for assessing marine mammal stocks: report of the GAMMS workshop, April 3–5, 1996, Seattle, WA. U.S. Department of Commerce, NOAA Technical Memorandum. NMFS-OPR-12. 93 pp.

Wagstaff, R. A. 1973. RANDI: Research ambient noise directionality model. Naval Undersea Center, Technical Publication.

Walker, C. H., and D. R. Livingstone. 1992. *Persistent Pollutants in Marine Ecosystems.* Pergamon Press Special Publication of SETAC (Society of Environmental Toxicology and Chemistry).

Wallace, J. M. 2000. North Atlantic Oscillation/Annular Mode: two paradigms—one phenomenon. *Quarterly Journal of the Royal Meterological Society* 126:791–805.

Walsh, D., B. Sjare, and E. Miller. 2001. Estimates of harp seal (*Phoca groenlandica*) by-catch in the Newfoundland lumpfish (*Cyclopterus lumpus*) fishery. Abstract, 14th Biennial Conference on the Biology of Marine Mammals, Vancouver, Canada.

Walsh, J. J., D. A. Dieterle, W. Maslowski, and T. E. Whitledge. 2004. Decadal shifts in biophysical forcing of Arctic marine food webs: numerical consequences. *Journal of Geophysical Research* 109:(C05031):1–31.

Walsh J. J., and K. A. Steidinger. 2001. Sarahan dust and Florida red tides: the cyannophyte connection. *Journal of Geophysical Research* 106:11597–11612.

Walsh, M. T., R. Y. Ewing, D. K. Odell, and G. D. Bossart. 2001. Mass strandings of cetaceans. Pp. 83–96 In *CRC Handbook of Marine Mammal Medicine* (L. A. Dierauf and F. M. D. Gulland, eds.). CRC Press, Boca Raton, FL.

Walters, C. J., V. Christensen, and D. Pauly. 1997. Structuring dynamic models of exploited ecosystems from trophic mass-balance assessments. *Reviews in Fish Biology and Fisheries* 7:139–172.

Walters, C. J., and R. Hilborn. 1976. Adaptive control of fishing systems. *Journal of the Fisheries Research Board of Canada* 33:145–159.

Walters, C. J., and C. S. Holling. 1990. Large-scale management experiments and learning by doing. *Ecology* 71:2060–2068.

Walters, C., and J. F. Kitchell. 2001. Cultivation / depensation effects on juvenile survival and recruitment: implications for the theory of fishing. *Canadian Journal of Fisheries and Aquatic Sciences* 58:39–50.

Walters, C., D. Pauly, and V. Christensen. 1999. Ecospace: prediction of mesoscale spatial patterns in trophic relationships of exploited ecosystems, with emphasis on the impacts of marine protected areas. *Ecosystems* 2:539–554.

Walters, C. J., D. Pauly, V. Christensen, and J. F. Kitchell. 2000. Representing density dependent consequences of life history strategies in aquatic ecosystems: ECOSIM II. *Ecosystems* 3:70–83.

Wang, G., N. T. Hobbs, K. M. Giesen, H. Galbraith, D. S. Ojima, and C. E. Braun. 2002. Relationships between climate and population dynamics of white-tailed ptarmigan *Lagopus leucurus* in Rocky Mountain National Park, Colorado, USA. *Climate Research* 23:81–87.

Waples, R. S. 1991. Definition of "species" under the Endangered Species Act: application to Pacific salmon. National Oceanic and Atmospheric Administration U.S. Department of Commerce, NOAA Technical Memorandum, NMFS-F / NWC-194. 29 pp.

Warner, R. E. 1968. The role of introduced diseases in the extinction of the endemic Hawaiian avifauna. *The Condor* 70:101–120.

Wartzok, D., and D. R. Ketten. 1999. Marine mammal sensory systems. Pp. 117–175 In *Biology of Marine Mammals* (J. E. Reynolds, III, and S. A. Rommel, eds.). Smithsonian Institution Press, Washington, DC.

Wartzok, D., S. Sayegh, H. Stone, J. Barchak, and W. Barnes. 1992. Acoustic tracking system for monitoring under-ice movements of polar seals. *Journal of the Acoustical Society of America* 92(2):682–687.

Wartzok, D., W. A. Watkins, B. Wursig, and C. I. Malme. 1989. Movements and behaviors of bowhead whales in response to repeated exposures to noises associated with industrial activities in the Beaufort Sea. Report from Purdue University for Amoco Production Company, Anchorage, AK.

Watkins, W. A. 1977. Acoustic behavior of sperm whales. *Oceanus* 20(2):50–58.

Watkins, W. A. 1986. Whale reactions to human activities in Cape Cod waters. *Marine Mammal Science* 2:251–262.

Watkins, W. A., K. E. Moore, and P. Tyack. 1985. Sperm whale acoustic behaviors in the southeast Caribbean. *Cetology* 49:1–15.

Watkins, W. A., and G. C. Ray. 1985. In-air and underwater sounds of the Ross seal, *Ossmatophoca rossi*. *Journal of the Acoustical Society of America* 77(4):1598–1600.

Watkins, W. A., and W. E. Schevill. 1974. Listening to Hawaiian spinner porpoises (*Stenella* cf. *longirostris*) with a three-dimensional hydrophone array. *Journal of Mammology* 55:319–328.

Watkins, W. A., and W. E. Schevill. 1975. Sperm whale codas. *Journal of the Acoustical Society of America* 26:1485–1490 + phono record.

Watkins, W. A., and D. Wartzok. 1985. Sensory biophysics of marine mammals. *Marine Mammal Science* 1:219–260.

Watt, J., D. B. Siniff, and J. A. Estes. 2000. Inter-decadal patterns of population and dietary change in sea otters at Amchitka island, Alaska. *Oecologia (Berlin)* 124:289–298.

Watts, A. J. 2003. *Jane's Underwater Warfare Systems* (15th Edition). Jane's, London.

Weber, M. L. 2002. *From Abundance to Scarcity: A History of U.S. Marine Fisheries Policy.* Island Press, Washington, DC. 245 pp.

Weir, C. R., C. Pollock, C. Cronin, and C. Taylor. 2001. Cetaceans of the Atlantic Frontier, north and west of Scotland. *Continental Shelf Research* 21:1047–1071.

Weisbrod, A. V., D. Shea, M. J. Moore, and J. J. Stegeman. 2000. Organochlorine exposure and bioaccumulation in the endangered northwest Atlantic right whale (*Eubalaena glacialis*) population. *Environmental Toxicology and Chemistry* 19: 654–666.

Weller, D. W. 2002. Predation on marine mammals. Pp. 985–994 In *Encyclopedia of Marine Mammals* (W. F. Perrin, B. Wursig, and J. G. M. Thewissen, eds.). Academic Press, San Diego, CA.

Weller, D. W., A. M. Burdin, A. L. Bradford, G. A. Tsidulko, Y. V. Ivashchenko, and R. L. Brownell, Jr. 2002. Gray whales off Sakhalin Island, Russia: June–September 2001, a joint U.S.-Russia Scientific Investigation. Final Report for Contract No. T03323111 with the Marine Mammal Commission.

Wells, R. A., D. M. Scott, and A. B. Irvine. 1987. The social structure of free-ranging bottlenose dolphins. In *Current Mammalogy*, Vol. 1 (H. H. Genoways, ed.). Plenum Press, New York, NY.

Wells, R. S., V. Tornero, A. Borrell, A. Aguilar, T. K. Rowles, H. L. Rhinehart, S. Hofmann, W. M. Jarman, A. A. Hohn, and J. C. Sweeney. 2004. Integrating life history and reproductive success data to examine potential relationships with organochlorine compounds for bottlenose dolphins (*Tursiops truncatus*) in Sarasota Bay, Florida. SC / 56 / E19. 52nd Meeting of the International Whaling Commission, Sorrento, Italy. Available from the IWC.

Wenz, G. M. 1962. Acoustic ambient noise in the ocean: spectra and sources. *Journal of the Acoustical Society of America* 34:1936–1956.

Wenz, G. M. 1969. Low-frequency deep-water ambient noise along the Pacific Coast of the United States. *Journal of Underwater Acoustics* 19:423–444 (declassified).

West, I. F., J. Molloy, M. F. Donoghue, and C. Pugsley. 1999. Recovery of environmental investigation costs attributable to commercial fishing: the New Zealand Conservation Services Levy Program. *Marine Technology Society Journal* 33(2):13–18.

White, W. B., and R. G. Peterson. 1996. An Antarctic circumpolar wave in surface pressure, wind, temperature and sea-ice extent. *Nature* 380:699–702.

Whitehead, H., and J. E. Carscadden. 1985. Predicting inshore whale abundance: whales and capelin off the Newfoundland coast. *Canadian Journal of Fisheries and Aquatic Sciences* 42(5): 976–981.

Wiberg, K., R. J. Letcher, C. D. Sandau, R. J. Norstrom, M. Tysklind and T. F. Bidleman. 2000. The enantioselective bioaccumulation of chiral chlordane and a-HCH contaminants in the polar bear food chain. *Environmental Science and Technology* 34:2668–2674.

Wiberg, K., M. Oehme, P. Haglund, H. Karlsson, M. Olsson, and C. Rappe. 1998. Enantioselective analysis of organochlorine pesticides in herring and seal from the Swedish marine environment. *Marine Pollution Bulletin* 36:345–353.

Wickens, P. 1995. A review of operational interactions between pinnipeds and fisheries. FAO Fisheries Technical Paper 346. 86 pp.

Willetto, C. E., T. M. O'Hara, and T. Rowles (eds.). 2002a. *Bowhead Whale Health and Physiology Workshop Report* (1–4 October 2001). Barrow, AK.

Willetto, C. E., T. M. O'Hara, and T. Rowles. 2002b. *Bowhead Whales Health and Physiology Workshop 2001:* Summary for the International Whaling Commission (IWC) Scientific Committee. Shimonoseki, Japan, Report No. SC/54/BRG1.

Wilson, B., H. Arnold, G. Bearzi, C. M. Fortuna, R. Gaspar, S. Ingram, C. Liret, S. Pribanic, A. J. Read, V. Ridoux, K. Schneider, K. W. Urian, R. S. Wells, C. Wood, C. Thompson, and P. S. Hamond. 1999. Epidermal lesions in bottlenose dolphins: impacts of natural and anthropogenic factors. *Proceedings of the Royal Society of London. Series B. Biological Sciences* 226: 1077–1083.

Wilson, E. O., and W. H. Bossart. 1971. *A Primer of Population Biology.* Sinauer Associates, Inc. Publishers, Sunderland, MA.

Wilson, K., O. N. Bjornstad, A. P. Dobson, S. Merler, G. Poglayan, S. E. Randolph, A. F. Read, and A. Skorping. 2002. Heterogeneities in macroparasite infections: patterns and processes. In *The Ecology of Wildlife Diseases* (P. J. Hudson, A. Rizzoli, B. T. Grenfell, H. Heesterbeek, and A. P. Dobson, eds.). Oxford University Press, Oxford, U.K.

Wilson, M. T., and P. J. Jokiel. 1986. Ciguatera at Midway: an assessment using the Hokama "stick test" for ciguatoxin. NOAA Technical Report NOAA-SWFSC-86-1.

Wilson, O. B. J., S. N. Wolf, and F. Ingenito. 1985. Measurements of acoustic ambient noise in shallow water due to breaking surf. *Journal of the Acoustical Society of America* 78(1):190–195.

Winship, A. J., and A. W. Trites. 2003. Prey consumption of Steller sea lions (*Eumetopias jubatus*) off Alaska: how much prey do they require? *Fisheries Bulletin* 101:147–167.

Witherell, D., C. Pautzke, and D. Fluharty. 2000. An ecosystem-based approach for Alaska groundfish fisheries. *ICES Journal of Marine Science* 57:771–777.

Wolf, R. J. 2001. The subsistence harvest of harbor seals and sea lions by Alaska natives in 2000. Alaska Department of Fish and Game, Division of Subsistence Technical Paper No. 266. Juneau, AK.

Wong, C. S., A. W. Garrison, P. D. Smith, and W. T. Foreman. 2001. Enantiomeric composition of chiral polychlorinated biphenyl atropisomers in aquatic and riparian biota. *Environmental Science and Technology* 35:2448–2454.

Work, T. M., B. Barr, A. M. Beale, L. Fritz, M. A. Quilliam, and J. L. C. Wright. 1993. Epidemiology of domoic acid poisoning in brown pelicans (*Pelicanus occidentalis*) and Brandt's cormorants (*Phalacrocorax penicillatus*) in California. *Journal of Zoo and Wildlife Medicine* 24:54–62.

World Council for Whalers. www.worldcouncilofwhalers.com/faqsframe.htm.

Worldwide Fund for Nature. 2003. Cetacean By-Catch Resource Center. www.cetaceanbycatch.org/. Accessed on 8 July 2003.

Wormworth, J. 1995. Toxins and tradition: the impact of food-chain contamination on the Inuit of northern Quebec. *Journal of the Canadian Medical Association* 152:1237–1240.

Woshner, V. M. 2000. Concentrations and Interactions of Selected Elements in Tissues of Four Marine Mammal Species Harvested by Inuit Hunters in Arctic Alaska, with an Intensive Histologic Assessment, Emphasizing the Beluga Whale. Ph.D. Thesis, University of Illinois, Urbana, IL.

Woshner, V. M., T. M. O'Hara, G. R. Bratton, and V. R. Beasley. 2001b. Concentrations and interactions of selected essential and non-essential elements in ringed seals and polar bears of Arctic Alaska. *Journal of Wildlife Diseases* 37:711–721.

Woshner, V. M., T. M. O'Hara, G. R. Bratton, R. S. Suydam, and V. R. Beasley. 2001a. Concentrations and interactions of selected essential and non-essential elements in bowhead and beluga whales of arctic Alaska. *Journal of Wildlife Diseases* 37:693–710.

Woshner, V. M., T. M. O'Hara, J. A. Eurell, M. A. Wallig, G. R. Bratton, V. R. Beasley, and R. S. Suydam. 2002. Distribution of inorganic mercury in liver and kidney of beluga whales, compared to bowhead whales, through autometallographic development of light microscopic tissue sections. *Toxicologic Pathology* 30:209–215.

Yano, K., and M. Dahlheim. 1995. Killer whale, *Orcinus orca,* depredation on longline catches of bottom fish in the southeastern Bering Sea and adjacent waters. *Fishery Bulletin* 93: 355–372.

Ylitalo, G. M., C. O. Matkin, J. Buzitis, M. M. Krahn, L. L. Jones, T. Rowles, and J. E. Stein. 2001. Influence of life-history parameters on organochlorine concentrations in free-ranging killer whales *Orcinus orca* from Prince William Sound, AK. *The Science of the Total Environment* 281:183–203.

Yodzis, P. 1988. The indeterminacy of ecological interactions as perceived through perturbation experiments. *Ecology* 69:508–515.

Yodzis, P. 1998. Local trophodynamics and the interaction of marine mammals and fisheries in the Benguela ecosystem. *Journal of Animal Ecology* 67:635–658.

Yodzis, P. 2000. Diffuse effects in food webs. *Ecology* 81(1):261–266.

Young, N. M. 2001. The conservation of marine mammals using a multi-party approach: an evaluation of the take reduction team process. *Ocean and Coastal Law Journal* 6:293–346.

Yu, D. P. 2002. Investigation of health status of Yangtze finless porpoise in the field. *Chinese Journal of Zoology* 37(5):70–73.

Yu, Y., G. A. Maykut, and D. A. Rothrock. 2004. Changes in the thickness distribution of Arctic sea ice between 1958–1970 and 1993–1997. *Journal of Geophysical Research* 109:1–13.

Yung, Y. K., C. K. Wong, M. J. Broom, J. A. Ogden, S. C. M. Chan, and Y. Leung. 1997. Long-term changes in hydrography, nutrients, and phytoplankton in Tolo Harbor, Hong Kong. *Hydrobiologia* 352:107–115.

Zarnke, R. L., T. C. Harder, H. W. Vos, J. M. V. Hoef, and A. D. M. E. Osterhaus. 1997. Serologic survey for phocid herpesvirus-1 and -2 in marine mammals from Alaska and Russia. *Journal of Wildlife Diseases* 33:459–465.

Zeppelin, T. K., D. J. Tollit, K. A. Call, T. J. Orchard, and C. J. Gudmundson. 2004. Sizes of walleye pollock (*Theragra chalcogramma*) and Atka mackerel (*Pleurogrammus monopterygius*) consumed by the western stock of Steller sea lions (*Eumetopias jubatus*) in Alaska from 1998 to 2000. *Fishery Bulletin* 102:509–521.

Contributors

Doug S. Butterworth
University of Cape Town
Cape Town, South Africa

Daniel Goodman
Montana State University
Bozeman, Montana, USA

Frances M. D. Gulland
The Marine Mammal Center
Sausalito, California, USA

Ailsa J. Hall
University of St Andrews
St. Andrews, Fife, Scotland

John Hildebrand
Scripps Institution of Oceanography and
University of California at San Diego
La Jolla, California, USA

Sue E. Moore
National Marine Fisheries Service and
University of Washington
Seattle, Washington, USA

Todd M. O'Hara
University of Alaska at Fairbanks
Fairbanks, Alaska, USA

Thomas J. O'Shea
U.S. Geological Survey
Fort Collins, Colorado, USA

William F. Perrin
National Marine Fisheries Service
La Jolla, California, USA

Éva E. Plagányi
University of Cape Town
Cape Town, South Africa

Timothy J. Ragen
Marine Mammal Commission
Bethesda, Maryland, USA

Andrew J. Read
Duke University Marine Laboratory
Beaufort, North Carolina, USA

Randall R. Reeves
Okapi Wildlife Associates
Hudson, Quebec, Canada

John E. Reynolds III
Mote Marine Laboratory
Sarasota, Florida, USA, and
Marine Mammal Commission
Bethesda, Maryland, USA

Barbara L. Taylor
National Marine Fisheries Service
La Jolla, California, USA

Frances M. Van Dolah
National Ocean Service
Charleston, South Carolina, USA

Index